Analyzing US Census Data

Census data are widely used by practitioners to understand demographic change, allocate resources, address inequalities, and make sound business decisions. Until recently, projects using US Census data have required proficiency with multiple web interfaces and software platforms to prepare, map, and present data products. This book introduces readers to tools in the R programming language for accessing and analyzing Census data, helping analysts manage these types of projects in a single computing environment. Chapters in this book cover the following key topics:

- Rapidly acquiring data from the decennial US Census and American Community Survey using R, then analyzing these datasets using **tidyverse** tools;
- Visualizing US Census data with a wide range of methods including charts in **ggplot2** as well as both static and interactive maps;
- Using R as a geographic information system (GIS) to manage, analyze, and model spatial demographic data from the US Census;
- Working with and modeling individual-level microdata from the American Community Survey's PUMS datasets;
- Applying these tools and workflows to the analysis of historical Census data, other US government datasets, and international Census data from countries like Canada, Brazil, Kenya, and Mexico.

Kyle Walker is an associate professor of geography at Texas Christian University, director of TCU's Center for Urban Studies, and a spatial data science consultant. His research focuses on demographic trends in the United States, demographic data visualization, and software tools for open spatial data science. He is the lead author of a number of R packages including **tigris**, **tidycensus**, and **mapboxapi**.

Analyzing US Census Data
Methods, Maps, and Models in R

Kyle Walker

CRC Press
Taylor & Francis Group
Boca Raton London New York

CRC Press is an imprint of the
Taylor & Francis Group, an **informa** business

A CHAPMAN & HALL BOOK

The front cover image shows the percent foreign-born by Census tract in NYC in 1910, mapped with ggplot2. Data source: NHGIS

First edition published 2023
by CRC Press
6000 Broken Sound Parkway NW, Suite 300, Boca Raton, FL 33487-2742

and by CRC Press
4 Park Square, Milton Park, Abingdon, Oxon, OX14 4RN

CRC Press is an imprint of Taylor & Francis Group, LLC

© 2023 Taylor & Francis Group, LLC

Reasonable efforts have been made to publish reliable data and information, but the author and publisher cannot assume responsibility for the validity of all materials or the consequences of their use. The authors and publishers have attempted to trace the copyright holders of all material reproduced in this publication and apologize to copyright holders if permission to publish in this form has not been obtained. If any copyright material has not been acknowledged, please write and let us know so we may rectify in any future reprint.

Except as permitted under U.S. Copyright Law, no part of this book may be reprinted, reproduced, transmitted, or utilized in any form by any electronic, mechanical, or other means, now known or hereafter invented, including photocopying, microfilming, and recording, or in any information storage or retrieval system, without written permission from the publishers.

For permission to photocopy or use material electronically from this work, access www.copyright.com or contact the Copyright Clearance Center, Inc. (CCC), 222 Rosewood Drive, Danvers, MA 01923, 978-750-8400. For works that are not available on CCC, please contact mpkbookspermissions@tandf.co.uk

Trademark notice: Product or corporate names may be trademarks or registered trademarks and are used only for identification and explanation without intent to infringe.

Library of Congress Cataloging-in-Publication Data
Names: Walker, Kyle E., author. Title: Analyzing US census data : methods, maps, and models in R / Kyle Walker. Description: First edition | Boca Raton : CRC Press, 2023. | Series: Chapman & Hall/CRC the R series | Includes bibliographical references and index. Identifiers: LCCN 2022021374 (print) | LCCN 2022021375 (ebook) | ISBN 9781032366449 (paperback) | ISBN 9781138560789 (hardback) | ISBN 9780203711415 (ebook) Subjects: LCSH: United States--Census. | R (Computer program language) Classification: LCC HA181 .W35 2023 (print) | LCC HA181 (ebook) | DDC 317.3--dc23/eng/20220912 LC record available at https://lccn.loc.gov/2022021374 LC ebook record available at https://lccn.loc.gov/2022021375

ISBN: 978-1-138-56078-9 (hbk)
ISBN: 978-1-032-36644-9 (pbk)
ISBN: 978-0-203-71141-5 (ebk)

DOI: 10.1201/9780203711415

Typeset in Latin Modern font
by KnowledgeWorks Global Ltd.

Publisher's note: This book has been prepared from camera-ready copy provided by the authors.

To Molly, Michaela, Landry, and Clara

Contents

Preface xiii

1 The US Census and the R programming language 1
  1.1 Census data: an overview 1
  1.2 Census hierarchies 2
  1.3 How to find US Census data 3
      1.3.1 Data downloads from the Census Bureau 5
      1.3.2 The Census API 5
      1.3.3 Third-party data distributors 7
  1.4 What is R? 8
      1.4.1 Getting started with R 8
      1.4.2 Basic data structures in R 8
      1.4.3 Functions and packages 9
      1.4.4 Package ecosystems in R 10
  1.5 Analyses using R and US Census data 11
      1.5.1 Census data packages in R: a brief summary 11
      1.5.2 Health resource access 12
      1.5.3 COVID-19 and pandemic response 12
      1.5.4 Politics and gerrymandering 12
      1.5.5 Social equity research 15
      1.5.6 Census data visualization 15

2 An introduction to tidycensus 17
  2.1 Getting started with tidycensus 17
      2.1.1 Decennial Census 18
      2.1.2 American Community Survey 20
  2.2 Geography and variables in tidycensus 22
      2.2.1 Geographic subsets 24
  2.3 Searching for variables in tidycensus 27
  2.4 Data structure in tidycensus 29
      2.4.1 Understanding GEOIDs 30
      2.4.2 Renaming variable IDs 32
  2.5 Other Census Bureau datasets in tidycensus 33
      2.5.1 Using `get_estimates()` 33
      2.5.2 Using `get_flows()` 35
  2.6 Debugging tidycensus errors 36
  2.7 Exercises 37

3 Wrangling Census data with tidyverse tools 39
  3.1 The tidyverse 39
  3.2 Exploring Census data with tidyverse tools 40
      3.2.1 Sorting and filtering data 40

		3.2.2	Using summary variables and calculating new columns	43
	3.3	\multicolumn{2}{l}{Group-wise Census data analysis}	45
		3.3.1	Making group-wise comparisons	46
		3.3.2	Tabulating new groups	47
	3.4	\multicolumn{2}{l}{Comparing ACS estimates over time}	49
		3.4.1	Time-series analysis: some cautions	50
		3.4.2	Preparing time-series ACS estimates	52
	3.5	\multicolumn{2}{l}{Handling margins of error in the American Community Survey with tidycensus}	56
		3.5.1	Calculating derived margins of error in tidycensus	57
		3.5.2	Calculating group-wise margins of error	59
	3.6	Exercises		60

4 Exploring US Census data with visualization — 61

- 4.1 Basic Census visualization with ggplot2 — 61
  - 4.1.1 Getting started with ggplot2 — 62
  - 4.1.2 Visualizing multivariate relationships with scatter plots — 64
- 4.2 Customizing ggplot2 visualizations — 66
  - 4.2.1 Improving plot legibility — 67
  - 4.2.2 Custom styling of ggplot2 charts — 69
  - 4.2.3 Exporting data visualizations from R — 71
- 4.3 Visualizing margins of error — 72
  - 4.3.1 Data setup — 72
  - 4.3.2 Using error bars for margins of error — 74
- 4.4 Visualizing ACS estimates over time — 76
- 4.5 Exploring age and sex structure with population pyramids — 78
  - 4.5.1 Preparing data from the Population Estimates API — 78
  - 4.5.2 Designing and styling the population pyramid — 80
- 4.6 Visualizing group-wise comparisons — 82
- 4.7 Advanced visualization with ggplot2 extensions — 86
  - 4.7.1 ggridges — 86
  - 4.7.2 ggbeeswarm — 87
  - 4.7.3 Geofaceted plots — 88
  - 4.7.4 Interactive visualization with plotly — 91
- 4.8 Learning more about visualization — 92
- 4.9 Exercises — 92

5 Census geographic data and applications in R — 93

- 5.1 Basic usage of tigris — 93
  - 5.1.1 Understanding tigris and simple features — 97
  - 5.1.2 Data availability in tigris — 100
- 5.2 Plotting geographic data — 101
  - 5.2.1 ggplot2 and `geom_sf()` — 101
  - 5.2.2 Interactive viewing with mapview — 103
- 5.3 tigris workflows — 105
  - 5.3.1 TIGER/Line and cartographic boundary shapefiles — 105
  - 5.3.2 Caching tigris data — 106
  - 5.3.3 Understanding yearly differences in TIGER/Line files — 107
  - 5.3.4 Combining tigris datasets — 108
- 5.4 Coordinate reference systems — 109
  - 5.4.1 Using the crsuggest package — 110
  - 5.4.2 Plotting with `coord_sf()` — 113

	5.5	Working with geometries	115
		5.5.1 Shifting and rescaling geometry for national US mapping	115
		5.5.2 Converting polygons to points	117
		5.5.3 Exploding multipolygon geometries to single parts	119
	5.6	Exercises	122
6	**Mapping Census data with R**		**123**
	6.1	Using geometry in tidycensus	123
		6.1.1 Basic mapping of sf objects with `plot()`	125
	6.2	Map-making with ggplot2 and geom_sf	126
		6.2.1 Choropleth mapping	126
		6.2.2 Customizing ggplot2 maps	127
	6.3	Map-making with tmap	128
		6.3.1 Choropleth maps with tmap	129
		6.3.2 Adding reference elements to a map	133
		6.3.3 Choosing a color palette	136
		6.3.4 Alternative map types with tmap	137
	6.4	Cartographic workflows with non-Census data	141
		6.4.1 National election mapping with tigris shapes	142
		6.4.2 Understanding and working with ZCTAs	143
	6.5	Interactive mapping	147
		6.5.1 Interactive mapping with Leaflet	147
		6.5.2 Alternative approaches to interactive mapping	151
	6.6	Advanced examples	154
		6.6.1 Mapping migration flows	155
		6.6.2 Linking maps and charts	156
		6.6.3 Reactive mapping with Shiny	158
	6.7	Working with software outside of R for cartographic projects	161
		6.7.1 Exporting maps from R	162
		6.7.2 Interoperability with other visualization software	163
	6.8	Exercises	165
7	**Spatial analysis with US Census data**		**167**
	7.1	Spatial overlay	167
		7.1.1 Note: aligning coordinate reference systems	168
		7.1.2 Identifying geometries within a metropolitan area	169
		7.1.3 Spatial subsets and spatial predicates	170
	7.2	Spatial joins	171
		7.2.1 Point-in-polygon spatial joins	172
		7.2.2 Spatial joins and group-wise spatial analysis	176
	7.3	Small area time-series analysis	180
		7.3.1 Area-weighted areal interpolation	182
		7.3.2 Population-weighted areal interpolation	183
		7.3.3 Making small-area comparisons	185
	7.4	Distance and proximity analysis	187
		7.4.1 Calculating distances	188
		7.4.2 Calculating travel times	190
		7.4.3 Catchment areas with buffers and isochrones	192
		7.4.4 Computing demographic estimates for zones with areal interpolation	194
	7.5	Better cartography with spatial overlay	196
		7.5.1 "Erasing" areas from Census polygons	196

	7.6	Spatial neighborhoods and spatial weights matrices	198
		7.6.1 Understanding spatial neighborhoods	199
		7.6.2 Generating the spatial weights matrix	201
	7.7	Global and local spatial autocorrelation	202
		7.7.1 Spatial lags and Moran's I	203
		7.7.2 Local spatial autocorrelation	204
		7.7.3 Identifying clusters and spatial outliers with local indicators of spatial association (LISA)	206
	7.8	Exercises	211

8 Modeling US Census data — 213

	8.1	Indices of segregation and diversity	213
		8.1.1 Data setup with spatial analysis	213
		8.1.2 The dissimilarity index	215
		8.1.3 Multi-group segregation indices	217
		8.1.4 Visualizing the diversity gradient	218
	8.2	Regression modeling with US Census data	220
		8.2.1 Data setup and exploratory data analysis	222
		8.2.2 Inspecting the outcome variable with visualization	223
		8.2.3 "Feature engineering"	225
		8.2.4 A first regression model	225
		8.2.5 Dimension reduction with principal components analysis	230
	8.3	Spatial regression	234
		8.3.1 Methods for spatial regression	236
		8.3.2 Choosing between spatial lag and spatial error models	239
	8.4	Geographically weighted regression	241
		8.4.1 Choosing a bandwidth for GWR	242
		8.4.2 Fitting and evaluating the GWR model	243
		8.4.3 Limitations of GWR	246
	8.5	Classification and clustering of ACS data	247
		8.5.1 Geodemographic classification	248
		8.5.2 Spatial clustering & regionalization	250
	8.6	Exercises	253

9 Introduction to Census microdata — 255

	9.1	What is "microdata?"	255
		9.1.1 Microdata resources: IPUMS	256
		9.1.2 Microdata and the Census API	256
	9.2	Using microdata in tidycensus	257
		9.2.1 Basic usage of `get_pums()`	257
		9.2.2 Understanding default data from `get_pums()`	258
	9.3	Working with PUMS variables	260
		9.3.1 Variables available in the ACS PUMS	261
		9.3.2 Recoding PUMS variables	261
		9.3.3 Using variables filters	262
	9.4	Public Use Microdata Areas (PUMAs)	263
		9.4.1 What is a PUMA?	263
		9.4.2 Working with PUMAs in PUMS data	265
	9.5	Exercises	267

10 Analyzing Census microdata — 269
- 10.1 PUMS data and the tidyverse — 269
  - 10.1.1 Basic tabulation of weights with tidyverse tools — 269
  - 10.1.2 Group-wise data tabulation — 272
- 10.2 Mapping PUMS data — 274
- 10.3 Survey design and the ACS PUMS — 275
  - 10.3.1 Getting replicate weights — 275
  - 10.3.2 Creating a survey object — 277
  - 10.3.3 Calculating estimates and errors with srvyr — 278
  - 10.3.4 Converting standard errors to margins of error — 279
- 10.4 Modeling with PUMS data — 280
  - 10.4.1 Data preparation — 281
  - 10.4.2 Fitting and evaluating the model — 282
- 10.5 Exercises — 284

11 Other Census and government data resources — 285
- 11.1 Mapping historical geographies of New York City with NHGIS — 285
  - 11.1.1 Getting started with NHGIS — 286
  - 11.1.2 Working with NHGIS data in R — 287
  - 11.1.3 Mapping NHGIS data in R — 289
- 11.2 Analyzing complete-count historical microdata with IPUMS and R — 292
  - 11.2.1 Getting microdata from IPUMS — 294
  - 11.2.2 Loading microdata into a database — 296
  - 11.2.3 Accessing your microdata database with R — 297
  - 11.2.4 Analyzing big Census microdata in R — 300
- 11.3 Other US government datasets — 302
  - 11.3.1 Accessing Census data resources with censusapi — 302
  - 11.3.2 Analyzing labor markets with lehdr — 306
  - 11.3.3 Bureau of Labor Statistics data with blscrapeR — 308
  - 11.3.4 Working with agricultural data with tidyUSDA — 310
- 11.4 Getting government data without R packages — 312
  - 11.4.1 Making requests to APIs with httr — 312
  - 11.4.2 Writing your own data access functions — 313
- 11.5 Exercises — 315

12 Working with Census data outside the United States — 317
- 12.1 The International Data Base and the idbr R package — 317
  - 12.1.1 Visualizing IDB data — 319
  - 12.1.2 Interactive and animated visualization of global demographic data — 322
- 12.2 Country-specific Census data packages — 325
  - 12.2.1 Canada: cancensus — 325
  - 12.2.2 Kenya: rKenyaCensus — 327
  - 12.2.3 Mexico: combining mxmaps and inegiR — 331
  - 12.2.4 Brazil: aligning the geobr R package with raw Census data files for spatial analysis — 333
- 12.3 Other international data resources — 341
- 12.4 Exercises — 341

Conclusion — 343

Bibliography — 345

Index — 353

Preface

Census data are widely used in the United States across numerous research and applied fields, including education, business, journalism, and many others. Until recently, the process of working with US Census data has required the use of a wide array of web interfaces and software platforms to prepare, map, and present data products. The goal of this book is to illustrate the utility of the R programming language for handling these tasks, allowing Census data users to manage their projects in a single computing environment.

This book focuses on two types of Census data products commonly used by analysts:

- **Aggregate data**, which involve counts or estimates released by the Census Bureau that are aggregated to some geographic unit of interest, such as a state;
- **Microdata**, which are individual, anonymized Census records that represent a sample of a given Census dataset.

Readers new to R and Census data should read this book in order, as each chapter includes concepts that build upon examples introduced in previous chapters. More experienced analysts might use chapters as standalone manuals tailored to specific tasks and topics of interest to them. A brief overview of each chapter follows below.

- Chapter 1 is a general overview of US Census data terms and definitions and gives some brief background about the R language and why R is an excellent environment for working with US Census data. It ends with motivating examples of excellent applied projects that use R and Census data.
- Chapter 2 introduces **tidycensus**, an R package for working with US Census Bureau data in a tidy format. Readers will learn how to make basic data requests with the package and understand various options in the package.
- Chapter 3 covers the analysis of US Census data using the **tidyverse**, an integrated framework of packages for data preparation and wrangling. It includes both simple and more complex examples of common **tidyverse** data wrangling tasks, and discusses workflows for handling margins of error in the American Community Survey.
- Chapter 4 introduces workflows for Census data visualization with a focus on the **ggplot2** package. Examples focus on best practices for preparing data for visualization and building charts well-suited for presenting Census data analyses.
- Chapter 5 introduces the **tigris** package for working with US Census Bureau geographic data in R. It includes an overview of spatial data structures in R with the **sf** package and covers key geospatial data topics like coordinate reference systems.
- Chapter 6 is all about mapping in R. Readers learn how to make simple shaded maps with US Census data using packages like **ggplot2** and **tmap**; maps with more complex data requirements like dot-density maps; and interactive geographic visualizations and apps.
- Chapter 7 covers spatial data analysis. Topics include geographic data overlay; distance and proximity analysis; and exploratory spatial data analysis with the **spdep** package.

- Chapter 8's topic is a modeling of geographic data. Readers learn how to compute indices of segregation and diversity with Census data; fit linear, spatial, and geographically weighted regression models; and develop workflows for geodemographic segmentation and regionalization.
- Chapter 9 focuses on individual-level *microdata* from the US Census Bureau's Public Use Microdata Sample. Readers will learn how to acquire these datasets with **tidycensus** and use them to generate unique insights.
- Chapter 10 covers the analysis of microdata with a focus on methods for analyzing and modeling complex survey samples. Topics include estimation of standard errors with replicate weights, mapping microdata, and modeling microdata appropriately with the **survey** and **srvyr** packages.
- Chapter 11's focus is Census datasets beyond the decennial US Census and American Community Survey. The first part of the chapter focuses on historical mapping with the National Historical Geographic Information System (NHGIS) and historical microdata analysis with IPUMS-USA; readers learn how to use the **ipumsr** R package and R's database tools to assist with these tasks. The second part of the chapter covers the wide range of datasets available from the US Census Bureau, and R packages like **censusapi** and **lehdr** that help analysts access those datasets.
- Chapter 12 covers Census data resources for regions outside the United States. It covers global demographic analysis with the US Census Bureau's International Data Base as well as country-specific examples from Canada, Mexico, Brazil, and Kenya.

Who this book is for

This book is designed for practitioners interested in working efficiently with data from the United States Census Bureau. It defines "practitioners" quite broadly, and analysts at all levels of expertise should find topics within this book useful. While this book focuses on examples from the United States, the topics covered are designed to be more general and applicable to use cases outside the United States and also outside the domain of working with Census data.

- Students and analysts newer to R will likely want to start with Chapters 2 through 4, which cover the basics of **tidycensus** and give examples of several tidyverse-centric workflows.
- Chapters 5 through 7 will likely appeal to analysts from a Geographic Information Systems (GIS) background, as these chapters introduce methods for handling and mapping spatial data that might alternatively be done in desktop GIS software like ArcGIS or QGIS.
- More experienced social science researchers will find Chapters 9 through 11 useful as they cover ACS microdata, an essential resource for many social science disciplines. Analysts coming to R from Stata or SAS will also learn how to use R's survey data modeling tools in these chapters.
- Data scientists interested in integrating spatial analysis into their work may want to focus on Chapters 7 and 8, which cover a range of methods that can be incorporated into business intelligence workflows.

- Journalists will find value throughout this book, though Chapters 2, 5–6, 9, and 12 may prove especially useful as they focus on rapid retrieval of US and international Census data that can be incorporated into articles and reports.

Of course, there are many other use cases for Census data that are not covered in this overview. If you are using the examples in this book for unique applications, please reach out!

About the author

I (Kyle Walker) work as an associate professor of Geography at Texas Christian University and as a spatial data science consultant. I'm a proud graduate of the University of Oregon (Go Ducks!), and I hold a Ph.D. in Geography from the University of Minnesota. I do research in the field of population geography, focusing on metropolitan demographic trends, and I consult broadly in areas such as commercial real estate, the health sciences, and general R training/software development. For consulting inquiries, please reach out to kyle@walker-data.com[1].

I live in Fort Worth, Texas, with my wife Molly and our three children, Michaela, Landry, and Clara, to whom this book is dedicated. While I enjoy developing open-source software, my true passions are exploring the country with my family and earnestly (though not always successfully) coaching my kids' sports teams.

Funding for this book comes in part from National Science Foundation grant BCS-1739662, "Demographic Inversion in US Metropolitan Areas".

---

[1] mailto:kyle@walker-data.com

1

The US Census and the R programming language

The main focus of this book is applied social data analysis in the R programming language, with a focus on data from the US Census Bureau. This chapter introduces both topics. The first part of the chapter covers the US Census and the US Census Bureau and gives an overview of how Census data can be accessed and used by analysts. The second part of the chapter is an introduction to the R programming language for readers who are new to R. This chapter wraps up with some examples of applied social data analysis projects that have used R and US Census data, setting the stage for the topics covered in the remainder of this book.

1.1 Census data: an overview

The US Constitution mandates in Article I, Sections 2 and 9 that a complete enumeration of the US population be taken every 10 years. The language from the Constitution is as follows (see https://www.census.gov/programs-surveys/decennial-census/about.html):

> The actual enumeration shall be made within three years after the first meeting of the Congress of the United States, and within every subsequent term of ten years, in such manner as they shall by law direct.

The government agency tasked with completing the enumeration of the US population is the US Census Bureau[1], part of the US Department of Commerce[2]. The first US Census was conducted in 1790, with enumerations taking place every 10 years since then. By convention, "Census day" is April 1 of the Census year.

The decennial US Census is intended to be a complete enumeration of the US population to assist with *apportionment*, which refers to the balanced arrangement of Congressional districts to ensure appropriate representation in the US House of Representatives. It asks a limited set of questions on race, ethnicity, age, sex, and housing tenure.

Before the 2010 decennial Census, one in six Americans also received the Census *long form*, which asked a wider range of demographic questions on income, education, language, housing,

---

[1] https://www.census.gov/en.html
[2] https://www.commerce.gov/

DOI: 10.1201/9780203711415-1

and more. The Census long form has since been replaced by the **American Community Survey**, which is now the premier source of detailed demographic information about the US population. The ACS is mailed to approximately 3.5 million households per year (representing around 3 percent of the US population), allowing for annual data updates. The Census Bureau releases two ACS datasets to the public: the **1-year ACS**, which covers areas of population 65,000 and greater, and the **5-year ACS**, which is a moving average of data over a 5-year period that covers geographies down to the Census block group. ACS data are distinct from decennial Census data in that data represent *estimates* rather than precise counts, and in turn are characterized by *margins of error* around those estimates. This topic is covered in more depth in Section 3.5. Due to data collection problems resulting from the COVID-19 pandemic, 2020 1-year ACS data were not released, replaced by experimental estimates for that year. The regular 1-year ACS resumed in 2021.

While the decennial US Census and American Community Survey are the most popular and widely used datasets produced by the US Census Bureau, the Bureau conducts hundreds of other surveys and disseminates data on a wide range of subjects to the public. These datasets include economic and business surveys, housing surveys, international data, population estimates and projections, and much more; a full listing is available on the Census website[3].

1.2 Census hierarchies

Aggregate data from the decennial US Census, American Community Survey, and other Census surveys are made available to the public at different *enumeration units*. Enumeration units are geographies at which Census data are tabulated. They include both *legal entities* such as states and counties and *statistical entities* that are not official jurisdictions but used to standardize data tabulation. The smallest unit at which data are made available from the decennial US Census is the *block,* and the smallest unit available in the ACS is the *block group*, which represents a collection of blocks. Other surveys are generally available at higher levels of aggregation.

Enumeration units represent different levels of the *Census hierarchy.* This hierarchy is summarized in Figure 1.1 (from https://www.census.gov/programs-surveys/geography/guidance/hierarchy.html).

The central axis of the diagram represents the central Census hierarchy of enumeration units, as each geography from Census blocks all the way up to the nation *nests* within its parent unit. This means that block groups are fully composed of Census blocks, Census tracts are fully composed of block groups, and so forth. An example of nesting is shown in Figure 1.2 for 2020 Census tracts in Benton County, Oregon.

The plot illustrates how Census tracts in Benton County neatly nest within its parent geography, the county. This means that the sum of Census data for Census tracts in Benton County will also equal Benton County's published total at the county level.

Reviewing the diagram shows that some Census geographies, like congressional districts, only nest within states, and that some other geographies do not nest within any parent

---

[3]https://www.census.gov/programs-surveys/surveys-programs.html

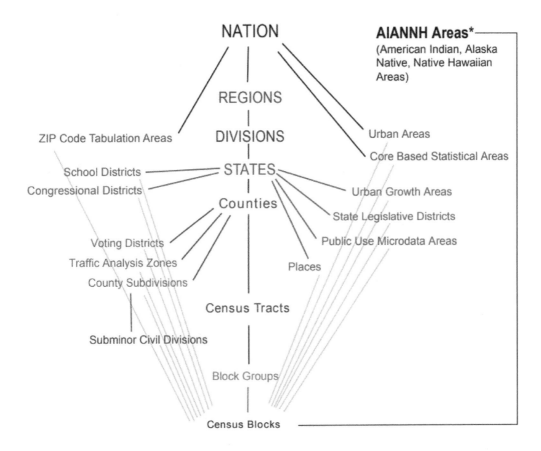

FIGURE 1.1 Census hierarchy of enumeration units

geography at all. A good example of this is the Zip Code Tabulation Area (ZCTA), which is used by the Census Bureau to represent postal code geographies in the United States. A more in-depth discussion of ZCTAs (and some of their pitfalls) is found in Section 6.4.2; a brief illustration is represented by Figure 1.3.

As the graphic illustrates, while some ZCTAs do fit entirely within Benton County, others overlap the county boundaries. In turn, "ZCTAs in Benton County" does not have the same meaning as "Census tracts in Benton County," as the former will extend into neighboring counties whereas the latter will not.

1.3 How to find US Census data

US Census data are available from a variety of sources, both directly from the Census Bureau and also from third-party distributors. This section gives an overview of some of these sources.

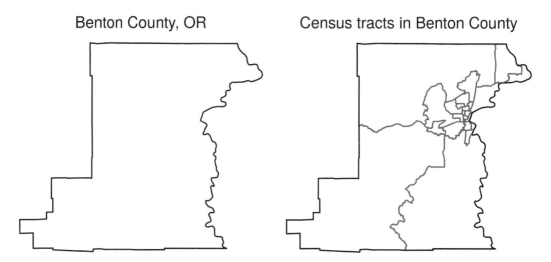

FIGURE 1.2 Benton County, OR Census tracts in relationship to the county boundary

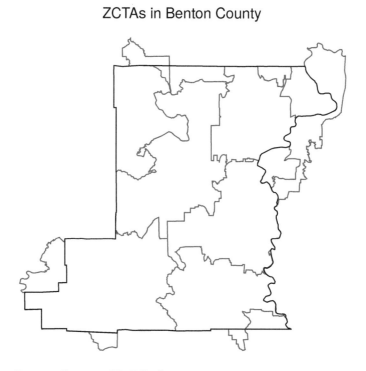

FIGURE 1.3 Benton County, OR ZCTAs in relationship to the county boundary

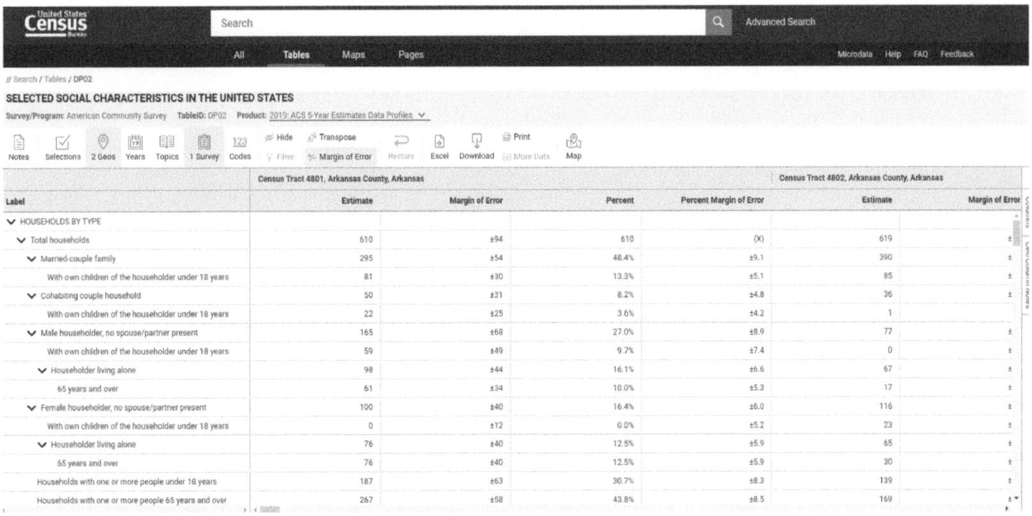

1.3 How to find US Census data

FIGURE 1.4 View of the data.census.gov interface

1.3.1 Data downloads from the Census Bureau

For years, researchers would visit the US Census Bureau's American FactFinder site[4] to download Census data for custom geographies. American FactFinder was decommissioned in 2020, giving way to a new data download interface, data.census.gov[5]. Users can interactively search several Census Bureau datasets (the decennial Census & ACS, along with the Economic Census, Population Estimates Program, and others), generate custom queries by geography, and download data extracts.

Figure 1.4 shows ACS table DP02 from the ACS Data Profile for Census tracts in Arkansas.

Users who are comfortable dealing with data in bulk and want to download the raw data will instead prefer the US Census Bureau's FTP site, `https://www2.census.gov/programs-surveys/`. This site includes a directory of Census surveys that can be navigated and downloaded. Figure 1.5 shows the directory structure for the 2019 American Community Survey[6], which is available by state or alternatively for the entire country.

National files are very large (the full 5-year file for all geographies is over 10GB of data zipped), so users will require dedicated software and computing workflows to interact with this data.

1.3.2 The Census API

The Obama Administration's Digital Government Strategy[7] prioritized widespread dissemination of government data resources to the public, highlighted by the data.gov[8] data portal. As part of this strategy, the Census Bureau released the Census

---

[4] https://www.census.gov/acs/www/data/data-tables-and-tools/american-factfinder/
[5] https://data.census.gov/cedsci/
[6] https://www2.census.gov/programs-surveys/acs/summary_file/2019/data/
[7] https://obamawhitehouse.archives.gov/digitalgov/apis
[8] https://www.data.gov/

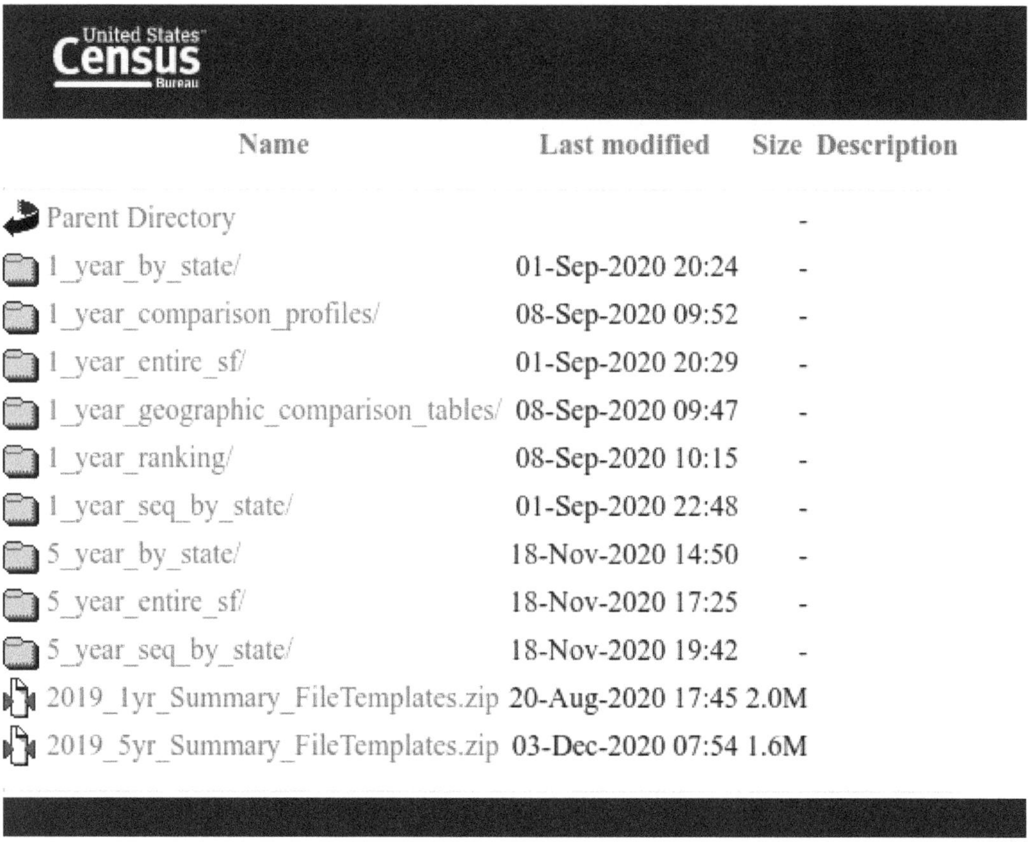

Name	Last modified	Size	Description
Parent Directory		-	
1_year_by_state/	01-Sep-2020 20:24	-	
1_year_comparison_profiles/	08-Sep-2020 09:52	-	
1_year_entire_sf/	01-Sep-2020 20:29	-	
1_year_geographic_comparison_tables/	08-Sep-2020 09:47	-	
1_year_ranking/	08-Sep-2020 10:15	-	
1_year_seq_by_state/	01-Sep-2020 22:48	-	
5_year_by_state/	18-Nov-2020 14:50	-	
5_year_entire_sf/	18-Nov-2020 17:25	-	
5_year_seq_by_state/	18-Nov-2020 19:42	-	
2019_1yr_Summary_FileTemplates.zip	20-Aug-2020 17:45	2.0M	
2019_5yr_Summary_FileTemplates.zip	03-Dec-2020 07:54	1.6M	

FIGURE 1.5 View of the Census FTP download site

Application Programming Interface, or API, in 2012[9]. This interface, available at https://www.census.gov/data/developers/data-sets.html, has grown to provide developers programmatic access to hundreds of data resources from the Census Bureau.

Census APIs are characterized by an *API endpoint*, which is a base web address for a given Census dataset, and a *query*, which customizes the data returned from the API. For example, the API endpoint for the 2010 Decennial US Census is https://api.census.gov/data/2010/dec/sf1; an example query that requests total population data for all counties in California, ?get=P001001,NAME&for=county:*&in=state:06, would be appended to the endpoint in a request to the Census API. The result of the query can be viewed here[10], returning data in JavaScript Object Notation (JSON) format as shown below.

```
[["P001001","NAME","state","county"],
["21419","Colusa County, California","06","011"],
["220000","Butte County, California","06","007"],
["1510271","Alameda County, California","06","001"],
["1175","Alpine County, California","06","003"],
```

---

[9]https://www.census.gov/newsroom/releases/archives/miscellaneous/cb12-135.html
[10]https://api.census.gov/data/2010/dec/sf1?get=P001001,NAME&for=county:*&in=state:06

```
["38091","Amador County, California","06","005"],
["45578","Calaveras County, California","06","009"],
["1049025","Contra Costa County, California","06","013"],
["28610","Del Norte County, California","06","015"],
["152982","Kings County, California","06","031"],
["28122","Glenn County, California","06","021"],
["134623","Humboldt County, California","06","023"],
["174528","Imperial County, California","06","025"],
["181058","El Dorado County, California","06","017"],
["930450","Fresno County, California","06","019"],
...
```

In most cases, users will interact with the Census API through *software libraries* that offer simplified programmatic access to the API's data resources. The example covered extensively in this book is the R package **tidycensus** (Walker and Herman, 2021), introduced in Chapter 2. Many other libraries exist for accessing the Census API; some of these resources are covered in Chapter 11, and readers will learn how to write their own data access functions in Section 11.4.2.

Users of the Census API through these software libraries will require a Census API key[11], which is free and fast to acquire. Getting an API key is covered in more detail in Section 2.1. Users may also want to join the US Census Bureau's Slack Community[12], where developers interact and answer each others' questions about using the API and associated software libraries.

1.3.3 Third-party data distributors

As the US Census Bureau provides data resources that are free and available for redistribution, several third-party data distributors have developed streamlined interfaces to use Census data. One of the most comprehensive of these resources is the University of Minnesota's National Historical Geographic Information System, or NHGIS[13] (Manson et al., 2021), which provides access to ACS data as well as decennial Census data back to 1790. NHGIS is covered in more depth in Section 11.1.1.

Two other recommended third-party Census data distributors are Census Reporter[14] and Social Explorer[15]. Census Reporter, a project based at Northwestern University's Knight Lab[16], is targeted toward journalists but offers a web interface that can help anyone explore tables available in the ACS and download ACS data. Social Explorer is a commercial product that offers both table-based and a map-based interface for exploring and visualizing Census data and makes mapping of Census data straightforward for users who aren't experienced with data analysis or mapping software.

Most readers of this book will want to learn how to use Census data resources to produce unique insights in their field of interest. This requires identifying a workflow to access and download custom data extracts relevant to their topics and study areas, then set up a software environment to help them wrangle, visualize, and model those data extracts. This

---

[11]https://api.census.gov/data/key_signup.html
[12]https://uscensusbureau.slack.com/
[13]https://www.nhgis.org/
[14]https://censusreporter.org/
[15]https://www.socialexplorer.com/
[16]https://knightlab.northwestern.edu/

1.4 What is R?

R (R Core Team, 2021) is one of the most popular programming languages and software environments for statistical computing and is the focus of this book with respect to software applications. This section introduces some basics of working with R and covers some terminology that will help readers work through the sections of this book. If you are an experienced R user, you can safely skip this section; however, readers new to R will find this information helpful before getting started with the applied examples in the book.

1.4.1 Getting started with R

To get started with R, visit the CRAN (Comprehensive R Archive Network) website at https://cloud.r-project.org/ and download the appropriate version of R for your operating system, then install the software. At the time of this writing, the most recent version of R is 4.1.1; it is a good idea to make sure you have the most recent version of R installed on your computer.

Once R is installed, I strongly recommend that you install **RStudio** (RStudio Team, 2021), the premier integrated development environment (IDE) for R. While you can run R without RStudio, RStudio offers a wide variety of utilities to make analysts' work with R easier and more streamlined. In fact, this entire book was written inside RStudio! RStudio can be installed from http://www.rstudio.com/download.

Once RStudio is installed, open it up and find the **Console** pane. This is an interactive console that allows you to type or copy-paste R commands and get results back.

1.4.2 Basic data structures in R

On a basic level, R can function as a calculator, computing everything from simple arithmetic to advanced math:

```
2 + 3
```

```
## [1] 5
```

Often, you will want to *assign* analytic results like this to an *object* (also commonly called a *variable*). Objects in R are created with an *assignment operator* (either <- or =) like this:

```
x <- 2 + 3
```

Above, we have *assigned* the result of the mathematical operation 2 + 3 to the object x:

```
x
```

```
## [1] 5
```

1.4 What is R?

Object names can be composed of any unquoted combination of letters and numbers so long as the first character is a letter. Our object, which stores the value 5, is characterized by a class:

```
class(x)
```

```
## [1] "numeric"
```

x is an object of class `numeric`, which is a general class indicating that we can perform mathematical operations on our object. Numeric objects can be contrasted with objects of class `"character"`, which represent character strings or textual information. Objects of class `"character"` are defined by either single- or double-quotes around a block of text.

```
y <- "census"
class(y)
```

```
## [1] "character"
```

There are *many* other classes of objects you'll encounter in this book; however, the distinction between objects of class `"numeric"` and `"character"` will come up frequently.

Data analysts will commonly encounter another class of object: the data frame and its derivatives (class `"data.frame"`). Data frames are rectangular objects characterized by *rows*, which generally represent individual observations, and *columns*, which represent characteristics or attributes common to those rows.

```
df <- data.frame(
  v1 = c(2, 5, 1, 7, 4),
  v2 = c(10, 2, 4, 2, 1),
  v3 = c("a", "b", "c", "d", "e")
)

df
```

```
##   v1 v2 v3
## 1  2 10  a
## 2  5  2  b
## 3  1  4  c
## 4  7  2  d
## 5  4  1  e
```

1.4.3 Functions and packages

The code that generates the data frame in the previous section uses two built-in *functions*: `data.frame()`, which creates columns from one or more *vectors* (defined as sequences of objects), and `c()`, which was used to create the vectors for the data frame. You can think of functions as "wrappers" that condense longer code-based workflows into simpler representations. R users can define their own functions as follows:

```
multiply <- function(x, y) {
  x * y
}
```

```
multiply(232, 7)
```

```
## [1] 1624
```

In this basic example, a function named `multiply()` is defined with `function`. `x` and `y` are *parameters*, which are locally-varying elements of the function. When the function is called, a user supplies *arguments*, which are passed to the parameters for some series of calculations. In this example, `x` takes on the value of 232, and `y` takes on the value of 7; the result is then returned by the `multiply()` function.

You can do quite a bit in R without ever having to write your own functions; however, you will almost certainly use functions written by others. In R, functions are generally available in *packages*, which are libraries of code designed to complete a related set of tasks. For example, the main focus of Chapter 2 is the **tidycensus** package, which includes functions to help users access Census data. Packages can be installed from CRAN with the `install.packages()` function:

```
install.packages("tidycensus")
```

Once installed, functions from a package can be loaded into a user's R environment with the `library()` command, e.g. `library(tidycensus)`. Alternatively, they can be used with the `package_name::function_name()` notation, e.g. `tidycensus::get_acs()`. Both notations are used at times in this book.

While "official" versions of R packages are usually published to CRAN and installable with `install.packages()`, more experimental or in-development R packages may be available on GitHub[17] instead. These packages should be installed with the `install_github()` function in the **remotes** package (Hester et al., 2021), referencing both the user name and the package name.

```
library(remotes)
install_github("Shelmith-Kariuki/rKenyaCensus")
```

While most packages used in this book are available on CRAN, some are only available on GitHub and should be installed accordingly.

1.4.4 Package ecosystems in R

R is *free and open source software (FOSS)*[18], which means that R is free to download and install, and its source code is open for anyone to view. This brings the substantial benefit of encouraging innovation from the user community, as anyone can create new packages and either submit them for publication to the official CRAN repository or host them on their personal GitHub page. In turn, new methodological innovations are often quickly accessible to the R user community. However, this can make R feel fragmented, especially for users

---

[17]https://github.com/
[18]https://en.wikipedia.org/wiki/Free_and_open_source_software

coming from commercial software designed to have a consistent interface. Package syntax will sometimes represent idiosyncratic choices of the developer, which can make R confusing to beginners.

The **tidyverse** ecosystem developed by RStudio (Wickham et al., 2019) is one of the most popular frameworks for data analysis in R and attempts to respond to problems introduced by package fragmentation. The tidyverse consists of a series of R packages designed to address common data analysis tasks (data wrangling, data reshaping, and data visualization, among many others) using a consistent syntax. Many R packages are now developed with integration within the tidyverse in mind. A good example of this is the **sf** package (Pebesma, 2018) which integrates spatial data analysis and the tidyverse. This book is largely written with the tidyverse and sf ecosystems in mind; **tidyverse** is covered in greater depth in Chapter 3, and **sf** is introduced in Chapter 5.

Other ecosystems exist that may be preferable to R users. R's core functionality, commonly termed "base R," consists of the original syntax of the language, which R users should get to know independent of their preferred analytic framework. Some R users prefer to maintain their analysis in base R as it does not require *dependencies*, meaning that it can run without installing external libraries. Another popular framework is **data.table** (Dowle and Srinivasan, 2021) and its associated packages, which extend base R's `data.frame` and allow for fast and high-performance data wrangling and analysis.

1.5 Analyses using R and US Census data

A large ecosystem of R packages exists to help analysts work with US Census data. A good summary of this ecosystem is found in Ari Lamstein and Logan Powell's *A Guide to Working with US Census Data in R* (Lamstein and Powell, 2018), and the ecosystem has grown further since their report was published. Below is a non-comprehensive summary of some R packages that will be of interest to readers; others will be covered throughout the book.

1.5.1 Census data packages in R: a brief summary

For users who prefer to work with the raw Census data files, the **totalcensus** package (Li, 2021) helps download Census data in bulk from the Census FTP server and loads it into R. For election analysts, the **PL94171** package (McCartan and Kenny, 2021) processes and loads PL-94171 redistricting files, including the most recent data from the 2020 Census.

Users who want to make more custom queries will be interested in R packages that interact with the Census APIs. The pioneering package in this area is the **acs** package (Glenn, 2019), which uses a custom interface and class system to return data extracts from various Census APIs. This package has informed a variety of other convenient Census data packages, such as **choroplethr** (Lamstein, 2020), which automates map production with data from the Census API. The **censusapi** package (Recht, 2021) also offers a comprehensive interface to the hundreds of datasets available from the Census Bureau via API. The examples in this book largely focus on the **tidycensus** and **tigris** packages created by the author, which interact with several Census API endpoints and return geographic data for mapping and spatial analysis.

R interfaces to third-party Census data resources have emerged as well. A good example is the **ipumsr** R package (Ellis and Burk, 2020), which helps users interact with datasets from the Minnesota Population Center like NHGIS. The aforementioned R packages – along with R's rich ecosystem for data analysis – have contributed to a wide range of projects using Census data in a variety of fields. A few such examples are highlighted below.

1.5.2 Health resource access

R and Census data have widespread applications in the analysis of health care and resource access. An excellent example is **censusapi** developer Hannah Recht's study of stroke care access in the Mississippi Delta and Appalachia, published in KHN in 2021 (Pattani et al., 2021). Figure 1.6 illustrates Recht's work integrating travel-time analytics with Census data to identify populations with limited access to stroke care across the US South.

Recht published her analysis code in a corresponding GitHub repository[19], allowing readers and developers to understand the methodology and reproduce the analysis. R packages used in the analysis include **censusapi** for Census data, **sf** for spatial analysis, and the **tidyverse** framework for data preparation and wrangling.

1.5.3 COVID-19 and pandemic response

R and Census data can also be used together to generate applications to the benefit of public health initiatives. A great example of this is the Texas COVID-19 Vaccine Tracker[20], developed by Matt Worthington at the University of Texas LBJ School of Public Affairs.

The application, illustrated in Figure 1.7, provides a comprehensive set of information about vaccine uptake and access around Texas. The example map shown visualizes vaccine doses administered per 1,000 persons in Austin-area ZCTAs. Census data are used throughout the application to help visitors understand differential vaccine update by regions and by demographics.

Source code for Worthington's application can be explored at its corresponding GitHub repository[21]. The site was built with a variety of R frameworks, including the Shiny framework for interactive dashboarding, and includes a range of static graphics, interactive graphics, and interactive maps generated from R.

1.5.4 Politics and gerrymandering

As discussed above, one of the primary purposes of the US Census is to determine congressional apportionment. This process requires the re-drawing of congressional districts every 10 years, termed *redistricting*. The redistricting process is politically fraught due to *gerrymandering*, which refers to the drawing of districts in a way that gives one political party (in the United States, Republican or Democratic) a built-in advantage over the other, potentially disenfranchising voters in the process.

Harvard University's ALARM project[22] uses R to analyze redistricting and gerrymandering and contribute to equitable solutions. The ALARM project has developed a veritable ecosystem of R packages that incorporate Census data and make Census data accessible to redistricting analysts. This includes the **PL94171** package mentioned above to get 2020 redistricting data, the **geomander** package to prepare data for redistricting analysis (Kenny

---

[19] https://github.com/khnews/2021-delta-appalachia-stroke-access
[20] https://texasvaccinetracker.com/
[21] https://github.com/utexas-lbjp-data/tx_vaccine_site
[22] https://alarm-redist.github.io/

1.5 Analyses using R and US Census data

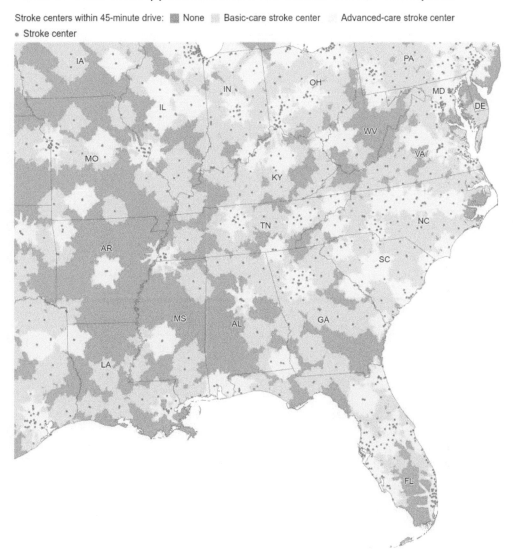

FIGURE 1.6 Accessibility to stroke care across the US South from Pattani et al. (2021). Image reprinted with permission from the publisher.

et al., 2021), and the **redist** package to algorithmically derive and evaluate redistricting solutions (Kenny et al., 2021). The example below, which is generated with modified code from the **redist** documentation, shows a basic example of potential redistricting solutions based on Census data for Iowa.

As court battles and contentious discussions around redistricting ramp up following the release of the 2020 Census redistricting files, R users can use ALARM's family of packages to analyze and produce potential solutions.

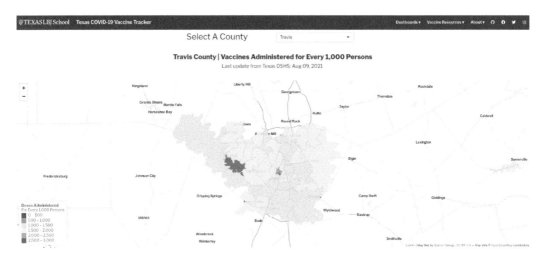

FIGURE 1.7 Screenshot of the Texas COVID-19 vaccine tracker website

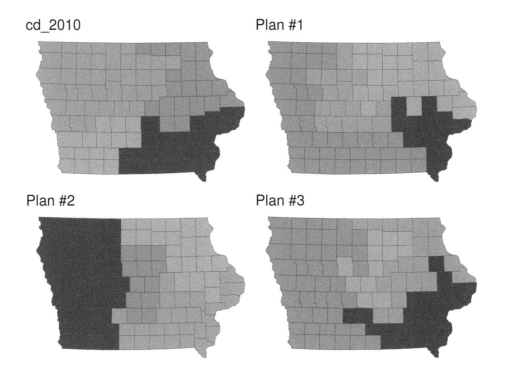

FIGURE 1.8 Example redistricting solutions using the redist package

1.5 Analyses using R and US Census data

FIGURE 1.9 Screenshot of the Mapping Immigrant America interactive map

1.5.5 Social equity research

Census data is a core resource for a large body of research in the social sciences as it speaks directly to issues of inequality and opportunity. Jerry Shannon's study of dollar store geography (Shannon, 2020) uses Census data with R in a compelling way for this purpose. His analysis examines the growth of dollar stores across the United States in relationship to patterns of racial segregation in the United States. Shannon finds that dollar stores are more likely to be found nearer to predominantly Black and Latino neighborhoods as opposed to predominantly white neighborhoods, even after controlling for the structural and economic characteristics of those neighborhoods. Shannon's analysis was completed in R, and his analysis code is available in the corresponding GitHub repository[23]. The demographic analysis uses American Community Survey data obtained from NHGIS.

1.5.6 Census data visualization

One of my favorite projects that I have worked on is *Mapping Immigrant America*, an interactive map of the US foreign-born population. The map scatters dots within Census tracts to proportionally represent the residences of immigrants based on data from the American Community Survey. The map is viewable at this link[24].

While version 1 of the map used a variety of tools to process the data including ArcGIS and QGIS, the data processing for version 2 was completed entirely in R, with data then uploaded to the Mapbox Studio[25] platform for hosting and visualization. The data preparation code can be viewed in the map's GitHub repository[26].

---

[23] https://github.com/jshannon75/metrodollars
[24] https://personal.tcu.edu/kylewalker/immigrant-america/#11.08/41.872/-87.7436
[25] https://studio.mapbox.com/
[26] https://github.com/walkerke/mb-immigrants

The map uses a *dasymetric dot-density* methodology (Walker, 2018) implemented in R using a series of techniques covered in this book. ACS data and their corresponding Census tract boundaries were acquired using tools learned in Chapter 2 and Section 6.1; areas with no population were removed from Census tracts using a spatial analysis technique covered in Section 7.5.1; and dots were generated for mapping using a method introduced in Section 6.3.4.3.

These examples are only a small sampling of the volumes of work completed by analysts using R and US Census Bureau data. In the next chapter, you will get started using R to access Census data for yourselves, with a focus on the **tidycensus** package.

2
An introduction to tidycensus

The **tidycensus** package (Walker and Herman, 2021), first released in 2017, is an R package designed to facilitate the process of acquiring and working with US Census Bureau population data in the R environment. The package has two distinct goals. First, **tidycensus** aims to make Census data available to R users in a **tidyverse**-friendly format, helping kick-start the process of generating insights from US Census data. Second, the package is designed to streamline the data wrangling process for spatial Census data analysts. With **tidycensus**, R users can request *geometry* along with attributes for their Census data, helping facilitate mapping and spatial analysis. This functionality of **tidycensus** is covered in more depth in Chapters 6, 7, and 8.

As discussed in the previous chapter, the US Census Bureau makes a wide range of datasets available to the user community through their APIs and other data download resources. **tidycensus** is not a comprehensive portal to these data resources; instead, it focuses on a select number of datasets implemented in a series of core functions. These core functions in **tidycensus** include:

- `get_decennial()`, which requests data from the US Decennial Census APIs for 2000, 2010, and 2020.
- `get_acs()`, which requests data from the 1-year and 5-year American Community Survey samples. Data are available from the 1-year ACS back to 2005 and the 5-year ACS back to 2005-2009.
- `get_estimates()`, an interface to the Population Estimates APIs. These datasets include yearly estimates of population characteristics by state, county, and metropolitan area, along with components of change demographic estimates like births, deaths, and migration rates.
- `get_pums()`, which accesses data from the ACS Public Use Microdata Sample APIs. These samples include anonymized individual-level records from the ACS organized by household and are highly useful for many different social science analyses. `get_pums()` is covered in more depth in Chapters 9 and 10.
- `get_flows()`, an interface to the ACS Migration Flows APIs. Includes information on in- and out-flows from various geographies for the 5-year ACS samples, enabling origin-destination analyses.

2.1 Getting started with tidycensus

To get started with **tidycensus**, users should install the package with `install.packages("tidycensus")` if not yet installed; load the package with

TABLE 2.1 Total population by state, 2010 Census

GEOID	NAME	variable	value
01	Alabama	P001001	4779736
02	Alaska	P001001	710231
04	Arizona	P001001	6392017
05	Arkansas	P001001	2915918
06	California	P001001	37253956
22	Louisiana	P001001	4533372
21	Kentucky	P001001	4339367
08	Colorado	P001001	5029196
09	Connecticut	P001001	3574097
10	Delaware	P001001	897934

library("tidycensus"); and set their Census API key with the census_api_key() function. API keys can be obtained at https://api.census.gov/data/key_signup.html. After you've signed up for an API key, be sure to activate the key from the email you receive from the Census Bureau so it works correctly. Declaring install = TRUE when calling census_api_key() will install the key for use in future R sessions, which may be convenient for many users.

```
library(tidycensus)
# census_api_key("YOUR KEY GOES HERE", install = TRUE)
```

2.1.1 Decennial Census

Once an API key is installed, users can obtain decennial Census or ACS data with a single function call. Let's start with get_decennial(), which is used to access decennial Census data from 2000, 2010, and 2020 decennial US Censuses.

To get data from the decennial US Census, users must specify a string representing the requested geography; a vector of Census variable IDs, represented by variable; or optionally a Census table ID, passed to table. The code below gets data on total population by state from the 2010 decennial Census.

```
total_population_10 <- get_decennial(
  geography = "state",
  variables = "P001001",
  year = 2010
)
```

The function returns a tibble of data from the 2010 US Census (the function default year) with information on total population by state, and assigns it to the object total_population_10. Data for 2000 or 2020 can also be obtained by supplying the appropriate year to the year parameter.

2.1.1.1 Summary files in the decennial Census

By default, get_decennial() uses the argument sumfile = "sf1", which fetches data from the decennial Census Summary File 1. This summary file exists for the 2000 and 2010 decennial

2.1 Getting started with tidycensus

TABLE 2.2 American Indian or Alaska Native alone population by state from the 2020 decennial Census

GEOID	NAME	variable	value
01	Alabama	P1_005N	33625
02	Alaska	P1_005N	111575
04	Arizona	P1_005N	319512
05	Arkansas	P1_005N	27177
06	California	P1_005N	631016
08	Colorado	P1_005N	74129

US Censuses, and includes core demographic characteristics for Census geographies. The 2000 and 2010 decennial Census data also include Summary File 2, which contains information on a range of population and housing unit characteristics and is specified as `"sf2"`. Detailed demographic information in the 2000 decennial Census such as income and occupation can be found in Summary Files 3 (`"sf3"`) and 4 (`"sf4"`). Data from the 2000 and 2010 Decennial Censuses for island territories other than Puerto Rico must be accessed at their corresponding summary files: `"as"` for American Samoa, `"mp"` for the Northern Mariana Islands, `"gu"` for Guam, and `"vi"` for the US Virgin Islands.

2020 Decennial Census data are available from the PL 94-171 Redistricting summary file, which is specified with `sumfile = "pl"` and is also available for 2010. The Redistricting summary files include a limited subset of variables from the decennial US Census to be used for legislative redistricting. These variables include total population and housing units, race and ethnicity, voting-age population, and group quarters population. For example, the code below retrieves information on the American Indian & Alaska Native population by state from the 2020 decennial Census.

```
aian_2020 <- get_decennial(
  geography = "state",
  variables = "P1_005N",
  year = 2020,
  sumfile = "pl"
)
```

The argument `sumfile = "pl"` is assumed (and in turn not required) when users request data for 2020 and will remain so until the main Demographic and Housing Characteristics File is released in 2023.

When users request data from the 2020 decennial Census for the first time in a given R session, `get_decennial()` prints out the following message:

```
Note: 2020 decennial Census data use differential privacy, a technique that
introduces errors into data to preserve respondent confidentiality.
? Small counts should be interpreted with caution.
? See https://www.census.gov/library/fact-sheets/2021/protecting-the-
confidentiality-of-the-2020-census-redistricting-data.html for additional guidance.
```

This message alerts users that 2020 decennial Census data use *differential privacy* as a method to preserve the confidentiality of individuals who responded to the Census. This can lead to inaccuracies in small area analyses using 2020 Census data and also can make

TABLE 2.3 Mexican-born population by state, 2016–2020 5-year ACS

GEOID	NAME	variable	estimate	moe
01	Alabama	B05006_150	46927	1846
02	Alaska	B05006_150	4181	709
04	Arizona	B05006_150	510639	8028
05	Arkansas	B05006_150	60236	2182
06	California	B05006_150	3962910	25353
08	Colorado	B05006_150	215778	4888
09	Connecticut	B05006_150	28086	2144
10	Delaware	B05006_150	14616	1065
11	District of Columbia	B05006_150	4026	761
12	Florida	B05006_150	257933	6418

comparisons of small counts across years difficult. A more in-depth discussion of differential privacy and the 2020 Census is found in the conclusion of this book.

2.1.2 American Community Survey

Similarly, `get_acs()` retrieves data from the American Community Survey. As discussed in the previous chapter, the ACS includes a wide variety of variables detailing characteristics of the US population not found in the decennial Census. The example below fetches data on the number of residents born in Mexico by state.

```
born_in_mexico <- get_acs(
  geography = "state",
  variables = "B05006_150",
  year = 2020
)
```

If the year is not specified, `get_acs()` defaults to the most recent 5-year ACS sample. The data returned is similar in structure to that returned by `get_decennial()`, but includes an `estimate` column (for the ACS estimate) and `moe` column (for the margin of error around that estimate) instead of a `value` column. Different years and different surveys are available by adjusting the `year` and `survey` parameters. `survey` defaults to the 5-year ACS; however, this can be changed to the 1-year ACS by using the argument `survey = "acs1"`. For example, the following code will fetch data from the 1-year ACS for 2019:

```
born_in_mexico_1yr <- get_acs(
  geography = "state",
  variables = "B05006_150",
  survey = "acs1",
  year = 2019
)
```

Note the differences between the 5-year ACS estimates and the 1-year ACS estimates shown. For states with larger Mexican-born populations like Arizona, California, and Colorado, the 1-year ACS data will represent the most up-to-date estimates, albeit characterized by

2.1 Getting started with tidycensus

TABLE 2.4 Mexican-born population by state, 2019 1-year ACS

GEOID	NAME	variable	estimate	moe
01	Alabama	B05006_150	NA	NA
02	Alaska	B05006_150	NA	NA
04	Arizona	B05006_150	516618	15863
05	Arkansas	B05006_150	NA	NA
06	California	B05006_150	3951224	40506
08	Colorado	B05006_150	209408	12214
09	Connecticut	B05006_150	26371	4816
10	Delaware	B05006_150	NA	NA
11	District of Columbia	B05006_150	NA	NA
12	Florida	B05006_150	261614	17571

larger margins of error relative to their estimates. For states with smaller Mexican-born populations like Alabama, Alaska, and Arkansas, however, the estimate returns NA, R's notation representing missing data. If you encounter this in your data's estimate column, it will generally mean that the estimate is too small for given geography to be deemed reliable by the Census Bureau. In this case, only the states with the largest Mexican-born populations have data available for that variable in the 1-year ACS, meaning that the 5-year ACS should be used to make full state-wise comparisons if desired.

If users try accessing data from the 2020 1-year ACS in **tidycensus**, they will encounter the following error:

```
Error: The regular 1-year ACS was not released in 2020 due to low response rates.
The Census Bureau released a set of experimental estimates for the 2020 1-year ACS
that are not available in tidycensus.
These estimates can be downloaded at https://www.census.gov/programs-
surveys/acs/data/experimental-data/1-year.html.
```

This means that for 1-year ACS data, **tidycensus** users will need to use older datasets (2019 and earlier) or access data for 2021 or later.

Variables from the ACS detailed tables, data profiles, summary tables, comparison profile, and supplemental estimates are available through **tidycensus**'s get_acs() function; the function will auto-detect from which dataset to look for variables based on their names. Alternatively, users can supply a table name to the table parameter in get_acs(); this will return data for every variable in that table. For example, to get all variables associated with table B01001, which covers sex broken down by age, from the 2016-2020 5-year ACS:

```
age_table <- get_acs(
  geography = "state",
  table = "B01001",
  year = 2020
)
```

To find all of the variables associated with a given ACS table, **tidycensus** downloads a dataset of variables from the Census Bureau website and looks up the variable codes for download. If the cache_table parameter is set to TRUE, the function instructs **tidycensus**

TABLE 2.5 Table B01001 by state from the 2016–2020 5-year ACS

GEOID	NAME	variable	estimate	moe
01	Alabama	B01001_001	4893186	NA
01	Alabama	B01001_002	2365734	1090
01	Alabama	B01001_003	149579	672
01	Alabama	B01001_004	150937	2202
01	Alabama	B01001_005	160287	2159
01	Alabama	B01001_006	96832	565
01	Alabama	B01001_007	65459	961
01	Alabama	B01001_008	36705	1467
01	Alabama	B01001_009	33089	1547
01	Alabama	B01001_010	93871	2045

to cache this dataset on the user's computer for faster future access. This only needs to be done once per ACS or Census dataset if the user would like to specify this option.

2.2 Geography and variables in tidycensus

The `geography` parameter in `get_acs()` and `get_decennial()` allows users to request data aggregated to common Census enumeration units. At the time of this writing, **tidycensus** accepts enumeration units nested within states and/or counties, when applicable. Census blocks are available in `get_decennial()` but not in `get_acs()` as block-level data are not available from the American Community Survey. To request data within states and/or counties, state and county names can be supplied to the `state` and `county` parameters, respectively. Arguments should be formatted in the way that they are accepted by the US Census Bureau API, specified in the table below. If an "Available by" geography is in bold, that argument is required for that geography.

The only geographies available in 2000 are `"state"`, `"county"`, `"county subdivision"`, `"tract"`, `"block group"`, and `"place"`. Some geographies available from the Census API are not available in tidycensus at the moment as they require more complex hierarchy specification than the package supports, and not all variables are available at every geography.

Geography	Definition	Available by	Available in
`"us"`	United States		`get_acs()`, `get_decennial()`, `get_estimates()`
`"region"`	Census region		`get_acs()`, `get_decennial()`, `get_estimates()`
`"division"`	Census division		`get_acs()`, `get_decennial()`, `get_estimates()`

2.2 Geography and variables in tidycensus

Geography	Definition	Available by	Available in
"state"	State or equivalent	state	get_acs(), get_decennial(), get_estimates(), get_flows()
"county"	County or equivalent	state, county	get_acs(), get_decennial(), get_estimates(), get_flows()
"county subdivision"	County subdivision	**state**, county	get_acs(), get_decennial(), get_estimates(), get_flows()
"tract"	Census tract	**state**, county	get_acs(), get_decennial()
"block group"	Census block group	**state**, county	get_acs() (2013-), get_decennial()
"block"	Census block	**state, county**	get_decennial()
"place"	Census-designated place	state	get_acs(), get_decennial(), get_estimates()
"alaska native regional corporation"	Alaska native regional corporation	state	get_acs(), get_decennial()
"american indian area/alaska native area/hawaiian home land"	Federal and state-recognized American Indian reservations and Hawaiian home lands	state	get_acs(), get_decennial()
"american indian area/alaska native area (reservation or statistical entity only)"	Only reservations and statistical entities	state	get_acs(), get_decennial()
"american indian area (off-reservation trust land only)/hawaiian home land"	Only off-reservation trust lands and Hawaiian home lands	state	get_acs(),
"metropolitan statistical area/micropolitan statistical area" OR "cbsa"	Core-based statistical area	state	get_acs(), get_decennial(), get_estimates(), get_flows()
"combined statistical area"	Combined statistical area	state	get_acs(), get_decennial(), get_estimates()
"new england city and town area"	New England city/town area	state	get_acs(), get_decennial()
"combined new england city and town area"	Combined New England area	state	get_acs(), get_decennial()
"urban area"	Census-defined urbanized areas		get_acs(), get_decennial()

Geography	Definition	Available by	Available in
`"congressional district"`	Congressional district for the year-appropriate Congress	state	`get_acs()`, `get_decennial()`
`"school district (elementary)"`	Elementary school district	**state**	`get_acs()`, `get_decennial()`
`"school district (secondary)"`	Secondary school district	**state**	`get_acs()`, `get_decennial()`
`"school district (unified)"`	Unified school district	**state**	`get_acs()`, `get_decennial()`
`"public use microdata area"`	PUMA (geography associated with Census microdata samples)	state	`get_acs()`
`"zip code tabulation area"` OR `"zcta"`	Zip code tabulation area	state	`get_acs()`, `get_decennial()`
`"state legislative district (upper chamber)"`	State senate districts	**state**	`get_acs()`, `get_decennial()`
`"state legislative district (lower chamber)"`	State house districts	**state**	`get_acs()`, `get_decennial()`
`"voting district"`	Voting districts (2020 only)	**state**	`get_decennial()`

The geography parameter must be typed exactly as in the table to request data correctly from the Census API; use the guide as a reference and copy-paste for longer strings. For core-based statistical areas and zip code tabulation areas, two heavily-requested geographies, the aliases `"cbsa"` and `"zcta"` can be used, respectively, to fetch data for those geographies.

```
cbsa_population <- get_acs(
  geography = "cbsa",
  variables = "B01003_001",
  year = 2020
)
```

2.2.1 Geographic subsets

For many geographies, **tidycensus** supports more granular requests that are subsetted by state or even by county, if supported by the API. This information is found in the "Available by" column in the guide above. If a geographic subset is in bold, it is required; if not, it is optional.

For example, an analyst might be interested in studying variations in household income in the state of Wisconsin. Although the analyst *can* request all counties in the United States, this is not necessary for this specific task. In turn, they can use the `state` parameter to subset the request for a specific state.

2.2 Geography and variables in tidycensus

TABLE 2.6 Population by CBSA

GEOID	NAME	variable	estimate	moe
10100	Aberdeen, SD Micro Area	B01003_001	42864	NA
10140	Aberdeen, WA Micro Area	B01003_001	73769	NA
10180	Abilene, TX Metro Area	B01003_001	171354	NA
10220	Ada, OK Micro Area	B01003_001	38385	NA
10300	Adrian, MI Micro Area	B01003_001	98310	NA
10380	Aguadilla-Isabela, PR Metro Area	B01003_001	295172	NA
10420	Akron, OH Metro Area	B01003_001	703286	NA
10460	Alamogordo, NM Micro Area	B01003_001	66804	NA
10500	Albany, GA Metro Area	B01003_001	147431	NA
10540	Albany-Lebanon, OR Metro Area	B01003_001	127216	NA

TABLE 2.7 Median household income by county in Wisconsin

GEOID	NAME	variable	estimate	moe
55001	Adams County, Wisconsin	B19013_001	48906	2387
55003	Ashland County, Wisconsin	B19013_001	47869	3190
55005	Barron County, Wisconsin	B19013_001	52346	2092
55007	Bayfield County, Wisconsin	B19013_001	57257	2496
55009	Brown County, Wisconsin	B19013_001	64728	1419
55011	Buffalo County, Wisconsin	B19013_001	58364	1871
55013	Burnett County, Wisconsin	B19013_001	53555	2513
55015	Calumet County, Wisconsin	B19013_001	76065	2314
55017	Chippewa County, Wisconsin	B19013_001	61215	2064
55019	Clark County, Wisconsin	B19013_001	54463	1089

```r
wi_income <- get_acs(
  geography = "county",
  variables = "B19013_001",
  state = "WI",
  year = 2020
)
```

tidycensus accepts state names (e.g. `"Wisconsin"`), state postal codes (e.g. `"WI"`), and state FIPS codes (e.g. `"55"`), so an analyst can use what they are most comfortable with.

Smaller geographies like Census tracts can also be subsetted by county. Given that Census tracts nest neatly within counties (and do not cross county boundaries), we can request all Census tracts for a given county by using the optional `county` parameter. Dane County, home to Wisconsin's capital city of Madison, is shown below. Note that the name of the county can be supplied as well as the FIPS code. If a state has two counties with similar names (e.g. "Collin" and "Collingsworth" in Texas), you'll need to spell out the full county string and type `"Collin County"`.

```r
dane_income <- get_acs(
  geography = "tract",
```

TABLE 2.8 Median household income in Dane County by Census tract

GEOID	NAME	variable	estimate	moe
55025000100	Census Tract 1, Dane County, Wisconsin	B19013_001	74054	15662
55025000201	Census Tract 2.01, Dane County, Wisconsin	B19013_001	92460	27067
55025000202	Census Tract 2.02, Dane County, Wisconsin	B19013_001	88092	5189
55025000204	Census Tract 2.04, Dane County, Wisconsin	B19013_001	82717	12175
55025000205	Census Tract 2.05, Dane County, Wisconsin	B19013_001	100000	17506
55025000301	Census Tract 3.01, Dane County, Wisconsin	B19013_001	37016	11524
55025000302	Census Tract 3.02, Dane County, Wisconsin	B19013_001	117321	28723
55025000401	Census Tract 4.01, Dane County, Wisconsin	B19013_001	100434	12108
55025000402	Census Tract 4.02, Dane County, Wisconsin	B19013_001	105850	12205
55025000406	Census Tract 4.06, Dane County, Wisconsin	B19013_001	74009	2811

```
  variables = "B19013_001",
  state = "WI",
  county = "Dane",
  year = 2020
)
```

With respect to geography and the American Community Survey, users should be aware that whereas the 5-year ACS covers geographies down to the block group, the 1-year ACS only returns data for geographies of population 65,000 and greater. This means that some geographies (e.g. Census tracts) will never be available in the 1-year ACS, and that other geographies such as counties are only partially available. To illustrate this, we can check the number of rows in the object wi_income:

```
nrow(wi_income)
```

```
## [1] 72
```

There are 72 rows in this dataset, one for each county in Wisconsin. However, if the same data were requested from the 2019 1-year ACS:

```
wi_income_1yr <- get_acs(
  geography = "county",
  variables = "B19013_001",
  state = "WI",
  year = 2019,
  survey = "acs1"
)
```

```
nrow(wi_income_1yr)
```

```
## [1] 23
```

There are only 23 rows in this dataset, representing the 23 counties that meet the "total population of 65,000 or greater" threshold required to be included in the 1-year ACS data.

2.3 Searching for variables in tidycensus

One additional challenge when searching for Census variables is understanding variable IDs, which are required to fetch data from the Census and ACS APIs. There are thousands of variables available across the different datasets and summary files. To make searching easier for R users, **tidycensus** offers the `load_variables()` function. This function obtains a dataset of variables from the Census Bureau website and formats it for fast searching, ideally in RStudio.

The function takes two required arguments: `year`, which takes the year or end-year of the Census dataset or ACS sample, and `dataset`, which references the dataset name. For the 2000 or 2010 Decennial Census, use `"sf1"` or `"sf2"` as the dataset name to access variables from Summary Files 1 and 2, respectively. The 2000 Decennial Census also accepts `"sf3"` and `"sf4"` for Summary Files 3 and 4. For 2020, the only dataset supported at the time of this writing is `"pl"` for the PL-94171 Redistricting dataset; more datasets will be supported as the 2020 Census data are released. An example request would look like `load_variables(year = 2020, dataset = "pl")` for variables from the 2020 Decennial Census Redistricting data.

For variables from the American Community Survey, users should specify the dataset as `"acs1"` for the 1-year ACS or `"acs5"` for the 5-year ACS. If no suffix to these dataset names is specified, users will retrieve data from the ACS Detailed Tables. Variables from the ACS Data Profile, Summary Tables, and Comparison Profile are also available by appending the suffixes `/profile`, `/summary`, or `/cprofile`, respectively. For example, a user requesting variables from the 2020 5-year ACS Detailed Tables would specify `load_variables(year = 2020, dataset = "acs5")`; a request for variables from the Data Profile then would be `load_variables(year = 2020, dataset = "acs5/profile")`. In addition to these datasets, the ACS Supplemental Estimates variables can be accessed with the dataset name `"acsse"`.

As this function requires processing thousands of variables from the Census Bureau, which may take a few moments depending on the user's internet connection, the user can specify `cache = TRUE` in the function call to store the data in the user's cache directory for future access. On subsequent calls of the `load_variables()` function, `cache = TRUE` will direct the function to look in the cache directory for the variables rather than the Census website.

An example of how `load_variables()` works is as follows:

```
v16 <- load_variables(2016, "acs5")
```

The returned data frame always has three columns: `name`, which refers to the Census variable ID; `label`, which is a descriptive data label for the variable; and `concept`, which refers to the topic of the data and often corresponds to a table of Census data. For the 5-year ACS detailed tables, the returned data frame also includes a fourth column, `geography`, which specifies the smallest geography at which a given variable is available from the Census API. As illustrated above, the data frame can be filtered using tidyverse tools for variable exploration. However, the RStudio integrated development environment includes an interactive data viewer, which is ideal for browsing this dataset and allows for interactive sorting and filtering. The data viewer can be accessed with the `View()` function:

```
View(v16)
```

TABLE 2.9 Variables in the 2012–2016 5-year ACS

name	label	concept	geography
B01001_001	Estimate!!Total	SEX BY AGE	block group
B01001_002	Estimate!!Total!!Male	SEX BY AGE	block group
B01001_003	Estimate!!Total!!Male!!Under 5 years	SEX BY AGE	block group
B01001_004	Estimate!!Total!!Male!!5 to 9 years	SEX BY AGE	block group
B01001_005	Estimate!!Total!!Male!!10 to 14 years	SEX BY AGE	block group
B01001_006	Estimate!!Total!!Male!!15 to 17 years	SEX BY AGE	block group
B01001_007	Estimate!!Total!!Male!!18 and 19 years	SEX BY AGE	block group
B01001_008	Estimate!!Total!!Male!!20 years	SEX BY AGE	block group
B01001_009	Estimate!!Total!!Male!!21 years	SEX BY AGE	block group
B01001_010	Estimate!!Total!!Male!!22 to 24 years	SEX BY AGE	block group

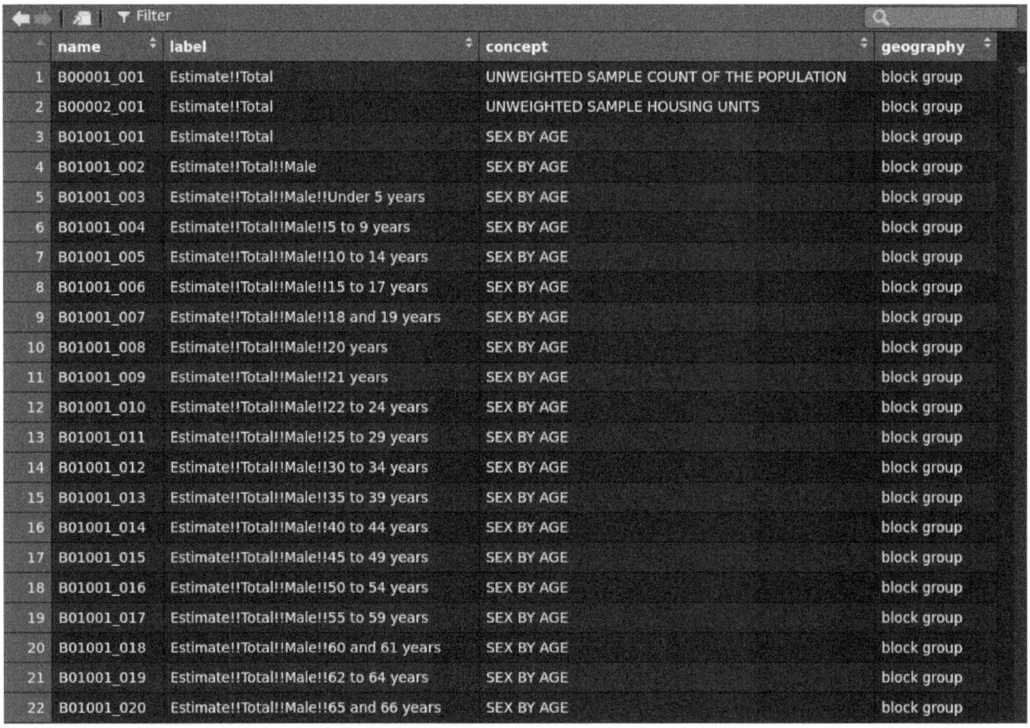

Filter

	name	label	concept	geography
1	B00001_001	Estimate!!Total	UNWEIGHTED SAMPLE COUNT OF THE POPULATION	block group
2	B00002_001	Estimate!!Total	UNWEIGHTED SAMPLE HOUSING UNITS	block group
3	B01001_001	Estimate!!Total	SEX BY AGE	block group
4	B01001_002	Estimate!!Total!!Male	SEX BY AGE	block group
5	B01001_003	Estimate!!Total!!Male!!Under 5 years	SEX BY AGE	block group
6	B01001_004	Estimate!!Total!!Male!!5 to 9 years	SEX BY AGE	block group
7	B01001_005	Estimate!!Total!!Male!!10 to 14 years	SEX BY AGE	block group
8	B01001_006	Estimate!!Total!!Male!!15 to 17 years	SEX BY AGE	block group
9	B01001_007	Estimate!!Total!!Male!!18 and 19 years	SEX BY AGE	block group
10	B01001_008	Estimate!!Total!!Male!!20 years	SEX BY AGE	block group
11	B01001_009	Estimate!!Total!!Male!!21 years	SEX BY AGE	block group
12	B01001_010	Estimate!!Total!!Male!!22 to 24 years	SEX BY AGE	block group
13	B01001_011	Estimate!!Total!!Male!!25 to 29 years	SEX BY AGE	block group
14	B01001_012	Estimate!!Total!!Male!!30 to 34 years	SEX BY AGE	block group
15	B01001_013	Estimate!!Total!!Male!!35 to 39 years	SEX BY AGE	block group
16	B01001_014	Estimate!!Total!!Male!!40 to 44 years	SEX BY AGE	block group
17	B01001_015	Estimate!!Total!!Male!!45 to 49 years	SEX BY AGE	block group
18	B01001_016	Estimate!!Total!!Male!!50 to 54 years	SEX BY AGE	block group
19	B01001_017	Estimate!!Total!!Male!!55 to 59 years	SEX BY AGE	block group
20	B01001_018	Estimate!!Total!!Male!!60 and 61 years	SEX BY AGE	block group
21	B01001_019	Estimate!!Total!!Male!!62 to 64 years	SEX BY AGE	block group
22	B01001_020	Estimate!!Total!!Male!!65 and 66 years	SEX BY AGE	block group

FIGURE 2.1 Variable viewer in RStudio

By browsing the table in this way, users can identify the appropriate variable IDs (found in the name column) that can be passed to the variables parameter in get_acs() or get_decennial(). Users may note that the raw variable IDs in the ACS, as consumed by the API, require a suffix of E or M. **tidycensus** does not require this suffix, as it will automatically return both the estimate and margin of error for a given requested variable. Additionally, if users desire an entire table of related variables from the ACS, the user should supply the characters prior to the underscore from a variable ID to the table parameter.

2.4 Data structure in tidycensus

Key to the design philosophy of **tidycensus** is its interpretation of tidy data. Following Wickham (2014), "tidy" data are defined as follows:

1. Each observation forms a row;
2. Each variable forms a column;
3. Each observational unit forms a table.

By default, **tidycensus** returns a tibble of ACS or decennial Census data in "tidy" format. For decennial Census data, this will include four columns:

- GEOID, representing the Census ID code that uniquely identifies the geographic unit;
- NAME, which represents a descriptive name of the unit;
- variable, which contains information on the Census variable name corresponding to that row;
- value, which contains the data values for each unit-variable combination. For ACS data, two columns replace the value column: estimate, which represents the ACS estimate, and moe, representing the margin of error around that estimate.

Given the terminology used by the Census Bureau to distinguish data, it is important to provide some clarifications of nomenclature here. Census or ACS *variables*, which are specific series of data available by enumeration unit, are interpreted in tidycensus as *characteristics* of those enumeration units. In turn, rows in datasets returned when output = "tidy", which is the default setting in the get_acs() and get_decennial() functions, represent data for unique unit-variable combinations. An example of this is illustrated below with income groups by state for the 2016 1-year American Community Survey.

```
hhinc <- get_acs(
  geography = "state",
  table = "B19001",
  survey = "acs1",
  year = 2016
)
```

In this example, each row represents state-characteristic combinations, consistent with the tidy data model. Alternatively, if a user desires the variables spread across the columns of the dataset, the setting output = "wide" will enable this. For ACS data, estimates and margins of error for each ACS variable will be found in their own columns. For example:

TABLE 2.10 Household income groups by state, 2016 1-year ACS

GEOID	NAME	variable	estimate	moe
01	Alabama	B19001_001	1852518	12189
01	Alabama	B19001_002	176641	6328
01	Alabama	B19001_003	120590	5347
01	Alabama	B19001_004	117332	5956
01	Alabama	B19001_005	108912	5308
01	Alabama	B19001_006	102080	4740
01	Alabama	B19001_007	103366	5246
01	Alabama	B19001_008	91011	4699
01	Alabama	B19001_009	86996	4418
01	Alabama	B19001_010	74864	4210

TABLE 2.11 Income table in wide form

GEOID	NAME	B19001_001E	B19001_001M	B19001_002E	B19001_002M	B19001_003E	B19001_003M
28	Mississippi	1091245	8803	113124	4835	87136	5004
29	Missouri	2372190	10844	160615	6705	122649	4654
30	Montana	416125	4426	26734	2183	24786	2391
31	Nebraska	747562	4452	45794	3116	33266	2466
32	Nevada	1055158	6433	68507	4886	42720	3071
33	New Hampshire	520643	5191	20890	2566	15933	1908
34	New Jersey	3194519	10274	170029	6836	118862	5855
35	New Mexico	758364	6296	66983	4439	48930	3220
36	New York	7209054	17665	543763	12132	352029	9607
37	North Carolina	3882423	16063	282491	7816	228088	7916

```
hhinc_wide <- get_acs(
  geography = "state",
  table = "B19001",
  survey = "acs1",
  year = 2016,
  output = "wide"
)
```

The wide-form dataset includes `GEOID` and `NAME` columns, as in the tidy dataset, but is also characterized by estimate/margin of error pairs across the columns for each Census variable in the table.

2.4.1 Understanding GEOIDs

The `GEOID` column returned by default in **tidycensus** can be used to uniquely identify geographic units in a given dataset. For geographies within the core Census hierarchy (Census block through state, as discussed in Section 1.2), GEOIDs can be used to uniquely identify specific units as well as units' parent geographies. Let's take the example of households by Census block from the 2020 Census in Cimarron County, Oklahoma.

2.4 Data structure in tidycensus

TABLE 2.12 Households by block in Cimarron County, Oklahoma

GEOID	NAME
400259503001110	Block 1110, Block Group 1, Census Tract 9503, Cimarron County, Oklahoma
400259503001157	Block 1157, Block Group 1, Census Tract 9503, Cimarron County, Oklahoma
400259501001837	Block 1837, Block Group 1, Census Tract 9501, Cimarron County, Oklahoma
400259501001751	Block 1751, Block Group 1, Census Tract 9501, Cimarron County, Oklahoma
400259501001906	Block 1906, Block Group 1, Census Tract 9501, Cimarron County, Oklahoma
400259501001996	Block 1996, Block Group 1, Census Tract 9501, Cimarron County, Oklahoma
400259503001208	Block 1208, Block Group 1, Census Tract 9503, Cimarron County, Oklahoma
400259501001888	Block 1888, Block Group 1, Census Tract 9501, Cimarron County, Oklahoma
400259503001047	Block 1047, Block Group 1, Census Tract 9503, Cimarron County, Oklahoma
400259501001313	Block 1313, Block Group 1, Census Tract 9501, Cimarron County, Oklahoma

```
cimarron_blocks <- get_decennial(
  geography = "block",
  variables = "H1_001N",
  state = "OK",
  county = "Cimarron",
  year = 2020,
  sumfile = "pl"
)
```

The mapping between the `GEOID` and `NAME` columns in the returned 2020 Census block data offers some insight into how GEOIDs work for geographies within the core Census hierarchy. Take the first block in the table, Block 1110, which has a GEOID of **400259503001110**. The GEOID value breaks down as follows:

- The first two digits, **40**, correspond to the Federal Information Processing Series (FIPS) code[1] for the state of Oklahoma. All states and US territories, along with other geographies at which the Census Bureau tabulates data, will have a FIPS code that can uniquely identify that geography.

- Digits 3 through 5, **025**, are representative of Cimarron County. These three digits will uniquely identify Cimarron County within Oklahoma. County codes are generally combined with their corresponding state codes to uniquely identify a county within the United States, as three-digit codes will be repeated across states. Cimarron County's code in this example would be **40025**.

- The next six digits, **950300**, represent the block's Census tract. The tract name in the `NAME` column is Census Tract 9503; the six-digit tract ID is right-padded with zeroes.

- The twelfth digit, **1**, represents the parent block group of the Census block. As there are no more than nine block groups in any Census tract, the block group name will not exceed 9.

- The last three digits, **110**, represent the individual Census block, though these digits are combined with the parent block group digit to form the block's name.

For geographies outside the core Census hierarchy, GEOIDs will uniquely identify geographic units but will only include IDs of parent geographies to the degree to which they nest within

---

[1] https://www.census.gov/library/reference/code-lists/ansi.html

TABLE 2.13 Multi-variable dataset for Georgia counties

GEOID	NAME	variable	estimate	moe
13001	Appling County, Georgia	medage	39.9	1.7
13001	Appling County, Georgia	medinc	37924.0	4761.0
13003	Atkinson County, Georgia	medage	35.9	1.5
13003	Atkinson County, Georgia	medinc	35703.0	5493.0
13005	Bacon County, Georgia	medage	36.5	1.0
13005	Bacon County, Georgia	medinc	36692.0	3774.0
13007	Baker County, Georgia	medage	52.2	4.8
13007	Baker County, Georgia	medinc	34034.0	9879.0
13009	Baldwin County, Georgia	medage	35.8	0.5
13009	Baldwin County, Georgia	medinc	46250.0	4707.0

them. For example, a geography that nests within states but may cross county boundaries like school districts will include the state GEOID in its GEOID but unique digits after that. Geographies like core-based statistical areas that do not nest within states will have fully unique GEOIDs, independent of any other geographic level of aggregation such as states.

2.4.2 Renaming variable IDs

Census variables IDs can be cumbersome to type and remember in the course of an R session. As such, **tidycensus** has built-in tools to automatically rename the variable IDs if requested by a user. For example, let's say that a user is requesting data on median household income (variable ID `B19013_001`) and median age (variable ID `B01002_001`). By passing a *named* vector to the `variables` parameter in `get_acs()` or `get_decennial()`, the functions will return the desired names rather than the Census variable IDs. Let's examine this for counties in Georgia from the 2016–2020 5-year ACS.

```
ga <- get_acs(
  geography = "county",
  state = "Georgia",
  variables = c(medinc = "B19013_001",
                medage = "B01002_001"),
  year = 2020
)
```

ACS variable IDs, which would be found in the `variable` column, are replaced by `medage` and `medinc`, as requested. When a wide-form dataset is requested, **tidycensus** will still append `E` and `M` to the specified column names, as illustrated below.

```
ga_wide <- get_acs(
  geography = "county",
  state = "Georgia",
  variables = c(medinc = "B19013_001",
                medage = "B01002_001"),
  output = "wide",
  year = 2020
)
```

2.5 Other Census Bureau datasets in tidycensus

TABLE 2.14 Georgia dataset in wide form

GEOID	NAME	medincE	medincM	medageE	medageM
13001	Appling County, Georgia	37924	4761	39.9	1.7
13003	Atkinson County, Georgia	35703	5493	35.9	1.5
13005	Bacon County, Georgia	36692	3774	36.5	1.0
13007	Baker County, Georgia	34034	9879	52.2	4.8
13011	Banks County, Georgia	50912	4278	41.5	1.1
13013	Barrow County, Georgia	62990	2562	36.0	0.3
13017	Ben Hill County, Georgia	32077	4008	39.5	1.4
13021	Bibb County, Georgia	41317	1220	36.3	0.3
13023	Bleckley County, Georgia	46992	6279	36.0	1.5
13027	Brooks County, Georgia	37516	4438	43.6	0.9

Median household income for each county is represented by `medincE`, for the estimate, and `medincM`, for the margin of error. At the time of this writing, custom variable names are only available for `variables` and not for `table`, as users will not always know the number of variables found in a table beforehand.

2.5 Other Census Bureau datasets in tidycensus

As mentioned earlier in this chapter, **tidycensus** does not grant access to all of the datasets available from the Census API; users should look at the **censusapi** package (Recht, 2021) for that functionality. However, the Population Estimates and ACS Migration Flows APIs are accessible with the `get_estimates()` and `get_flows()` functions, respectively. This section includes brief examples of each.

2.5.1 Using `get_estimates()`

The Population Estimates Program[2], or PEP, provides yearly estimates of the US population and its components between decennial Censuses. It differs from the ACS in that it is not directly based on a dedicated survey but rather projects forward data from the most recent decennial Census based on birth, death, and migration rates. In turn, estimates in the PEP will differ slightly from what you may see in data returned by `get_acs()`, as the estimates are produced using a different methodology.

One advantage of using the PEP to retrieve data is that it allows you to access the indicators used to produce the intercensal population estimates. These indicators can be specified as variables direction in the `get_estimates()` function in **tidycensus**, or requested in bulk by using the `product` argument. The products available include `"population"`, `"components"`, `"housing"`, and `"characteristics"`. For example, we can request all components of change population estimates for 2019 for a specific county:

```
library(tidycensus)
library(tidyverse)
```

---

[2] https://www.census.gov/programs-surveys/popest.html

TABLE 2.15 Components of change estimates for Queens County, NY

NAME	GEOID	variable	value
Queens County, New York	36081	BIRTHS	27453.000000
Queens County, New York	36081	DEATHS	16380.000000
Queens County, New York	36081	DOMESTICMIG	−41789.000000
Queens County, New York	36081	INTERNATIONALMIG	9883.000000
Queens County, New York	36081	NATURALINC	11073.000000
Queens County, New York	36081	NETMIG	−31906.000000
Queens County, New York	36081	RBIRTH	12.124644
Queens County, New York	36081	RDEATH	7.234243
Queens County, New York	36081	RDOMESTICMIG	−18.456152
Queens County, New York	36081	RINTERNATIONALMIG	4.364836

```
queens_components <- get_estimates(
  geography = "county",
  product = "components",
  state = "NY",
  county = "Queens",
  year = 2019
)
```

The returned variables include raw values for births and deaths (`BIRTHS` and `DEATHS`) during the previous 12 months, defined as mid-year 2018 (July 1) to mid-year 2019. Crude rates per 1000 people in Queens County are also available with `RBIRTH` and `RDEATH`. `NATURALINC`, the natural increase, then measures the number of births minus the number of deaths. Net domestic and international migration are also available as counts and rates, and the `NETMIG` variable accounts for the overall migration, domestic and international included. Alternatively, a single variable or vector of variables can be requested with the `variable` argument, and the `output = "wide"` argument can also be used to spread the variable names across the columns.

The `product = "characteristics"` argument also has some unique options. The argument `breakdown` lets users get breakdowns of population estimates for the US, states, and counties by `"AGEGROUP"`, `"RACE"`, `"SEX"`, or `"HISP"` (Hispanic origin). If set to `TRUE`, the `breakdown_labels` argument will return informative labels for the population estimates. For example, to get population estimates by sex and Hispanic origin for metropolitan areas, we can use the following code:

```
louisiana_sex_hisp <- get_estimates(
  geography = "state",
  product = "characteristics",
  breakdown = c("SEX", "HISP"),
  breakdown_labels = TRUE,
  state = "LA",
  year = 2019
)
```

2.6 Other Census Bureau datasets in tidycensus

TABLE 2.16 Population characteristics for Louisiana

GEOID	NAME	value	SEX	HISP
22	Louisiana	4648794	Both sexes	Both Hispanic Origins
22	Louisiana	4401822	Both sexes	Non-Hispanic
22	Louisiana	246972	Both sexes	Hispanic
22	Louisiana	2267050	Male	Both Hispanic Origins
22	Louisiana	2135979	Male	Non-Hispanic
22	Louisiana	131071	Male	Hispanic
22	Louisiana	2381744	Female	Both Hispanic Origins
22	Louisiana	2265843	Female	Non-Hispanic
22	Louisiana	115901	Female	Hispanic

TABLE 2.17 Migration flows data for Honolulu, HI

GEOID1	GEOID2	FULL1_NAME	FULL2_NAME	variable	estimate	moe
15003	NA	Honolulu County, Hawaii	Africa	MOVEDIN	152	156
15003	NA	Honolulu County, Hawaii	Africa	MOVEDOUT	NA	NA
15003	NA	Honolulu County, Hawaii	Africa	MOVEDNET	NA	NA
15003	NA	Honolulu County, Hawaii	Asia	MOVEDIN	7680	884
15003	NA	Honolulu County, Hawaii	Asia	MOVEDOUT	NA	NA
15003	NA	Honolulu County, Hawaii	Asia	MOVEDNET	NA	NA
15003	NA	Honolulu County, Hawaii	Central America	MOVEDIN	192	100
15003	NA	Honolulu County, Hawaii	Central America	MOVEDOUT	NA	NA
15003	NA	Honolulu County, Hawaii	Central America	MOVEDNET	NA	NA
15003	NA	Honolulu County, Hawaii	Caribbean	MOVEDIN	97	78

The `value` column gives the estimate characterized by the population labels in the `SEX` and `HISP` columns. For example, the estimated population value in 2019 for Hispanic males in Louisiana was 131,071.

2.5.2 Using `get_flows()`

As of version 1.0, **tidycensus** also includes support for the ACS Migration Flows API. The flows API returns information on both in- and out-migration for states, counties, and metropolitan areas. By default, the function allows for analysis of in-migrants, emigrants, and net migration for a given geography using data from a given 5-year ACS sample. In the example below, we request migration data for Honolulu County, Hawaii. In-migration for world regions is available along with out-migration and net migration for US locations.

```
honolulu_migration <- get_flows(
  geography = "county",
  state = "HI",
  county = "Honolulu",
  year = 2019
)
```

`get_flows()` also includes functionality for migration flow mapping; this advanced feature will be covered in Section 6.6.1.

2.6 Debugging tidycensus errors

At times, you may think that you've formatted your use of a **tidycensus** function correctly but the Census API doesn't return the data you expected. Whenever possible, **tidycensus** carries through the error message from the Census API or translates common errors for the user. In the example below, a user has mis-typed the variable ID:

```
state_pop <- get_decennial(
  geography = "state",
  variables = "P01001",
  year = 2010
)
```

```
Error : Your API call has errors.
The API message returned is error: error: unknown variable 'P01001'.
```

The "unknown variable" error message from the Census API is carried through to the user. In other instances, users might request geographies that are not available in a given dataset:

```
cbsa_ohio <- get_acs(
  geography = "cbsa",
  variables = "DP02_0068P",
  state = "OH",
  year = 2019
)
```

```
Error: Your API call has errors.
The API message returned is error: unknown/unsupported geography heirarchy.
```

The user above has attempted to get bachelor's degree attainment by CBSA in Ohio from the ACS Data Profile. However, CBSA geographies are not available by state given that many CBSAs cross state boundaries. In response, the API returns an "unsupported geography hierarchy" error.

To assist with debugging errors, or more generally to help users understand how **tidycensus** functions are being translated to Census API calls, **tidycensus** offers a parameter `show_call` that when set to `TRUE` prints out the actual API call that **tidycensus** is making to the Census API.

```
cbsa_bachelors <- get_acs(
  geography = "cbsa",
  variables = "DP02_0068P",
  year = 2019,
  show_call = TRUE
)
```

The printed URL[3] can be copy-pasted into a web browser where users can see the raw JSON returned by the Census API and inspect the results.

```
[["DP02_0068PE","DP02_0068PM","NAME","metropolitan
    statistical area/micropolitan statistical area"],
["15.7","1.5","Big Stone Gap, VA Micro Area","13720"],
["31.6","1.0","Billings, MT Metro Area","13740"],
["27.9","0.7","Binghamton, NY Metro Area","13780"],
["31.4","0.4","Birmingham-Hoover, AL Metro Area","13820"],
["33.3","1.0","Bismarck, ND Metro Area","13900"],
["21.2","2.0","Blackfoot, ID Micro Area","13940"],
["35.2","1.1","Blacksburg-Christiansburg, VA Metro Area","13980"],
["44.8","1.1","Bloomington, IL Metro Area","14010"],
["40.8","1.2","Bloomington, IN Metro Area","14020"],
["24.9","1.0","Bloomsburg-Berwick, PA Metro Area","14100"],
...
```

A common use-case for `show_call = TRUE` is to understand what data is available from the API, especially if functions in **tidycensus** are returning NA in certain rows. If the raw API call itself contains missing values for given variables, this will confirm that the requested data are not available from the API at a given geography.

2.7 Exercises

1. Review the available geographies in tidycensus from the geography table in this chapter. Acquire data on median age (variable B01002_001) for geography we have not yet used.

2. Use the `load_variables()` function to find a variable that interests you that we haven't used yet. Use `get_acs()` to fetch data from the 2016-2020 ACS for counties in the state where you live, where you have visited, or where you would like to visit.

---

[3] https://api.census.gov/data/2019/acs/acs5/profile?get=DP02_0068PE%2CDP02_0068PM%2CNAME&for=metropolitan%20statistical%20area%2Fmicropolitan%20statistical%20area%3A%2A

3

Wrangling Census data with tidyverse tools

One of the most popular frameworks for data analysis in R is the **tidyverse**, a suite of packages designed for integrated data wrangling, visualization, and modeling. The "tidy" or long-form data returned by default in **tidycensus** is designed to work well with tidyverse analytic workflows. This chapter provides an overview of how to use tidyverse tools to gain additional insights about US Census data retrieved with **tidycensus**. It concludes with discussion about margins of error (MOEs) in the American Community Survey and how to wrangle and interpret MOEs appropriately.

3.1 The tidyverse

The tidyverse[1] is a collection of R packages that are designed to work together in common data wrangling, analysis, and visualization projects. Many of these R packages, maintained by RStudio, are among the most popular R packages worldwide. Some of the key packages you'll use in the tidyverse include:

- **readr** (Wickham and Hester, 2021), which contains tools for importing and exporting datasets;
- **dplyr** (Wickham et al., 2021a), a powerful framework for data wrangling tasks;
- **tidyr** (Wickham, 2021b), a package for reshaping data;
- **purrr** (Henry and Wickham, 2020), a comprehensive framework for functional programming and iteration;
- **ggplot2** (Wickham, 2016), a data visualization package based on the Grammar of Graphics

The core data structure used in the tidyverse is the *tibble*, which is an R data frame with some small enhancements to improve the user experience. **tidycensus** returns tibbles by default.

A full treatment of the tidyverse and its functionality is beyond the scope of this book; however, the examples in this chapter will introduce you to several key tidyverse features using US Census Bureau data. For a more general and broader treatment of the tidyverse, I recommend the *R for Data Science* book (Wickham and Grolemund, 2017).

---

[1] https://www.tidyverse.org/

3.2 Exploring Census data with tidyverse tools

Census data queries using **tidycensus**, combined with core tidyverse functions, are excellent ways to explore downloaded Census data. Chapter 2 covered how to download data from various Census datasets using **tidycensus** and return the data in a desired format. A common next step in an analytic process will involve data exploration, which is handled by a wide range of tools in the tidyverse.

To get started, the **tidycensus** and **tidyverse** packages are loaded. "tidyverse" is not specifically a package itself but rather loads several core packages within the tidyverse. The package load message gives you more information:

```
library(tidycensus)
library(tidyverse)
```

```
## -- Attaching packages ------------------------------------ tidyverse 1.3.1 --

## v ggplot2 3.3.5      v purrr   0.3.4
## v tibble  3.1.6      v dplyr   1.0.8
## v tidyr   1.2.0      v stringr 1.4.0
## v readr   2.1.2      v forcats 0.5.1

## -- Conflicts --------------------------------------- tidyverse_conflicts() --
## x dplyr::filter() masks stats::filter()
## x dplyr::lag()    masks stats::lag()
```

Eight tidyverse packages are loaded: **ggplot2**, **tibble** (Müller and Wickham, 2021), **purrr**, **dplyr**, **readr**, and **tidyr** are included along with **stringr** (Wickham, 2019) for string manipulation and **forcats** (Wickham, 2021a) for working with factors. These tools collectively can be used for many core Census data analysis tasks.

3.2.1 Sorting and filtering data

For a first example, let's request data on median age from the 2016-2020 ACS with `get_acs()` for all counties in the United States. This requires specifying `geography = "county"` and leaving state set to `NULL`, the default.

```
median_age <- get_acs(
  geography = "county",
  variables = "B01002_001",
  year = 2020
)
```

The default method for printing data used by the **tibble** package shows the first 10 rows of the dataset, which in this case prints counties in Alabama. A first exploratory data analysis question might involve understanding which counties are the *youngest* and *oldest* in the United States as measured by median age. This task can be accomplished with the `arrange()` function found in the **dplyr** package. `arrange()` sorts a dataset by values in one or more columns and returns the sorted result. To view the dataset in ascending order of a given column, supply the data object and a column name to the `arrange()` function.

3.2 Exploring Census data with tidyverse tools   41

TABLE 3.1 Median age for US counties

GEOID	NAME	variable	estimate	moe
01001	Autauga County, Alabama	B01002_001	38.6	0.6
01003	Baldwin County, Alabama	B01002_001	43.2	0.4
01005	Barbour County, Alabama	B01002_001	40.1	0.6
01007	Bibb County, Alabama	B01002_001	39.9	1.2
01009	Blount County, Alabama	B01002_001	41.0	0.5
01011	Bullock County, Alabama	B01002_001	39.7	1.9
01013	Butler County, Alabama	B01002_001	41.2	0.6
01015	Calhoun County, Alabama	B01002_001	39.5	0.4
01017	Chambers County, Alabama	B01002_001	41.9	0.7
01019	Cherokee County, Alabama	B01002_001	46.8	0.5

TABLE 3.2 The youngest counties in the United States by median age

GEOID	NAME	variable	estimate	moe
35011	De Baca County, New Mexico	B01002_001	22.2	6.9
51678	Lexington city, Virginia	B01002_001	22.2	0.8
16065	Madison County, Idaho	B01002_001	23.5	0.2
46121	Todd County, South Dakota	B01002_001	23.6	0.6
51750	Radford city, Virginia	B01002_001	23.7	0.6
13053	Chattahoochee County, Georgia	B01002_001	24.0	0.7
02158	Kusilvak Census Area, Alaska	B01002_001	24.1	0.2
49049	Utah County, Utah	B01002_001	25.0	0.1
46027	Clay County, South Dakota	B01002_001	25.2	0.5
53075	Whitman County, Washington	B01002_001	25.2	0.2

```
arrange(median_age, estimate)
```

Per the 2016-2020 ACS, the youngest county is De Baca County, New Mexico. Two of the five youngest "counties" in the United States are independent cities in Virginia, which are treated as county-equivalents. Both Lexington and Radford are college towns; Lexington is home to both Washington & Lee University and the Virginia Military Institute, and Radford houses Radford University.

To retrieve the *oldest* counties in the United States by median age, an analyst can use the `desc()` function available in **dplyr** to sort the `estimate` column in descending order.

```
arrange(median_age, desc(estimate))
```

The oldest county in the United States is Sumter County, Florida. Sumter County is home to The Villages, a Census-designated place that includes a large age-restricted community also called The Villages[2].

The tidyverse includes several tools for parsing datasets that allow for exploration beyond sorting and browsing data. The `filter()` function in **dplyr** queries a dataset for rows where

---

[2]https://www.thevillages.com/

TABLE 3.3 The oldest counties in the United States by median age

GEOID	NAME	variable	estimate	moe
12119	Sumter County, Florida	B01002_001	68.0	0.3
48301	Loving County, Texas	B01002_001	62.2	37.8
48243	Jeff Davis County, Texas	B01002_001	61.3	36.8
08027	Custer County, Colorado	B01002_001	60.1	3.4
12015	Charlotte County, Florida	B01002_001	59.5	0.2
51091	Highland County, Virginia	B01002_001	59.5	4.8
35003	Catron County, New Mexico	B01002_001	59.4	2.6
51133	Northumberland County, Virginia	B01002_001	59.3	0.7
26131	Ontonagon County, Michigan	B01002_001	59.1	0.5
48443	Terrell County, Texas	B01002_001	59.1	9.4

TABLE 3.4 Counties with a median age of 50 or above

GEOID	NAME	variable	estimate	moe
02105	Hoonah-Angoon Census Area, Alaska	B01002_001	52.1	2.9
04007	Gila County, Arizona	B01002_001	50.4	0.2
04012	La Paz County, Arizona	B01002_001	57.4	0.6
04015	Mohave County, Arizona	B01002_001	52.3	0.2
04025	Yavapai County, Arizona	B01002_001	54.1	0.2
05005	Baxter County, Arkansas	B01002_001	52.3	0.5
05089	Marion County, Arkansas	B01002_001	52.1	0.8
05097	Montgomery County, Arkansas	B01002_001	50.6	0.8
05137	Stone County, Arkansas	B01002_001	50.0	0.5
06009	Calaveras County, California	B01002_001	52.8	0.6

a given condition evaluates to TRUE, and retains those rows only. For analysts who are familiar with databases and SQL, this is equivalent to a WHERE clause. This helps analysts subset their data for specific areas by their characteristics and answer questions like "how many counties in the United States have a median age of 50 or older?"

```
filter(median_age, estimate >= 50)
```

Functions like `arrange()` and `filter()` operate on row values and organize data by row. Other tidyverse functions, like **tidyr**'s `separate()`, operate on columns. The NAME column, returned by default by most **tidycensus** functions, contains a basic description of the location that can be more intuitive than the GEOID. For the 2016-2020 ACS, NAME is formatted as "X County, Y", where X is the county name and Y is the state name. `separate()` can split this column into two columns where one retains the county name and the other retains the state; this can be useful for analysts who need to complete a comparative analysis by state.

```
separate(
  median_age,
  NAME,
  into = c("county", "state"),
```

3.2 Exploring Census data with tidyverse tools

TABLE 3.5 Separate columns for county and state

GEOID	county	state	variable	estimate	moe
01001	Autauga County	Alabama	B01002_001	38.6	0.6
01003	Baldwin County	Alabama	B01002_001	43.2	0.4
01005	Barbour County	Alabama	B01002_001	40.1	0.6
01007	Bibb County	Alabama	B01002_001	39.9	1.2
01009	Blount County	Alabama	B01002_001	41.0	0.5
01011	Bullock County	Alabama	B01002_001	39.7	1.9
01013	Butler County	Alabama	B01002_001	41.2	0.6
01015	Calhoun County	Alabama	B01002_001	39.5	0.4
01017	Chambers County	Alabama	B01002_001	41.9	0.7
01019	Cherokee County	Alabama	B01002_001	46.8	0.5

```
  sep = ", "
)
```

You may have noticed above that existing variable names are unquoted when referenced in tidyverse functions. Many tidyverse functions use non-standard evaluation to refer to column names, which means that column names can be used as arguments directly without quotation marks. Non-standard evaluation makes interactive programming faster, especially for beginners; however, it can introduce some complications when writing your own functions or R packages. A full treatment of non-standard evaluation is beyond the scope of this book; Hadley Wickham's *Advanced R* (Wickham, 2019) is the best resource on the topic if you'd like to learn more.

3.2.2 Using summary variables and calculating new columns

Data in Census and ACS tables, as in the example above, are frequently comprised of variables that individually constitute sub-categories such as the numbers of households in different household income bands. One limitation of the approach above, however, is that the data and the resulting analysis return estimated counts, which are difficult to compare across geographies. For example, Maricopa County in Arizona is the state's most populous county with 4.3 million residents; the second-largest county, Pima, only has just over 1 million residents and six of the state's 15 counties have fewer than 100,000 residents. In turn, comparing Maricopa's estimates with those of smaller counties in the state would often be inappropriate.

A solution to this issue might involve **normalizing** the estimated count data by dividing it by the overall population from which the sub-group is derived. Appropriate denominators for ACS tables are frequently found in the tables themselves as variables. In ACS table B19001, which covers the number of households by income bands, the variable `B19001_001` represents the total number of households in a given enumeration unit, which we removed from our analysis earlier. Given that this variable is an appropriate denominator for the other variables in the table, it merits its own column to facilitate the calculation of proportions or percentages.

TABLE 3.6 Race and ethnicity in Arizona

GEOID	NAME	variable	estimate	moe	summary_est	summary_moe
04001	Apache County, Arizona	White	12993	56	71714	NA
04001	Apache County, Arizona	Black	544	56	71714	NA
04001	Apache County, Arizona	Native	51979	327	71714	NA
04001	Apache County, Arizona	Asian	262	76	71714	NA
04001	Apache County, Arizona	HIPI	49	14	71714	NA
04001	Apache County, Arizona	Hispanic	4751	NA	71714	NA
04003	Cochise County, Arizona	White	69095	350	126442	NA
04003	Cochise County, Arizona	Black	4512	378	126442	NA
04003	Cochise County, Arizona	Native	1058	176	126442	NA
04003	Cochise County, Arizona	Asian	2371	241	126442	NA

In **tidycensus**, this can be accomplished by supplying a variable ID to the `summary_var` parameter in both the `get_acs()` and `get_decennial()` functions. When using `get_decennial()`, doing so will create two new columns for the decennial Census datasets, `summary_var` and `summary_value`, representing the summary variable ID and the summary variable's value. When using `get_acs()`, using `summary_var` creates three new columns for the ACS datasets, `summary_var`, `summary_est`, and `summary_moe`, which include the ACS estimate and margin of error for the summary variable.

With this information in hand, normalizing data is straightforward. The following example uses the `summary_var` parameter to compare the population of counties in Arizona by race & Hispanic origin with their baseline populations, using data from the 2016-2020 ACS.

```
race_vars <- c(
  White = "B03002_003",
  Black = "B03002_004",
  Native = "B03002_005",
  Asian = "B03002_006",
  HIPI = "B03002_007",
  Hispanic = "B03002_012"
)

az_race <- get_acs(
  geography = "county",
  state = "AZ",
  variables = race_vars,
  summary_var = "B03002_001",
  year = 2020
)
```

By using dplyr's `mutate()` function, we calculate a new column, `percent`, representing the percentage of each Census tract's population that corresponds to each racial/ethnic group in 2016-2020. The `select()` function, also in dplyr, retains only those columns that we need to view.

```
az_race_percent <- az_race %>%
```

TABLE 3.7 Race and ethnicity in Arizona as percentages

NAME	variable	percent
Apache County, Arizona	White	18.1178013
Apache County, Arizona	Black	0.7585688
Apache County, Arizona	Native	72.4809661
Apache County, Arizona	Asian	0.3653401
Apache County, Arizona	HIPI	0.0683270
Apache County, Arizona	Hispanic	6.6249268
Cochise County, Arizona	White	54.6456083
Cochise County, Arizona	Black	3.5684345
Cochise County, Arizona	Native	0.8367473
Cochise County, Arizona	Asian	1.8751681

```
mutate(percent = 100 * (estimate / summary_est)) %>%
select(NAME, variable, percent)
```

The above example introduces some additional syntax common to tidyverse data analyses. The `%>%` operator from the **magrittr** R package (Bache and Wickham, 2020) is a *pipe* operator that allows for analysts to develop *analytic pipelines*, which are deeply embedded in tidyverse-centric data analytic workflows. The pipe operator passes the result of a given line of code as the first argument of the code on the next line. In turn, analysts can develop data analysis pipelines of related operations that fit together in a coherent way.

tidyverse developers recommend that the pipe operator be read as "then." The above code can in turn be interpreted as "Create a new data object `az_race_percent` by using the existing data object `az_race` THEN creating a new `percent` column THEN selecting the `NAME`, `variable`, and `percent` columns."

Since R version 4.1, the base installation of R also includes a pipe operator, `|>`. It works much the same way as the **magrittr** pipe `%>%`, though `%>%` has some small additional features that make it work well within tidyverse analysis pipelines. In turn, `%>%` will be used in the examples throughout this book.

3.3 Group-wise Census data analysis

The split-apply-combine model of data analysis, as discussed in Wickham (2011), is a powerful framework for analyzing demographic data. In general terms, an analyst will apply this framework as follows:

- The analyst identifies salient groups in a dataset between which they want to make comparisons. The dataset is then **split** into multiple pieces, one for each group.

- A function is then **applied** to each group in turn. This might be a simple summary function, such as taking the maximum or calculating the mean, or a custom function defined by the analyst.

TABLE 3.8 Largest group by county in Arizona

NAME	variable	percent
Apache County, Arizona	Native	72.48097
Cochise County, Arizona	White	54.64561
Coconino County, Arizona	White	53.80798
Gila County, Arizona	White	61.85232
Graham County, Arizona	White	50.88764
Greenlee County, Arizona	Hispanic	47.26889
La Paz County, Arizona	White	57.18089
Maricopa County, Arizona	White	54.55515
Mohave County, Arizona	White	76.69694
Navajo County, Arizona	Native	42.71386

- Finally, the results of the function applied to each group are **combined** back into a single dataset, allowing the analyst to compare the results by group.

Given the hierarchical nature of US Census Bureau data, "groups" across which analysts can make comparisons are found in just about every analytic tasks. In many cases, the split-apply-combine model of data analysis will be useful to analysts as they make sense of patterns and trends found in Census data.

In the tidyverse, split-apply-combine is implemented with the group_by() function in the dplyr package. group_by() does the work for the analyst of splitting a dataset into groups, allowing subsequent functions used by the analyst in an analytic pipeline to be applied to each group then combined back into a single dataset. The examples that follow illustrate some common group-wise analyses.

3.3.1 Making group-wise comparisons

The az_race_percent dataset created above is an example of a dataset suitable for group-wise data analysis. It includes two columns that could be used as group definitions: NAME, representing the county, and variable, representing the racial or ethnic group. Split-apply-combine could be used for either group definition to make comparisons for data in Arizona across these categories.

In a first example, we can deploy group-wise data analysis to identify the largest racial or ethnic group in each county in Arizona. This involves setting up a data analysis pipeline with the **magrittr** pipe and calculating a *grouped filter* where the filter() operation will be applied specific to each group. In this example, the filter condition will be specified as percent == max(percent). We can read the analytic pipeline then as "Create a new dataset, largest_group, by using the az_race_dataset THEN grouping the dataset by the NAME column THEN filtering for rows that are equal to the maximum value of percent for each group."

```
largest_group <- az_race_percent %>%
  group_by(NAME) %>%
  filter(percent == max(percent))
```

The result of the grouped filter allows us to review the most common racial or ethnic group in each Arizona County along with how their percentages vary. For example, in two Arizona

3.3 Group-wise Census data analysis

TABLE 3.9 Median percentage by group

variable	median_pct
Asian	0.9918415
Black	1.2857283
HIPI	0.1070384
Hispanic	30.4721836
Native	3.6344427
White	53.8079773

counties (Greenlee and Navajo), none of the racial or ethnic groups form a majority of the population.

`group_by()` is commonly paired with the `summarize()` function in data analysis pipelines. `summarize()` generates a new, condensed dataset that by default returns a column for the grouping variable(s) and columns representing the results of one or more functions applied to those groups. In the example below, the `median()` function is used to identify the median percentage for each of the racial & ethnic groups in the dataset across counties in Arizona. In turn, `variable` is passed to `group_by()` as the grouping variable.

```
az_race_percent %>%
  group_by(variable) %>%
  summarize(median_pct = median(percent))
```

The result of this operation tells us the median county percentage of each racial and ethnic group for the state of Arizona. A broader analysis might involve the calculation of these percentages hierarchically, finding the median county percentage of given attributes across states, for example.

3.3.2 Tabulating new groups

In the examples above, suitable groups in the `NAME` and `variable` columns were already found in the data retrieved with `get_acs()`. Commonly, analysts will also need to calculate new custom groups to address specific analytic questions. For example, variables in ACS table B19001 represent groups of households whose household incomes fall into a variety of categories: less than \$10,000/year, between \$10,000/year and \$19,999/year, and so forth. These categories may be more granular than needed by an analyst. As such, an analyst might take the following steps: 1) recode the ACS variables into wider income bands; 2) group the data by the wider income bands; 3) calculate grouped sums to generate new estimates.

Consider the following example, using household income data for Minnesota counties from the 2012-2016 ACS:

```
mn_hh_income <- get_acs(
  geography = "county",
  table = "B19001",
  state = "MN",
  year = 2016
)
```

TABLE 3.10 Table B19001 for counties in Minnesota, 2012–2016 ACS

GEOID	NAME	variable	estimate	moe
27001	Aitkin County, Minnesota	B19001_001	7640	262
27001	Aitkin County, Minnesota	B19001_002	562	77
27001	Aitkin County, Minnesota	B19001_003	544	72
27001	Aitkin County, Minnesota	B19001_004	472	69
27001	Aitkin County, Minnesota	B19001_005	508	68
27001	Aitkin County, Minnesota	B19001_006	522	92
27001	Aitkin County, Minnesota	B19001_007	447	61
27001	Aitkin County, Minnesota	B19001_008	390	49
27001	Aitkin County, Minnesota	B19001_009	426	64
27001	Aitkin County, Minnesota	B19001_010	415	65

Our data include household income categories for each county in the rows. However, let's say we only need three income categories for purposes of analysis: below $35,000/year, between $35,000/year and $75,000/year, and $75,000/year and up.

We first need to do some transformation of our data to recode the variables appropriately. First, we will remove variable `B19001_001`, which represents the total number of households for each county. Second, we use the `case_when()` function from the **dplyr** package to identify groups of variables that correspond to our desired groupings. Given that the variables are ordered in the ACS table in relationship to the household income values, the less than operator can be used to identify groups.

The syntax of `case_when()` can appear complex to beginners, so it is worth stepping through how the function works. Inside the `mutate()` function, which is used to create a new variable named `incgroup`, `case_when()` steps through a series of logical conditions that are evaluated in order similar to a series of if/else statements. The first condition is evaluated, telling the function to assign the value of `below35k` to all rows with a `variable` value that comes before `"B19001_008"` – which in this case will be `B19001_002` (income less than $10,000) through `B19001_007` (income between $30,000 and $34,999). The second condition is then evaluated *for all those rows not accounted for by the first condition*. This means that `case_when()` knows not to assign `"bw35kand75k"` to the income group of $10,000 and below even though its variable comes before `B19001_013`. The final condition in `case_when()` can be set to `TRUE`, which in this scenario translates as "all other values."

```
mn_hh_income_recode <- mn_hh_income %>%
  filter(variable != "B19001_001") %>%
  mutate(incgroup = case_when(
    variable < "B19001_008" ~ "below35k",
    variable < "B19001_013" ~ "bw35kand75k",
    TRUE ~ "above75k"
))
```

Our result illustrates how the different variable IDs are mapped to the new, recoded categories that we specified in `case_when()`. The `group_by() %>% summarize()` workflow can now be applied to the recoded categories by county to tabulate the data into a smaller number of groups.

TABLE 3.11 Recoded household income categories

GEOID	NAME	variable	estimate	moe	incgroup
27001	Aitkin County, Minnesota	B19001_002	562	77	below35k
27001	Aitkin County, Minnesota	B19001_003	544	72	below35k
27001	Aitkin County, Minnesota	B19001_004	472	69	below35k
27001	Aitkin County, Minnesota	B19001_005	508	68	below35k
27001	Aitkin County, Minnesota	B19001_006	522	92	below35k
27001	Aitkin County, Minnesota	B19001_007	447	61	below35k
27001	Aitkin County, Minnesota	B19001_008	390	49	bw35kand75k
27001	Aitkin County, Minnesota	B19001_009	426	64	bw35kand75k
27001	Aitkin County, Minnesota	B19001_010	415	65	bw35kand75k
27001	Aitkin County, Minnesota	B19001_011	706	81	bw35kand75k

TABLE 3.12 Grouped sums by income bands

GEOID	incgroup	estimate
27001	above75k	1706
27001	below35k	3055
27001	bw35kand75k	2879
27003	above75k	61403
27003	below35k	24546
27003	bw35kand75k	39311
27005	above75k	4390
27005	below35k	4528
27005	bw35kand75k	4577
27007	above75k	4491

```
mn_group_sums <- mn_hh_income_recode %>%
  group_by(GEOID, incgroup) %>%
  summarize(estimate = sum(estimate))
```

Our data now reflect the new estimates by group by county.

3.4 Comparing ACS estimates over time

A common task when working with Census data is to examine demographic change over time. Data from the Census API – and consequently **tidycensus** – only go back to the 2000 Decennial Census. For historical analysts who want to go even further back, decennial Census data are available since 1790 from the National Historical Geographic Information System[3], or NHGIS, which will be covered in detail in Chapter 11.

---

[3] https://www.nhgis.org/

TABLE 3.13 2016-2020 age table for Oglala Lakota County, SD

GEOID	NAME	variable	estimate	moe
46102	Oglala Lakota County, South Dakota	B01001_001	14277	NA
46102	Oglala Lakota County, South Dakota	B01001_002	6930	132
46102	Oglala Lakota County, South Dakota	B01001_003	761	66
46102	Oglala Lakota County, South Dakota	B01001_004	794	128
46102	Oglala Lakota County, South Dakota	B01001_005	707	123
46102	Oglala Lakota County, South Dakota	B01001_006	394	20
46102	Oglala Lakota County, South Dakota	B01001_007	227	15
46102	Oglala Lakota County, South Dakota	B01001_008	85	53
46102	Oglala Lakota County, South Dakota	B01001_009	165	70
46102	Oglala Lakota County, South Dakota	B01001_010	356	101

3.4.1 Time-series analysis: some cautions

Before engaging in any sort of time series analysis of Census data, analysts need to account for potential problems that can emerge when using Census data longitudinally. One major issue that can emerge is *geography changes* over time. For example, let's say we are interested in analyzing data on Oglala Lakota County, South Dakota. We can get recent data from the ACS using tools learned in Chapter 2:

```
oglala_lakota_age <- get_acs(
  geography = "county",
  state = "SD",
  county = "Oglala Lakota",
  table = "B01001",
  year = 2020
)
```

To understand how the age composition of the county has changed over the past 10 years, we may want to look at the 2006-2010 ACS for the county. Normally, we would just change the year argument to `2010`:

```
oglala_lakota_age_10 <- get_acs(
  geography = "county",
  state = "SD",
  county = "Oglala Lakota",
  table = "B01001",
  year = 2010
)
```

```
## Error: Your API call has errors.  The API message returned is .
```

The request errors, and we don't get an informative error message back from the API as was discussed in Section 2.6. The problem here is that Oglala Lakota County had a different name in 2010, Shannon County, meaning that the `county = "Oglala Lakota"` argument will not return any data. In turn, the equivalent code for the 2006-2010 ACS would use `county = "Shannon"`.

3.4 Comparing ACS estimates over time

TABLE 3.14 200-2010 age table for Oglala Lakota County, SD (then named Shannon County)

GEOID	NAME	variable	estimate	moe
46113	Shannon County, South Dakota	B01001_001	13437	NA
46113	Shannon County, South Dakota	B01001_002	6553	47
46113	Shannon County, South Dakota	B01001_003	770	99
46113	Shannon County, South Dakota	B01001_004	565	151
46113	Shannon County, South Dakota	B01001_005	833	151
46113	Shannon County, South Dakota	B01001_006	541	47
46113	Shannon County, South Dakota	B01001_007	275	99
46113	Shannon County, South Dakota	B01001_008	164	89
46113	Shannon County, South Dakota	B01001_009	143	54
46113	Shannon County, South Dakota	B01001_010	342	83

```
oglala_lakota_age_10 <- get_acs(
  geography = "county",
  state = "SD",
  county = "Shannon",
  table = "B01001",
  year = 2010
)
```

Note the differences in the GEOID column between the two tables of data. When a county or geographic entity changes its name, the Census Bureau assigns it a new GEOID, meaning that analysts need to take care when dealing with those changes. A full listing of geography changes is available on the Census website for each year[4].

In addition to changes in geographic identifiers, variable IDs can change over time as well. For example, the ACS Data Profile is commonly used for pre-computed normalized ACS estimates. Let's say that we are interested in analyzing the percentage of residents age 25 and up with a 4-year college degree for counties in Colorado from the 2019 1-year ACS. We'd first look up the appropriate variable ID with load_variables(2019, "acs1/profile") then use get_acs():

```
co_college19 <- get_acs(
  geography = "county",
  variables = "DP02_0068P",
  state = "CO",
  survey = "acs1",
  year = 2019
)
```

We get back data for counties of population 65,000 and greater as these are the geographies available in the 1-year ACS. The data make sense: Boulder County, home to the University

---

[4]https://www.census.gov/programs-surveys/acs/technical-documentation/table-and-geography-changes.2019.html

TABLE 3.15 ACS Data Profile data in 2019

GEOID	NAME	variable	estimate	moe
08001	Adams County, Colorado	DP02_0068P	25.4	1.3
08005	Arapahoe County, Colorado	DP02_0068P	43.8	1.3
08013	Boulder County, Colorado	DP02_0068P	64.8	1.8
08014	Broomfield County, Colorado	DP02_0068P	56.9	3.3
08031	Denver County, Colorado	DP02_0068P	53.1	1.1

TABLE 3.16 ACS Data Profile data in 2018

GEOID	NAME	variable	estimate	moe
08001	Adams County, Colorado	DP02_0068P	375798	NA
08005	Arapahoe County, Colorado	DP02_0068P	497198	NA
08013	Boulder County, Colorado	DP02_0068P	263938	NA
08014	Broomfield County, Colorado	DP02_0068P	53400	NA
08031	Denver County, Colorado	DP02_0068P	575870	NA

of Colorado, has a very high percentage of its population with a 4-year degree or higher. However, when we run the exact same query for the 2018 1-year ACS:

```
co_college18 <- get_acs(
  geography = "county",
  variables = "DP02_0068P",
  state = "CO",
  survey = "acs1",
  year = 2018
)
```

The values are completely different, and clearly not percentages! This is because variable IDs for the Data Profile **are unique to each year** and in turn should not be used for time-series analysis. The returned results above represent the civilian population age 18 and up, and have nothing to do with educational attainment.

3.4.2 Preparing time-series ACS estimates

The safest option for time-series analysis in the ACS is to use the Comparison Profile Tables. These tables are available for both the 1-year and 5-year ACS, and allow for comparison of demographic indicators over the past 5 years for a given year. Using the Comparison Profile tables also brings the benefit of additional variable harmonization, such as inflation-adjusted income estimates.

Data from the Comparison Profile are accessed just like other ACS variables using `get_acs()`. The example below illustrates how to get data from the ACS Comparison Profile on inflation-adjusted median household income for counties and county-equivalents in Alaska.

```
ak_income_compare <- get_acs(
  geography = "county",
  variables = c(
```

3.4 Comparing ACS estimates over time

TABLE 3.17 Comparative income data from the ACS CP tables

GEOID	NAME	variable	estimate
02016	Aleutians West Census Area, Alaska	income15	92500
02016	Aleutians West Census Area, Alaska	income20	87443
02020	Anchorage Municipality, Alaska	income15	85534
02020	Anchorage Municipality, Alaska	income20	84813
02050	Bethel Census Area, Alaska	income15	55692
02050	Bethel Census Area, Alaska	income20	54400
02090	Fairbanks North Star Borough, Alaska	income15	77590
02090	Fairbanks North Star Borough, Alaska	income20	76464

```
    income15 = "CP03_2015_062",
    income20 = "CP03_2020_062"
  ),
  state = "AK",
  year = 2020
)
```

For the 2016-2020 ACS, the "comparison year" is 2015, representing the closest non-overlapping 5-year dataset, which in this case is 2011-2015. We can examine the results, which are inflation-adjusted for appropriate comparison:

3.4.2.1 Iterating over ACS years with tidyverse tools

Using the Detailed Tables also represents a safer option than the Data Profile, as it ensures that variable IDs will remain consistent across years allowing for consistent and correct analysis. That said, there still are some potential pitfalls to account for when using the Detailed Tables. The Census Bureau will add and remove variables from survey to survey depending on data needs and data availability. For example, questions are sometimes added and removed from the ACS survey meaning that you won't always be able to get every data point for every year and geography combination. In turn, it is still important to check on data availability using `load_variables()` for the years you plan to analyze before carrying out your time-series analysis.

Let's re-engineer the analysis above on educational attainment in Colorado counties, which below will be computed for a time series from 2010 to 2019. Information on "bachelor's degree or higher" is split by sex and across different tiers of educational attainment in the detailed tables, found in ACS table 15002. Given that we only need a few variables (representing estimates of populations age 25+ who have finished a 4-year degree or graduate degrees, by sex), we'll request those variables directly rather than the entire B15002 table.

```
college_vars <- c("B15002_015",
                  "B15002_016",
                  "B15002_017",
                  "B15002_018",
                  "B15002_032",
                  "B15002_033",
```

```
        "B15002_034",
        "B15002_035")
```

We'll now use these variables to request data on college degree holders from the ACS for counties in Colorado for each of the 1-year ACS surveys from 2010 to 2019. In most cases, this process should be streamlined with *iteration*. Thus far, we are familiar with using the `year` argument in `get_acs()` to request data for a specific year. Writing out ten different calls to `get_acs()`, however – one for each year – would be tedious and would require a fair amount of repetitive code! Iteration helps us avoid repetitive coding as it allows us to carry out the same process over a sequence of values. Programmers familiar with iteration will likely know of "loop" operators like `for` and `while`, which are available in base R and most other programming languages in some variety. Base R also includes the `*apply()` family of functions (e.g. `lapply()`, `mapply()`, `sapply()`), which iterates over a sequence of values and applies a given function to each value.

The tidyverse approach to iteration is found in the **purrr** package. **purrr** includes a variety of functions that are designed to integrate well in workflows that require iteration and use other tidyverse tools. The `map_*()` family of functions iterate over values and try to return a desired result; `map()` returns a list, `map_int()` returns an integer vector, and `map_chr()` returns a character vector, for example. With tidycensus, the `map_dfr()` function is particularly useful. `map_dfr()` iterates over an input and applies it to a function or process defined by the user, then row-binds the result into a single data frame. The example below illustrates how this works for the years 2010 through 2019.

```
years <- 2010:2019
names(years) <- years

college_by_year <- map_dfr(years, ~{
  get_acs(
    geography = "county",
    variables = college_vars,
    state = "CO",
    summary_var = "B15002_001",
    survey = "acs1",
    year = .x
  )
}, .id = "year")
```

For users newer to R, iteration and purrr syntax can feel complex, so it is worth stepping through how the above code sample works.

- First, a numeric vector of years is defined with the syntax `2010:2019`. This will create a vector of years at 1-year intervals. These values are set as the names of the vector as well, as `map_dfr()` has additional functionality for working with named objects.
- `map_dfr()` then takes three arguments above.
    - The first argument is the object that `map_dfr()` will iterate over, which in this case is our `years` vector. This means that the process we set up will be run once for each element of `years`.
    - The second argument is a formula we specify with the tilde (~) operator and curly braces ({...}). The code inside the curly braces will be run once for each element of

3.5 Comparing ACS estimates over time

TABLE 3.18 Educational attainment over time

year	GEOID	NAME	variable	estimate	moe	summary_est	summary_moe
2010	08001	Adams County, Colorado	B15002_015	20501	1983	275849	790
2011	08001	Adams County, Colorado	B15002_015	21233	2124	281231	865
2012	08001	Adams County, Colorado	B15002_015	19238	2020	287924	693
2013	08001	Adams County, Colorado	B15002_015	23818	2445	295122	673
2014	08001	Adams County, Colorado	B15002_015	20255	1928	304394	541
2015	08001	Adams County, Colorado	B15002_015	22962	2018	312281	705
2016	08001	Adams County, Colorado	B15002_015	25744	2149	318077	525
2017	08001	Adams County, Colorado	B15002_015	26159	2320	324185	562
2018	08001	Adams County, Colorado	B15002_015	28113	2078	331247	955
2019	08001	Adams County, Colorado	B15002_015	27552	2070	336931	705

years. The local variable .x, used inside the formula, takes on each value of years sequentially. In turn, we are running the equivalent of get_acs() with year = 2010, year = 2011, and so forth. Once get_acs() is run for each year, the result is combined into a single output data frame.

- The .id argument, which is optional but used here, creates a new column in the output data frame that contains values equivalent to the names of the input object, which in this case is years. By setting .id = "year", we tell map_dfr() to name the new column that will contain these values year.

Let's review the result:

```
college_by_year %>%
  arrange(NAME, variable, year)
```

The result is a long-form dataset that contains a time series of each requested ACS variable for each county in Colorado that is available in the 1-year ACS. The code below outlines a group_by() %>% summarize() workflow for calculating the percentage of the population age 25 and up with a 4-year college degree, then uses the pivot_wider() function from the tidyr package to spread the years across the columns for tabular data display.

```
percent_college_by_year <- college_by_year %>%
  group_by(NAME, year) %>%
  summarize(numerator = sum(estimate),
            denominator = first(summary_est)) %>%
  mutate(pct_college = 100 * (numerator / denominator)) %>%
  pivot_wider(id_cols = NAME,
              names_from = year,
              values_from = pct_college)
```

This particular format is suitable for data display or writing to an Excel spreadsheet for colleagues who are not R-based. Methods for visualization of time-series estimates from the ACS will be covered in Section 4.4.

TABLE 3.19 Percent college by year

NAME	2010	2011	2012	2013	2014	2015	2016
Adams County, Colorado	20.57394	20.51801	20.64538	23.09384	22.16929	22.79742	22.95293
Arapahoe County, Colorado	37.03001	38.24506	39.28435	39.42478	40.94194	41.03578	41.48359
Boulder County, Colorado	57.50285	59.05601	57.88284	58.53214	58.04066	60.57147	60.63005
Broomfield County, Colorado	NA	NA	NA	NA	NA	56.07776	51.94338
Denver County, Colorado	40.87971	42.97122	44.65358	44.35340	44.25600	47.10820	47.39683
Douglas County, Colorado	54.96800	53.27936	55.09223	57.66999	56.48866	56.06928	59.42687
El Paso County, Colorado	34.11467	35.69184	34.91315	35.47612	36.49302	36.43089	38.67864
Jefferson County, Colorado	40.83113	39.54961	41.43825	41.04234	41.99768	43.20923	43.51953
Larimer County, Colorado	45.80197	42.83543	44.71423	43.33800	42.67180	46.16705	46.78871
Mesa County, Colorado	24.99285	25.82724	23.01511	27.63325	25.14875	30.27630	25.02980

3.5 Handling margins of error in the American Community Survey with tidycensus

A topic of critical importance when working with data from the American Community Survey is the *margin of error*. As opposed to the decennial US Census, which is based on a complete enumeration of the US population, the ACS is based on a sample with estimates characterized by margins of error. By default, MOEs are returned at a 90 percent confidence level. This can be translated roughly as "we are 90 percent sure that the true value falls within a range defined by the estimate plus or minus the margin of error."

As discussed in Chapter 2, **tidycensus** takes an opinionated approach to margins of error. When applicable, **tidycensus** will always return the margin of error associated with an estimate, and does not have an option available to return estimates only. For "tidy" or long-form data, these margins of error will be found in the moe column; for wide-form data, margins of error will be found in columns with an M suffix.

The confidence level of the MOE can be controlled with the moe_level argument in get_acs(). The default moe_level is 90, which is what the Census Bureau returns by default. tidycensus can also return MOEs at a confidence level of 95 or 99, which uses Census Bureau-recommended formulas to adjust the MOE. For example, we might look at data on median household income by county in Rhode Island using the default moe_level of 90:

```
get_acs(
  geography = "county",
  state = "Rhode Island",
  variables = "B19013_001",
  year = 2020
)
```

A stricter margin of error will increase the size of the MOE relative to its estimate.

```
get_acs(
  geography = "county",
  state = "Rhode Island",
  variables = "B19013_001",
```

3.5 Handling margins of error in the American Community Survey with tidycensus

TABLE 3.20 Default MOE at 90 percent confidence level

GEOID	NAME	variable	estimate	moe
44001	Bristol County, Rhode Island	B19013_001	85413	6122
44003	Kent County, Rhode Island	B19013_001	75857	2022
44005	Newport County, Rhode Island	B19013_001	84282	2629
44007	Providence County, Rhode Island	B19013_001	62323	1270
44009	Washington County, Rhode Island	B19013_001	86970	3651

TABLE 3.21 MOE at 99 percent confidence level

GEOID	NAME	variable	estimate	moe
44001	Bristol County, Rhode Island	B19013_001	85413	9527.246
44003	Kent County, Rhode Island	B19013_001	75857	3146.699
44005	Newport County, Rhode Island	B19013_001	84282	4091.331
44007	Providence County, Rhode Island	B19013_001	62323	1976.413
44009	Washington County, Rhode Island	B19013_001	86970	5681.799

```
  year = 2020,
  moe_level = 99
)
```

3.5.1 Calculating derived margins of error in tidycensus

For small geographies or small populations, margins of error can get quite large, in some cases exceeding their corresponding estimates. In the example below, we can examine data on age groups by sex for the population age 65 and older for Census tracts in Salt Lake County, Utah. We will first generate a vector of variable IDs for which we want to request data from the ACS using some base R functionality.

In this workflow, an analyst has used `load_variables()` to look up the variables that represent estimates for populations age 65 and up; this includes B01001_020 through B01001_025 for males, and B01001_044 through B01001_049 for females. Typing out each variable individually would be tedious, so an analyst can use string concatenation to generate the required vector of variable IDs as follows:

```
vars <- paste0("B01001_0", c(20:25, 44:49))

vars
```

```
## [1] "B01001_020" "B01001_021" "B01001_022" "B01001_023" "B01001_024"
## [6] "B01001_025" "B01001_044" "B01001_045" "B01001_046" "B01001_047"
## [11] "B01001_048" "B01001_049"
```

When R evaluates nested expressions like this, it starts with the inner-most expressions then evaluates them from the inside out. The steps taken to assemble the correct vector of variable IDs are as follows:

- First, the expressions 20:25 and 44:49 are evaluated. The colon operator : in base R, when used between two numbers, will generate a vector of numbers from the first to the second

TABLE 3.22 Example Census tract in Salt Lake City

GEOID	variable	estimate	moe
49035100100	B01001_020	11	13
49035100100	B01001_021	25	18
49035100100	B01001_022	7	10
49035100100	B01001_023	4	7
49035100100	B01001_024	0	12
49035100100	B01001_025	17	20
49035100100	B01001_044	0	12
49035100100	B01001_045	4	7
49035100100	B01001_046	21	17
49035100100	B01001_047	123	168

at intervals of 1. This creates vectors of the integers 20 through 25 and 44 through 49, which will serve as the suffixes for the variable IDs.

- Second, the `c()` function is used to combine the two vectors of integers into a single vector.
- Third, the `paste0()` function concatenates the string prefix `"B01001_0"` with each of the integers in the vector created with `c()` and returns a vector of variable IDs. `paste0()` is a convenient extension of the more flexible `paste()` function that concatenates strings with no spaces in between them by default.

The resulting variables object, named `vars`, can now be used to request variables in a call to `get_acs()`.

```
salt_lake <- get_acs(
  geography = "tract",
  variables = vars,
  state = "Utah",
  county = "Salt Lake",
  year = 2020
)
```

We will now want to examine the margins of error around the estimates in the returned data. Let's focus on a specific Census tract in Salt Lake County using `filter()`:

```
example_tract <- salt_lake %>%
  filter(GEOID == "49035100100")

example_tract %>%
  select(-NAME)
```

In many cases, the margins of error exceed their corresponding estimates. For example, the ACS data suggest that in Census tract 49035100100, for the male population age 85 and up (variable ID `B01001_0025`), there are anywhere between 0 and 45 people in that Census tract. This can make ACS data for small geographies problematic for planning and analysis purposes.

3.5 Handling margins of error in the American Community Survey with tidycensus

A potential solution to large margins of error for small estimates in the ACS is to aggregate data upward until a satisfactory margin of error to estimate ratio is reached. The US Census Bureau publishes formulas for appropriately calculating margins of error around such derived estimates[5], which are included in tidycensus with the following functions:

- `moe_sum()`: calculates a margin of error for a derived sum;
- `moe_product()`: calculates a margin of error for a derived product;
- `moe_ratio()`: calculates a margin of error for a derived ratio;
- `moe_prop()`: calculates a margin of error for a derived proportion.

In their most basic form, these functions can be used with constants. For example, let's say we had an ACS estimate of 25 with a margin of error of 5 around that estimate. The appropriate denominator for this estimate is 100 with a margin of error of 3. To determine the margin of error around the derived proportion of 0.25, we can use `moe_prop()`:

```
moe_prop(25, 100, 5, 3)
```

```
## [1] 0.0494343
```

Our margin of error around the derived estimate of 0.25 is approximately 0.049.

3.5.2 Calculating group-wise margins of error

These margin of error functions in **tidycensus** can in turn be integrated into tidyverse-centric analytic pipelines to handle large margins of error around estimates. Given that the smaller age bands in the Salt Lake City dataset are characterized by too much uncertainty for our analysis, we decide in this scenario to aggregate our data upward to represent populations aged 65 and older by sex.

In the code below, we use the `case_when()` function to create a new column, sex, that represents a mapping of the variables we pulled from the ACS to their sex categories. We then employ a familiar `group_by() %>% summarize()` method to aggregate our data by Census tract and sex. Notably, the call to `summarize()` includes a call to tidycensus's `moe_sum()` function, which will generate a new column that represents the margin of error around the derived sum.

```
salt_lake_grouped <- salt_lake %>%
  mutate(sex = case_when(
    str_sub(variable, start = -2) < "26" ~ "Male",
    TRUE ~ "Female"
  )) %>%
  group_by(GEOID, sex) %>%
  summarize(sum_est = sum(estimate),
            sum_moe = moe_sum(moe, estimate))
```

The margins of error relative to their estimates are now much more reasonable than in the disaggregated data.

---

[5]https://www2.census.gov/programs-surveys/acs/tech_docs/statistical_testing/2018_Instructions_for_Stat_Testing_ACS.pdf?

TABLE 3.23 Grouped margins of error

GEOID	sex	sum_est	sum_moe
49035100100	Female	165	170.72493
49035100100	Male	64	34.43835
49035100200	Female	170	57.26255
49035100200	Male	128	47.51842
49035100306	Female	155	77.48548
49035100306	Male	136	85.53362
49035100307	Female	207	82.03658
49035100307	Male	110	58.89822
49035100308	Female	157	110.97297
49035100308	Male	91	59.97499

That said, the Census Bureau issues a note of caution[6] (American Community Survey Office, 2020):

> All derived MOE methods are approximations and users should be cautious in using them. This is because these methods do not consider the correlation or covariance between the basic estimates. They may be overestimates or underestimates of the derived estimate's standard error depending on whether the two basic estimates are highly correlated in either the positive or negative direction. As a result, the approximated standard error may not match direct calculations of standard errors or calculations obtained through other methods.

This means that your "best bet" is to first search the ACS tables to see if your data are found in aggregated form elsewhere before doing the aggregation and MOE estimation yourself. In many cases, you'll find aggregated information in the ACS combined tables, Data Profile, or Subject Tables that will include pre-computed margins of error for you.

3.6 Exercises

- The ACS Data Profile includes a number of pre-computed percentages, which can reduce your data wrangling time. The variable in the 2015-2019 ACS for "percent of the population age 25 and up with a bachelor's degree" is `DP02_0068P`. For a state of your choosing, use this variable to determine:
    - The county with the highest percentage in the state;
    - The county with the lowest percentage in the state;
    - The median value for counties in your chosen state.

---

[6] https://www2.census.gov/programs-surveys/acs/tech_docs/statistical_testing/2019_Instructions_for_Stat_Testing_ACS.pdf?

4

Exploring US Census data with visualization

The core visualization package within the **tidyverse** suite of packages is **ggplot2** (Wickham, 2016). Originally developed by RStudio chief scientist Hadley Wickham, **ggplot2** is a widely-used visualization framework by R developers, accounting for 70,000 downloads per day in June 2021 from the RStudio CRAN mirror. **ggplot2** allows R users to visualize data using a *layered grammar of graphics* approach, in which plot objects are initialized upon which the R user layers plot elements.

ggplot2 is an ideal package for visualization of US Census data, especially when obtained in tidy format by the **tidycensus** package. It has powerful capacity for basic charts, group-wise comparisons, and advanced chart types such as maps.

This chapter includes several examples of how R users can visualize data from the US Census Bureau using **ggplot2**. Chart types explored in this chapter include basic chart types; faceted, or "small multiples" plots; population pyramids; margin of error plots for ACS data; and advanced visualizations using extensions to **ggplot2**. Finally, the chapter introduces the **plotly** package for interactive visualization, which can be used to convert **ggplot2** objects to interactive web graphics.

4.1 Basic Census visualization with ggplot2

A critical part of the Census data analysis process is *data visualization*, where an analyst examines patterns and trends found in their data graphically. In many cases, the exploratory analyses outlined in the previous two chapters would be augmented significantly with accompanying graphics. This first section illustrates some examples for getting started with exploratory Census data visualization with **ggplot2**.

To get started, we'll return to a dataset used in Section 2.4.2, which includes data on median household income and median age by county in the state of Georgia from the 2016-2020 ACS. We are requesting the data in wide format, which will spread the estimate and margin of error information across the columns.

```
library(tidycensus)

ga_wide <- get_acs(
  geography = "county",
  state = "Georgia",
  variables = c(medinc = "B19013_001",
                medage = "B01002_001"),
  output = "wide",
```

```
  year = 2020
)
```

4.1.1 Getting started with ggplot2

ggplot2 visualizations are initialized with the `ggplot()` function, to which a user commonly supplies a dataset and an *aesthetic*, defined with the `aes()` function. Within the `aes()` function, a user can specify a series of mappings onto either the data axes or other characteristics of the plot, such as element fill or color.

After initializing the ggplot object, users can layer plot elements onto the plot object. Essential to the plot is a `geom`, which specifies one of many chart types available in **ggplot2**. For example, `geom_bar()` will create a bar chart, `geom_line()` a line chart, `geom_point()` a point plot, and so forth. Layers are linked to the ggplot object by using the + operator.

One of the first exploratory graphics an analyst will want to produce when examining a new dataset is a *histogram*, which characterizes the distribution of values in a column through varying lengths of bars. This first example, shown in Figure 4.1, uses **ggplot2** and its `geom_histogram()` function to generate such a histogram of median household income by county in Georgia. The optional call to `options(scipen = 999)` instructs R to avoid using scientific notation in its output, including on the **ggplot2** tick labels.

```
library(tidyverse)
options(scipen = 999)

ggplot(ga_wide, aes(x = medincE)) + 
  geom_histogram()
```

The histogram shows that the modal median household income of Georgia counties is around $40,000 per year, with a longer tail of wealthier counties on the right-hand side of the plot. In the histogram, counties are organized into "bins", which are groups of equal width along the X-axis. The Y-axis then represents the number of counties that fall within each bin. By default, **ggplot2** organizes the data into 30 bins; this option can be changed with the `bins` parameter. For example, we can re-make the visualization with half the previous number of bins by including the argument `bins = 15` in our call to `geom_histogram()`, shown in Figure 4.2.

```
ggplot(ga_wide, aes(x = medincE)) + 
  geom_histogram(bins = 15)
```

Histograms are not the only options for visualizing univariate data distributions. A popular alternative is the *box-and-whisker plot*, which is implemented in **ggplot2** with `geom_boxplot()`. In this example, the column `medincE` is passed to the y parameter instead of x; this creates a vertical rather than horizontal box plot.

```
ggplot(ga_wide, aes(y = medincE)) + 
  geom_boxplot()
```

4.1 Basic Census visualization with ggplot2

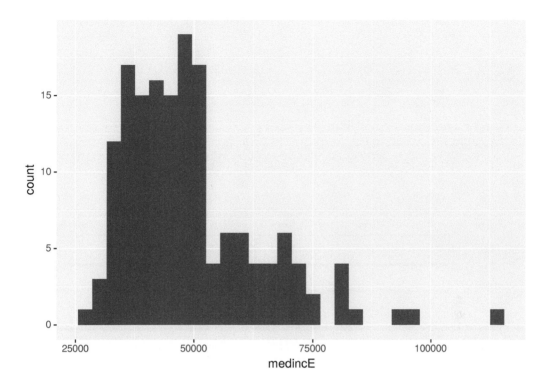

FIGURE 4.1 Histogram of median household income, Georgia counties

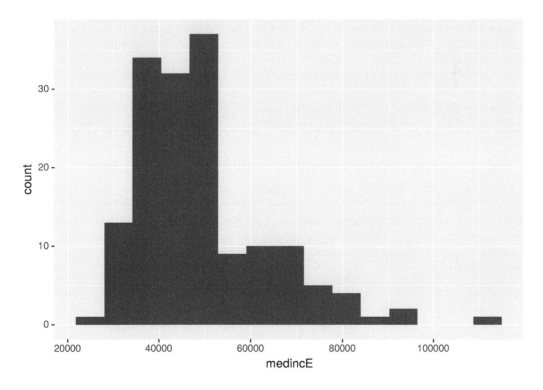

FIGURE 4.2 Histogram with the number of bins reduced to 15

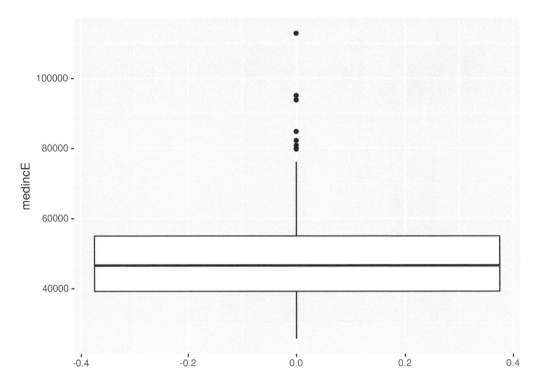

FIGURE 4.3 Box plot of median household income, Georgia counties

Figure 4.3 visualizes the distribution of median household incomes by county in Georgia with a number of different components. The central *box* covers the interquartile range (the IQR, representing the 25th to 75th percentile of values in the distribution) with a central line representing the value of the distribution's median. The *whiskers* then extend to either the minimum and maximum values of the distribution *or* 1.5 times the IQR. In this example, the lower whisker extends to the minimum value, and the upper whisker extends to 1.5 times the IQR. Values beyond the whiskers are represented as *outliers* on the plot with points.

4.1.2 Visualizing multivariate relationships with scatter plots

As part of the exploratory data analysis process, analysts will often want to visualize *interrelationships between Census variables* along with the univariate data distributions discussed above. For two numeric variables, a common exploratory chart is a *scatter plot*, which maps values in one column to the X-axis and values in another column to the Y-axis. The resulting plot then gives the analyst a sense of the nature of the relationship between the two variables.

Scatter plots are implemented in **ggplot2** with the `geom_point()` function, which plots points on a chart relative to X and Y values for observations in a dataset. This requires specification of two columns in the call to `aes()` as opposed to the single column used in the univariate distribution visualization examples. The example that follows generates a scatter plot to visualize the relationship between county median age and county median household income in Georgia.

4.2 Basic Census visualization with ggplot2

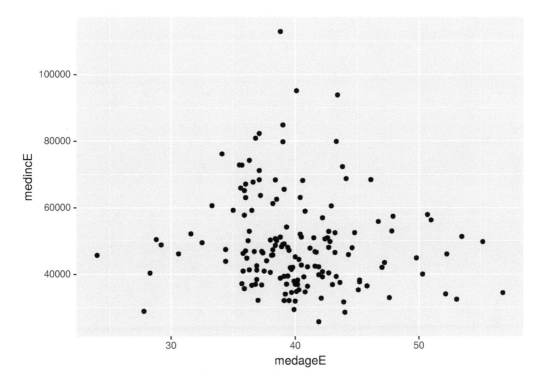

FIGURE 4.4 Scatter plot of median age and median household income, counties in Georgia

```
ggplot(ga_wide, aes(x = medageE, y = medincE)) + 
  geom_point()
```

Figure 4.4 shows a cloud of points that in some cases can suggest the nature of the correlation between the two columns. In this example, however, the correlation is not immediately clear from the distribution of points. Fortunately, **ggplot2** includes the ability to "layer on" additional chart elements to help clarify the nature of the relationship between the two columns. The geom_smooth() function draws a fitted line representing the relationship between the two columns on the plot. The argument method = "lm" draws a straight line based on a linear model fit; smoothed relationships can be visualized as well with method = "loess".

```
ggplot(ga_wide, aes(x = medageE, y = medincE)) + 
  geom_point() + 
  geom_smooth(method = "lm")
```

The regression line in Figure 4.5 suggests a modest negative relationship between the two columns, showing that county median household income in Georgia tends to decline slightly as median age increases.

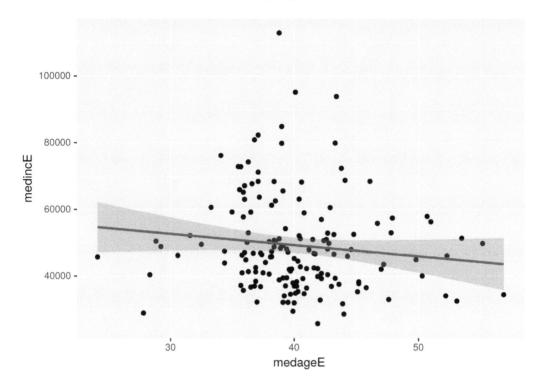

FIGURE 4.5 Scatter plot with linear relationship superimposed on the graphic

4.2 Customizing ggplot2 visualizations

The attractive defaults of **ggplot2** visualizations allow for the creation of legible graphics with little to no customization. This helps greatly with exploratory data analysis tasks where the primary audience is the analyst exploring the dataset. Analysts planning to present their work to an external audience, however, will want to customize the appearance of their plots beyond the defaults to maximize interpretability. This section covers how to take a Census data visualization that is relatively illegible by default and polish it up for eventual presentation and export from R.

In this example, we will create a visualization that illustrates the percent of commuters that take public transportation to work for the largest metropolitan areas in the United States. The data come from the 2019 1-year American Community Survey Data Profile, variable `DP03_0021P`. To determine this information, we can use **tidyverse** tools to sort our data by descending order of a summary variable representing total population and then retaining the 20 largest metropolitan areas by population using the `slice_max()` function.

```
library(tidycensus)
library(tidyverse)

metros <-  get_acs(
  geography = "cbsa",
```

4.2 Customizing ggplot2 visualizations

TABLE 4.1 Large metro areas by public transit commuting share

GEOID	NAME	variable	estimate	moe	summary_est	summary_moe
35620	New York-Newark-Jersey City, NY-NJ-PA Metro Area	DP03_0021P	31.6	0.2	19216182	NA
31080	Los Angeles-Long Beach-Anaheim, CA Metro Area	DP03_0021P	4.8	0.1	13214799	NA
16980	Chicago-Naperville-Elgin, IL-IN-WI Metro Area	DP03_0021P	12.4	0.3	9457867	1469
19100	Dallas-Fort Worth-Arlington, TX Metro Area	DP03_0021P	1.3	0.1	7573136	NA
26420	Houston-The Woodlands-Sugar Land, TX Metro Area	DP03_0021P	2.0	0.2	7066140	NA

```
  variables = "DP03_0021P",
  summary_var = "B01003_001",
  survey = "acs1",
  year = 2019
) %>%
  slice_max(summary_est, n = 20)
```

The returned data frame has 7 columns, as is standard for `get_acs()` with a summary variable, but has 20 rows as specified by the `slice_max()` command. While the data can be filtered and sorted further to facilitate comparative analysis, it also can be represented succinctly with a visualization. The tidy format returned by `get_acs()` is well-suited for visualization with **ggplot2**.

In the basic example below, we can create a bar chart comparing public transportation as commute share for the most populous metropolitan areas in the United States with a minimum of code. The first argument to `ggplot()` in the example below is the name of our dataset; the second argument is an aesthetic mapping of columns to plot elements, specified inside the `aes()` function. This plot initialization is then linked with the + operator to the `geom_col()` function to create a bar chart.

```
ggplot(metros, aes(x = NAME, y = estimate)) + 
  geom_col()
```

While the chart in Figure 4.6 is a visualization of the `metros` dataset, it tells us little about the data given the lack of necessary formatting. The x-axis labels are so lengthy that they overlap and are impossible to read; the axis titles are not intuitive; and the data are not sorted, making it difficult to compare similar observations.

4.2.1 Improving plot legibility

Fortunately, the plot can be made more legible by cleaning up the metropolitan area name, re-ordering the data in descending order, then adding layers to the plot definition. Additionally, **ggplot2** visualization can be used in combination with **magrittr** piping and **tidyverse** functions, allowing analysts to string together data manipulation and visualization processes.

Our first step will be to format the NAME column in a more intuitive way. The NAME column by default provides a description of each geography as formatted by the US Census Bureau. However, a detailed description like "Atlanta-Sandy Springs-Roswell, GA Metro Area" is likely unnecessary for our chart, as the same metropolitan area can be represented on the chart by the name of its first principal city, which in this case would be "Atlanta". To accomplish this, we can overwrite the NAME column by using the tidyverse function `str_remove()`, found in the **stringr** package. The example uses *regular expressions* to first remove all the text after the first dash, then remove the text after the first comma if no dash

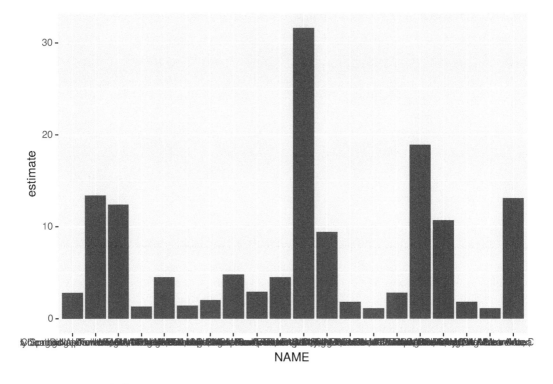

FIGURE 4.6 A first bar chart with ggplot2

was originally present. These two subsequent calls to `mutate()` will account for the various ways that metropolitan area names are specified.

On the chart, the legibility can be further improved by mapping the metro name to the y-axis and the ACS estimate to the x-axis, and plotting the points in descending order of their estimate values. The ordering of points in this way is accomplished with the `reorder()` function, used inside the call to `aes()`. As the result of the `mutate()` operations is piped to the `ggplot()` function in this example with the `%>%` operator, the dataset argument to `ggplot()` is inferred by the function.

```
metros %>%
  mutate(NAME = str_remove(NAME, "-.*$")) %>%
  mutate(NAME = str_remove(NAME, ",.*$")) %>%
  ggplot(aes(y = reorder(NAME, estimate), x = estimate)) +
  geom_col()
```

The plot shown in Figure 4.7 is much more legible after our modifications. Metropolitan areas can be directly compared with one another, and the metro area labels convey enough information about the different places without overwhelming the plot with long axis labels. However, the plot still lacks information to inform the viewer about the plot's content. This can be accomplished by specifying *labels* inside the `labs()` function. In the example below, we'll specify a title and subtitle, and modify the X and Y axis labels from their defaults.

4.2 Customizing ggplot2 visualizations

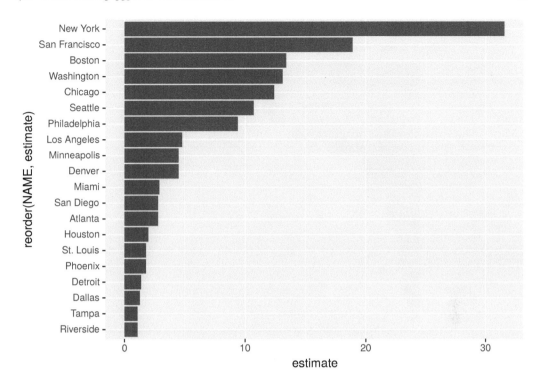

FIGURE 4.7 An improved bar chart with ggplot2

```
metros %>%
  mutate(NAME = str_remove(NAME, "-.*$")) %>%
  mutate(NAME = str_remove(NAME, ",.*$")) %>%
  ggplot(aes(y = reorder(NAME, estimate), x = estimate)) +
  geom_col() +
  theme_minimal() +
  labs(title = "Public transit commute share",
       subtitle = "2019 1-year ACS estimates",
       y = "",
       x = "ACS estimate",
       caption = "Source: ACS Data Profile; tidycensus R package")
```

The inclusion of labels provides key information about the contents of the plot and also gives it a more polished look for presentation as evidenced in Figure 4.8.

4.2.2 Custom styling of ggplot2 charts

While an analyst may be comfortable with the plot as-is, **ggplot2** allows for significant customization with respect to stylistic presentation. The example below makes a few such modifications. This includes styling the bars on the plot with a different color and internal transparency; changing the font; and customizing the axis tick labels.

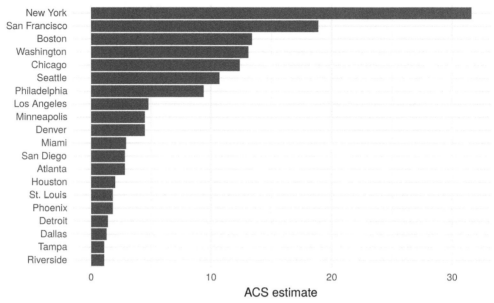

FIGURE 4.8 A cleaned-up bar chart with ggplot2

```
library(scales)

metros %>%
  mutate(NAME = str_remove(NAME, "-.*$")) %>%
  mutate(NAME = str_remove(NAME, ",.*$")) %>%
  ggplot(aes(y = reorder(NAME, estimate), x = estimate)) +
  geom_col(color = "navy", fill = "navy",
           alpha = 0.5, width = 0.85) +
  theme_minimal(base_size = 12, base_family = "Verdana") +
  scale_x_continuous(labels = label_percent(scale = 1)) +
  labs(title = "Public transit commute share",
       subtitle = "2019 1-year ACS estimates",
       y = "",
       x = "ACS estimate",
       caption = "Source: ACS Data Profile variable; tidycensus R package")
```

The code used to produced the styled graphic shown in Figure 4.9 uses the following modifications:

- While aesthetic mappings relative to a column in the input dataset will be specified in a call to aes(), **ggplot2** geoms can be styled directly in their corresponding functions. Bars in **ggplot2** are characterized by both a *color*, which is the outline of the bar, and a *fill*. The above code sets both to "navy" then modifies the internal transparency of the

4.2 Customizing ggplot2 visualizations

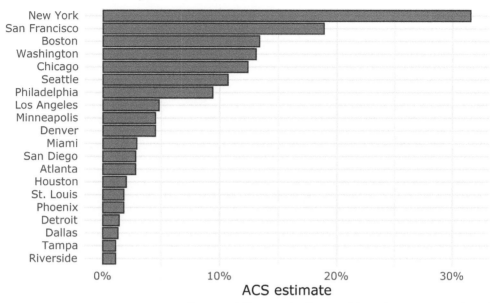

FIGURE 4.9 A ggplot2 bar chart with custom styling

bar with the `alpha` argument. Finally, `width = 0.85` slightly increases the spacing between bars.

- In the call to `theme_minimal()`, `base_size` and `base_family` parameters are available. `base_size` specifies the base font size to which plot text elements will be drawn; this defaults to 11. In many cases, you will want to increase `base_size` to improve plot legibility. `base_family` allows you to change the font family used on your plot. In this example, `base_family` is set to `"Verdana"`, but you can use any font families accessible to R from your operating system. To check this information, use the `system_fonts()` function in the **systemfonts** package (Pedersen et al., 2021).

- The `scale_x_continuous()` function is used to customize the X-axis of the plot. The `labels` parameter can accept a range of values including a function (used here) or formula that operates over the tick labels. The **scales** (Wickham and Seidel, 2022) package contains many useful formatting functions to neatly present tick labels, such as `label_percent()`, `label_dollar()`, and `label_date()`. The functions also accept arguments to modify the presentation.

4.2.3 Exporting data visualizations from R

Once an analyst has settled on a visualization design, they may want to export their image from R to display on a website, in a blog post, or in a report. Plots generated in RStudio can be exported with the **Export > Save as Image** command; however, analysts who want more programmatic control over their image exports can script this with **ggplot2**. The `ggsave()` function in **ggplot2** will save the last plot generated to an image file in the user's current working directory by default. The specified file extension will control the output image format, e.g. .png.

```
ggsave("metro_transit.png")
```

ggsave() includes a variety of options for more fine-grained control over the image output. Common options used by analysts will be width and height to control the image size; dpi to control the image resolution (in dots per inch); and path to specify the directory in which the image will be located. For example, the code below would write the most recent plot generated to an 8 inch by 5 inch image file in a custom location with a resolution of 300 dpi.

```
ggsave(
  filename = "metro_transit.png",
  path = "~/images",
  width = 8,
  height = 5,
  units = "in",
  dpi = 300
)
```

4.3 Visualizing margins of error

As discussed in Chapter 3, handling margins of error appropriately is of significant importance for analysts working with ACS data. While **tidycensus** has tools available for working with margins of error in a data wrangling workflow, it is also often useful to visualize those margins of error to illustrate the degree of uncertainty around estimates, especially when making comparisons between those estimates.

In the above example visualization of public transportation mode share by metropolitan area for the largest metros in the United States, estimates are associated with margins of error; however, these margins of error are relatively small given the large population size of the geographic units represented in the plot. However, if studying demographic trends for geographies of smaller population size – like counties, Census tracts, or block groups – comparisons can be subject to a considerable degree of uncertainty.

4.3.1 Data setup

In the example below, we will compare the median household incomes of counties in the US state of Maine from the 2016-2020 ACS. Before doing so, it is helpful to understand some basic information about counties in Maine, such as the number of counties and their total population. We can retrieve this information with **tidycensus** and 2020 decennial Census data.

```
maine <- get_decennial(
  state = "Maine",
  geography = "county",
```

4.3 Visualizing margins of error

TABLE 4.2 Population sizes of counties in Maine

GEOID	NAME	variable	value
23005	Cumberland County, Maine	totalpop	303069
23031	York County, Maine	totalpop	211972
23019	Penobscot County, Maine	totalpop	152199
23011	Kennebec County, Maine	totalpop	123642
23001	Androscoggin County, Maine	totalpop	111139
23003	Aroostook County, Maine	totalpop	67105
23017	Oxford County, Maine	totalpop	57777
23009	Hancock County, Maine	totalpop	55478
23025	Somerset County, Maine	totalpop	50477
23013	Knox County, Maine	totalpop	40607
23027	Waldo County, Maine	totalpop	39607
23023	Sagadahoc County, Maine	totalpop	36699
23015	Lincoln County, Maine	totalpop	35237
23029	Washington County, Maine	totalpop	31095
23007	Franklin County, Maine	totalpop	29456
23021	Piscataquis County, Maine	totalpop	16800

```
  variables = c(totalpop = "P1_001N"),
  year = 2020
) %>%
  arrange(desc(value))
```

There are sixteen counties in Maine, ranging in population from a maximum of 303,069 to a minimum of 16,800. In turn, estimates for the counties with smaller population sizes are likely to be subject to a larger margin of error than those with larger baseline populations.

Comparing median household incomes of these sixteen counties illustrates this point. Let's first obtain this data with **tidycensus** then clean up the NAME column with str_remove() to remove redundant information.

```
maine_income <- get_acs(
  state = "Maine",
  geography = "county",
  variables = c(hhincome = "B19013_001"),
  year = 2020
) %>%
  mutate(NAME = str_remove(NAME, " County, Maine"))
```

Using some of the tips covered in the previous visualization section, we can produce a plot with appropriate styling and formatting to rank the counties.

```
ggplot(maine_income, aes(x = estimate, y = reorder(NAME, estimate))) +
  geom_point(size = 3, color = "darkgreen") +
  labs(title = "Median household income",
       subtitle = "Counties in Maine",
```

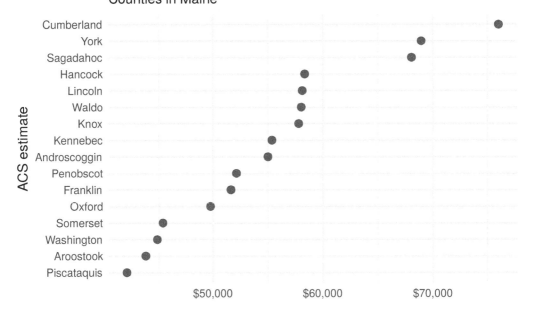

FIGURE 4.10 A dot plot of median household income by county in Maine

```
    x = "",
    y = "ACS estimate") + 
  theme_minimal(base_size = 12.5) + 
  scale_x_continuous(labels = label_dollar())
```

Figure 4.10 suggests a ranking of counties from the wealthiest (Cumberland) to the poorest (Piscataquis). However, the data used to generate this chart is significantly different from the metropolitan area data used in the previous example. In our first example, ACS estimates covered the top 20 US metros by population – areas that all have populations exceeding 2.8 million. For these areas, margins of error are small enough that they do not meaningfully change the interpretation of the estimates given the large sample sizes used to generate them. However, as discussed in Section 3.5, smaller geographies may have much larger margins of error relative to their ACS estimates.

4.3.2 Using error bars for margins of error

Several county estimates on the chart are quite close to one another, which may mean that the ranking of counties is misleading given the margin of error around those estimates. We can explore this by looking directly at the data.

```
maine_income %>% 
  arrange(desc(moe))
```

4.3 Visualizing margins of error

TABLE 4.3 Margins of error in Maine

GEOID	NAME	variable	estimate	moe
23023	Sagadahoc	hhincome	68039	4616
23015	Lincoln	hhincome	58125	3974
23027	Waldo	hhincome	58034	3482
23007	Franklin	hhincome	51630	2948
23021	Piscataquis	hhincome	42083	2883
23025	Somerset	hhincome	45382	2694
23009	Hancock	hhincome	58345	2593
23013	Knox	hhincome	57794	2528
23017	Oxford	hhincome	49761	2380
23003	Aroostook	hhincome	43791	2306
23029	Washington	hhincome	44847	2292
23031	York	hhincome	68932	2239
23011	Kennebec	hhincome	55368	2112
23001	Androscoggin	hhincome	55002	2003
23019	Penobscot	hhincome	52128	1836
23005	Cumberland	hhincome	76014	1563

Specifically, margins of error around the estimated median household incomes vary from a low of $1563 (Cumberland County) to a high of $4616 (Sagadahoc County). In many cases, the margins of error around estimated county household income exceed the differences between counties of neighboring ranks, suggesting uncertainty in the ranks themselves.

In turn, a dot plot like the one shown in Figure 4.10 intended to visualize a ranking of county household incomes in Maine may be misleading. However, using visualization tools in **ggplot2**, we can visualize the uncertainty around each estimate, giving chart readers a sense of the uncertainty in the ranking. This is accomplished with the `geom_errorbar()` function, which will plot horizontal error bars around each dot that stretch to a given value around each estimate. In this instance, we will use the `moe` column to determine the lengths of the error bars. The result is shown in Figure 4.11.

```
ggplot(maine_income, aes(x = estimate, y = reorder(NAME, estimate))) +
  geom_errorbarh(aes(xmin = estimate - moe, xmax = estimate + moe)) +
  geom_point(size = 3, color = "darkgreen") +
  theme_minimal(base_size = 12.5) +
  labs(title = "Median household income",
       subtitle = "Counties in Maine",
       x = "2016-2020 ACS estimate",
       y = "") +
  scale_x_continuous(labels = label_dollar())
```

Adding the horizontal error bars around each point gives us critical information to help us understand how our ranking of Maine counties by median household income. For example, while the ACS estimate suggests that Piscataquis County has the lowest median household income in Maine, the large margin of error around the estimate for Piscataquis County suggests that either Aroostook or Washington Counties *could* conceivably have lower median household incomes. Additionally, while Hancock County has a higher estimated median

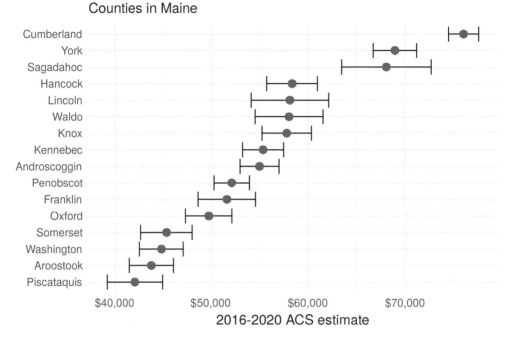

FIGURE 4.11 Median household income by county in Maine with error bars shown

household income than Lincoln, Waldo, and Knox Counties, the margin of error plot shows us that this ranking is subject to considerable uncertainty.

4.4 Visualizing ACS estimates over time

Section 3.4.2 covered how to obtain a time series of ACS estimates to explore temporal demographic shifts. While the output table usefully represented the time series of educational attainment in Colorado counties, data visualization is also commonly used to illustrate change over time. Arguably the most common chart type chosen for time-series visualization is the line chart, which **ggplot2** handles capably with the `geom_line()` function.

For an illustrative example, we'll obtain 1-year ACS data from 2005 through 2019 on median home value for Deschutes County, Oregon, home to the city of Bend and large numbers of in-migrants in recent years from the Bay Area in California. As in Chapter 3, `map_dfr()` is used to iterate over a named vector of years, creating a time-series dataset of median home value in Deschutes County since 2005, and we use the formula specification for anonymous functions so that `~ .x` translates to `function(x) x`.

```
years <- 2005:2019
names(years) <- years
```

4.4 Visualizing ACS estimates over time

TABLE 4.4 Time series of median home values in Deschutes County, OR

year	GEOID	NAME	variable	estimate	moe
2005	41017	Deschutes County, Oregon	B25077_001	236100	13444
2006	41017	Deschutes County, Oregon	B25077_001	336600	11101
2007	41017	Deschutes County, Oregon	B25077_001	356700	16765
2008	41017	Deschutes County, Oregon	B25077_001	331600	17104
2009	41017	Deschutes County, Oregon	B25077_001	284300	12652
2010	41017	Deschutes County, Oregon	B25077_001	260700	18197
2011	41017	Deschutes County, Oregon	B25077_001	216200	18065
2012	41017	Deschutes County, Oregon	B25077_001	235300	19016
2013	41017	Deschutes County, Oregon	B25077_001	240000	16955
2014	41017	Deschutes County, Oregon	B25077_001	257200	18488

```r
deschutes_value <- map_dfr(years, ~{
  get_acs(
    geography = "county",
    variables = "B25077_001",
    state = "OR",
    county = "Deschutes",
    year = .x,
    survey = "acs1"
  )
}, .id = "year")
```

This information can be visualized with familiar **ggplot2** syntax. `deschutes_value` is specified as the input dataset, with `year` mapped to the X-axis and `estimate` mapped to the y-axis. The argument `group = 1` is used to help **ggplot2** understand how to connect the yearly data points with lines given that only one county is being visualized. `geom_line()` then draws the lines, and we layer points on top of the lines as well to highlight the actual ACS estimates.

```r
ggplot(deschutes_value, aes(x = year, y = estimate, group = 1)) +
  geom_line() +
  geom_point()
```

The chart shows rising home values prior to the 2008 recession; a notable drop after the housing market crash; and rising values since 2011, reflecting increased demand from wealthy in-migrants from locations like the Bay Area. Given what we have learned in previous sections, there are also several opportunities for chart cleanup. This can include more intuitive tick and axis labels; a re-designed visual scheme; and a title and caption. We can also build the margin of error information into the line chart like we did in the previous section. We'll use the **ggplot2** function `geom_ribbon()` to draw the margin of error interval around the line, helping represent uncertainty in the ACS estimates.

```r
ggplot(deschutes_value, aes(x = year, y = estimate, group = 1)) +
  geom_ribbon(aes(ymax = estimate + moe, ymin = estimate - moe),
              fill = "navy",
              alpha = 0.4) +
```

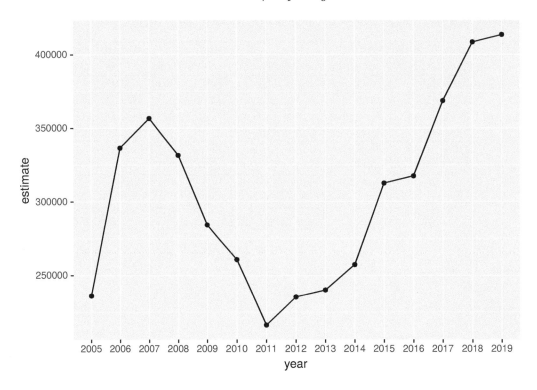

FIGURE 4.12 A time series chart of median home values in Deschutes County, OR

```
geom_line(color = "navy") +
geom_point(color = "navy", size = 2) +
theme_minimal(base_size = 12) +
scale_y_continuous(labels = label_dollar(scale = .001, suffix = "k")) +
labs(title = "Median home value in Deschutes County, OR",
     x = "Year",
     y = "ACS estimate",
     caption = "Shaded area represents margin of error around the ACS estimate")
```

4.5 Exploring age and sex structure with population pyramids

A common method for visualizing the demographic structure of a particular area is the *population pyramid*. Population pyramids are typically constructed by visualizing population size or proportion on the x-axis; age cohort on the y-axis; and sex is represented categorically with male and female bars mirrored around a central axis.

4.5.1 Preparing data from the Population Estimates API

We can illustrate this type of visualization using data from the Population Estimates API for the state of Utah. We first obtain data using the get_estimates() function in **tidycensus**

4.5 Exploring age and sex structure with population pyramids

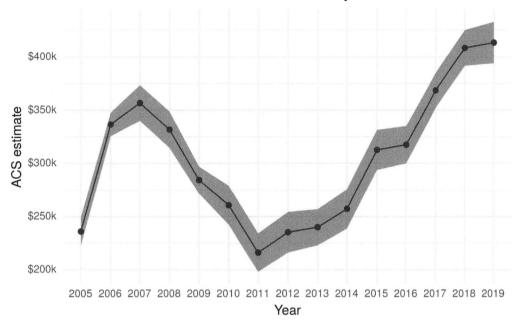

FIGURE 4.13 The Deschutes County home value line chart with error ranges shown

for 2019 population estimates from the Census Bureau's Population Estimates API. An extract of the returned data is shown in Table 4.5.

```
utah <- get_estimates(
  geography = "state",
  state = "UT",
  product = "characteristics",
  breakdown = c("SEX", "AGEGROUP"),
  breakdown_labels = TRUE,
  year = 2019
)
```

The function returns a long-form dataset in which each row represents population values broken down by age and sex for the state of Utah. However, there are some key issues with this dataset that must be addressed before constructing a population pyramid. First, several rows represent values that we don't need for our population pyramid visualization. For example, the first few rows in the dataset represent population values for `"Both sexes"` or for `"All ages"`. In turn, it will be necessary to isolate those rows that represent 5-year age bands by sex, and remove the rows that do not. This can be resolved with some data wrangling using tidyverse tools.

In the dataset returned by `get_estimates()`, 5-year age bands are identified in the `AGEGROUP` column beginning with the word `"Age"`. We can filter this dataset for rows that match this pattern, and remove those rows that represent both sexes. This leaves us with rows that represent 5-year age bands by sex. However, to achieve the desired visual effect, data for one

TABLE 4.5 Age and sex data for Utah from the PEP API

GEOID	NAME	value	SEX	AGEGROUP
49	Utah	3205958	Both sexes	All ages
49	Utah	247803	Both sexes	Age 0 to 4 years
49	Utah	258976	Both sexes	Age 5 to 9 years
49	Utah	267985	Both sexes	Age 10 to 14 years
49	Utah	253847	Both sexes	Age 15 to 19 years
49	Utah	264652	Both sexes	Age 20 to 24 years
49	Utah	251376	Both sexes	Age 25 to 29 years
49	Utah	220430	Both sexes	Age 30 to 34 years
49	Utah	231242	Both sexes	Age 35 to 39 years
49	Utah	212211	Both sexes	Age 40 to 44 years

TABLE 4.6 Filtered and transformed Utah population data

GEOID	NAME	value	SEX	AGEGROUP
49	Utah	-127060	Male	Age 0 to 4 years
49	Utah	-132868	Male	Age 5 to 9 years
49	Utah	-137940	Male	Age 10 to 14 years
49	Utah	-129312	Male	Age 15 to 19 years
49	Utah	-135806	Male	Age 20 to 24 years
49	Utah	-129179	Male	Age 25 to 29 years
49	Utah	-111776	Male	Age 30 to 34 years
49	Utah	-117335	Male	Age 35 to 39 years
49	Utah	-108090	Male	Age 40 to 44 years
49	Utah	-89984	Male	Age 45 to 49 years

sex must mirror another, split by a central vertical axis. To accomplish this, we can set the values for all `Male` values to negative. This transformation is shown in Table 4.6.

```
utah_filtered <- filter(utah, str_detect(AGEGROUP, "^Age"),
              SEX != "Both sexes") %>%
  mutate(value = ifelse(SEX == "Male", -value, value))
```

4.5.2 Designing and styling the population pyramid

The data are now ready for visualization. The core components of the pyramid visualization require mapping the population value and the age group to the chart axes. Sex can be mapped to the `fill` aesthetic allowing for the plotting of these categories by color.

```
ggplot(utah_filtered, aes(x = value, y = AGEGROUP, fill = SEX)) +
  geom_col()
```

The visualization shown in Figure 4.14 represents a functional population pyramid that is nonetheless in need of some cleanup. In particular, the axis labels are not informative; the y-axis tick labels have redundant information ("Age" and "years"); and the x-axis tick labels are difficult to parse. Cleaning up the plot allows us to use some additional visualization

4.5 Exploring age and sex structure with population pyramids

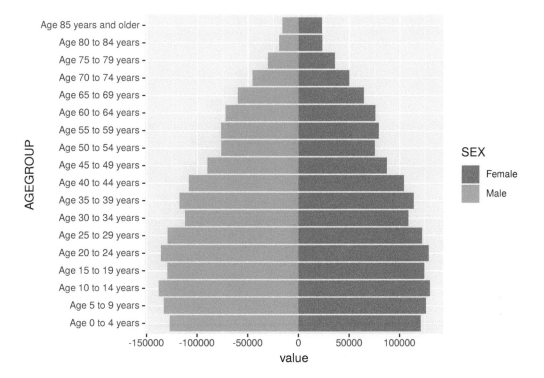

FIGURE 4.14 A first population pyramid

options in **ggplot2**. In addition to specifying appropriate chart labels, we can format the axis tick labels by using appropriate scale_* functions in **ggplot2** and setting the X-axis limits to show both sides of 0 equally. In particular, this involves the use of custom absolute values to represent population sizes, and the removal of redundant age group information. We'll also make use of an alternative **ggplot2** theme, theme_minimal(), which uses a white background with muted gridlines. The result is shown in Figure 4.15.

```
utah_pyramid <- ggplot(utah_filtered, 
                      aes(x = value, 
                          y = AGEGROUP,
                          fill = SEX)) + 
  geom_col(width = 0.95, alpha = 0.75) + 
  theme_minimal(base_family = "Verdana", 
                base_size = 12) + 
  scale_x_continuous(
    labels = ~ number_format(scale = .001, suffix = "k")(abs(.x)),
    limits = 140000 * c(-1,1)
  ) + 
  scale_y_discrete(labels = ~ str_remove_all(.x, "Age\\s|\\syears")) + 
  scale_fill_manual(values = c("darkred", "navy")) + 
  labs(x = "",
       y = "2019 Census Bureau population estimate",
       title = "Population structure in Utah",
       fill = "",
```

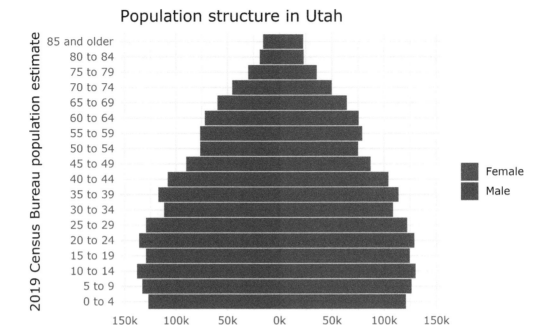

FIGURE 4.15 A formatted population pyramid of Utah

```
  caption = "Sources: US Census Bureau PEP, tidycensus R package")

utah_pyramid
```

4.6 Visualizing group-wise comparisons

One of the most powerful features of **ggplot2** is its ability to generate *faceted plots*, which are also commonly referred to as *small multiples*. Faceted plots allow for the sub-division of a dataset into groups, which are then plotted side-by-side to facilitate comparisons between those groups. This is particularly useful when examining how distributions of values vary across different geographies. An example shown below involves a comparison of median home values by Census tract for three counties in the Portland, Oregon area: Multnomah, which contains the city of Portland, and the suburban counties of Clackamas and Washington.

```
housing_val <- get_acs(
  geography = "tract",
  variables = "B25077_001",
  state = "OR",
  county = c(
```

4.6 Visualizing group-wise comparisons

TABLE 4.7 Median home values by Census tract in the Portland, OR area, 2016–2020 ACS

GEOID	NAME	variable	estimate	moe
41005020101	Census Tract 201.01, Clackamas County, Oregon	B25077_001	666700	131453
41005020102	Census Tract 201.02, Clackamas County, Oregon	B25077_001	909000	130787
41005020201	Census Tract 202.01, Clackamas County, Oregon	B25077_001	897400	97893
41005020202	Census Tract 202.02, Clackamas County, Oregon	B25077_001	821200	93103
41005020302	Census Tract 203.02, Clackamas County, Oregon	B25077_001	565600	32555

TABLE 4.8 Data with NAME column split by comma

GEOID	tract	county	state	variable	estimate	moe
41005020101	Census Tract 201.01	Clackamas County	Oregon	B25077_001	666700	131453
41005020102	Census Tract 201.02	Clackamas County	Oregon	B25077_001	909000	130787
41005020201	Census Tract 202.01	Clackamas County	Oregon	B25077_001	897400	97893
41005020202	Census Tract 202.02	Clackamas County	Oregon	B25077_001	821200	93103
41005020302	Census Tract 203.02	Clackamas County	Oregon	B25077_001	565600	32555

```r
    "Multnomah",
    "Clackamas",
    "Washington",
    "Yamhill",
    "Marion",
    "Columbia"
  ),
  year = 2020
)
```

As with other datasets obtained with **tidycensus**, the `NAME` column contains descriptive information that can be parsed to make comparisons. In this case, Census tract ID, county, and state are separated with commas; in turn the tidyverse `separate()` function can split this column into three columns accordingly.

```r
housing_val2 <- separate(
  housing_val,
  NAME,
  into = c("tract", "county", "state"),
  sep = ", "
)
```

As explored in previous chapters, a major strength of the tidyverse is its ability to perform group-wise data analysis. The dimensions of median home values by Census tract in each of the three counties can be explored in this way. For example, a call to `group_by()` followed by `summarize()` facilitates the calculation of county minimums, means, medians, and maximums.

```r
housing_val2 %>%
  group_by(county) %>%
  summarize(min = min(estimate, na.rm = TRUE),
```

TABLE 4.9 Summary statistics for Census tracts in Portland-area counties

county	min	mean	median	max
Clackamas County	62800	449940.7	426700	909000
Columbia County	218100	277590.9	275900	362200
Marion County	48700	270969.2	261200	483500
Multnomah County	192900	455706.2	425950	1033500
Washington County	221900	419618.3	406100	769700
Yamhill County	230000	333815.8	291100	545500

```
            mean = mean(estimate, na.rm = TRUE),
            median = median(estimate, na.rm = TRUE),
            max = max(estimate, na.rm = TRUE))
```

While these basic summary statistics offer some insights into comparisons between the three counties, they are limited in their ability to help us understand the dynamics of the overall distribution of values. This task can in turn be augmented through visualization, which allows for quick visual comparison of these distributions. Group-wise visualization in **ggplot2** can be accomplished with the `facet_wrap()` function added onto any existing **ggplot2** code that has salient groups to visualize. For example, a kernel density plot (shown in Figure 4.16) can show the overall shape of the distribution of median home values in our dataset:

```
ggplot(housing_val2, aes(x = estimate)) + 
  geom_density()
```

Mapping the `county` column onto the `fill` aesthetic will then draw superimposed density plots by county on the chart, as shown in Figure 4.17:

```
ggplot(housing_val2, aes(x = estimate, fill = county)) + 
  geom_density(alpha = 0.3)
```

Alternatively, adding the `facet_wrap()` function, and specifying `county` as the column used to group the data, splits this visualization into side-by-side graphics based on the counties to which each Census tract belongs.

```
ggplot(housing_val2, aes(x = estimate)) + 
  geom_density(fill = "darkgreen", color = "darkgreen", alpha = 0.5) + 
  facet_wrap(~county) + 
  scale_x_continuous(labels = dollar_format(scale = 0.000001, 
                                            suffix = "m")) + 
  theme_minimal(base_size = 14) + 
  theme(axis.text.y = element_blank(),
        axis.text.x = element_text(angle = 45)) + 
  labs(x = "ACS estimate",
       y = "",
       title = "Median home values by Census tract, 2015-2019 ACS")
```

4.6 Visualizing group-wise comparisons

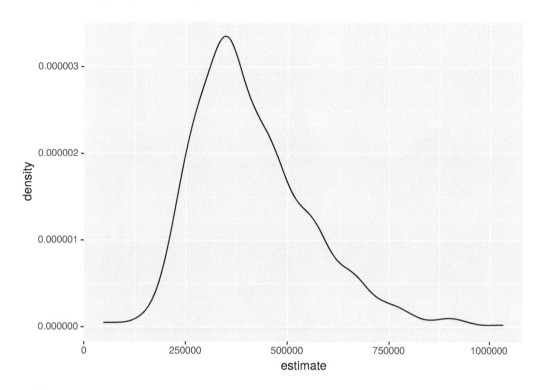

FIGURE 4.16 A density plot using all values in the dataset

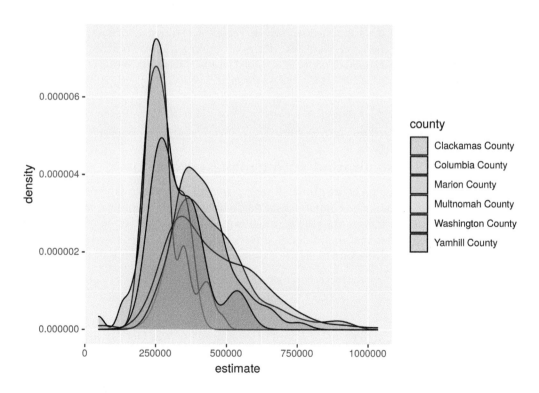

FIGURE 4.17 A density plot with separate curves for each county

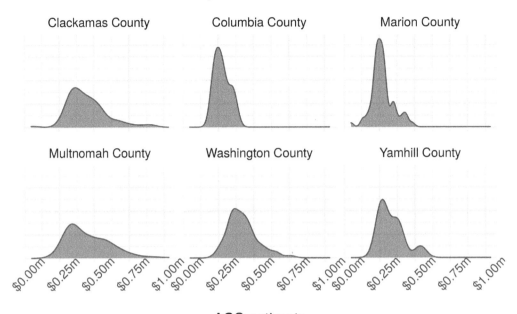

FIGURE 4.18 An example faceted density plot

The side-by-side comparative graphics shown in Figure 4.18 illustrate how the value distributions vary between the three counties. Home values in all three counties are common around $250,000, but Multnomah County has some Census tracts that represent the highest values in the dataset.

4.7 Advanced visualization with ggplot2 extensions

While the core functionality of **ggplot2** is very powerful, a notable advantage of using **ggplot2** for visualization is the contributions made by its user community in the form of *extensions*. ggplot2 extensions are packages developed by practitioners outside the core **ggplot2** development team that add functionality to the package. I encourage you to review the gallery of **ggplot2** extensions from the extensions website[1]; I highlight some notable examples below.

4.7.1 ggridges

The **ggridges** package (Wilke, 2021) adapts the concept of the faceted density plot to generate *ridgeline plots*, in which the densities overlap one another. The example in Figure 4.19 creates a ridgeline plot using the Portland-area home value data; `geom_density_ridges()` generates the ridgelines, and `theme_ridges()` styles the plot in an appropriate manner.

---

[1] https://exts.ggplot2.tidyverse.org/

4.7 Advanced visualization with ggplot2 extensions

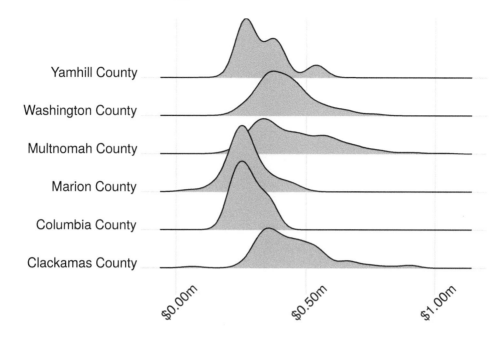

FIGURE 4.19 Median home values in Portland-area counties visualized with ggridges

```
library(ggridges)

ggplot(housing_val2, aes(x = estimate, y = county)) + 
  geom_density_ridges() + 
  theme_ridges() + 
  labs(x = "Median home value: 2016-2020 ACS estimate",
       y = "") + 
  scale_x_continuous(labels = label_dollar(scale = .000001, suffix = "m"),
                     breaks = c(0, 500000, 1000000)) + 
  theme(axis.text.x = element_text(angle = 45))
```

The overlapping density "ridges" offer both a pleasing aesthetic but also a practical way to compare the different data distributions. As **ggridges** extends **ggplot2**, analysts can style the different chart components to their liking using the methods introduced earlier in this chapter.

4.7.2 ggbeeswarm

The **ggbeeswarm** package (Clarke and Sherrill-Mix, 2017) extends ggplot2 by allowing users to generate *beeswarm plots*, in which clouds of points are jittered to show the overall density of a distribution of data values. Beeswarm plots can be compelling ways to visualize multiple data variables on a chart, such as the distributions of median household income by the racial and ethnic composition of neighborhoods. This is the motivating example for the chart below, which looks at household income by race/ethnicity in New York City. The data wrangling in the first part of the code chunk takes advantage of some of the skills covered in Chapters 2 and 3, allowing for visualization of household income distributions by the largest group in each Census tract.

```r
library(ggbeeswarm)

ny_race_income <- get_acs(
  geography = "tract",
  state = "NY",
  county = c("New York", "Bronx", "Queens", "Richmond", "Kings"),
  variables = c(White = "B03002_003",
                Black = "B03002_004",
                Asian = "B03002_006",
                Hispanic = "B03002_012"),
  summary_var = "B19013_001",
  year = 2020
) %>%
  group_by(GEOID) %>%
  filter(estimate == max(estimate, na.rm = TRUE)) %>%
  ungroup() %>%
  filter(estimate != 0)

ggplot(ny_race_income, aes(x = variable, y = summary_est, color = summary_est)) +
  geom_quasirandom(alpha = 0.5) +
  coord_flip() +
  theme_minimal(base_size = 13) +
  scale_color_viridis_c(guide = "none") +
  scale_y_continuous(labels = label_dollar()) +
  labs(x = "Largest group in Census tract",
       y = "Median household income",
       title = "Household income distribution by largest racial/ethnic group",
       subtitle = "Census tracts, New York City",
       caption = "Data source: 2016-2020 ACS")
```

The plot in Figure 4.20 shows that the wealthiest neighborhoods in New York City – those with median household incomes exceeding $150,000 – are all plurality or majority non-Hispanic white. However, the chart also illustrates that there are a range of values among neighborhoods with pluralities of the different racial and ethnic groups, suggesting a nuanced portrait of the intersections between race and income inequality in the city.

4.7.3 Geofaceted plots

The next four chapters of the book, Chapters 5 through 8, are all about spatial data, mapping, and spatial analysis. Geographic location can be incorporated into Census data visualizations without using geographic information explicitly by way of the **geofacet** package (Hafen, 2020). Geofaceted plots are enhanced versions of faceted visualizations that arrange subplots in relationship to their relative geographic location. The geofacet package has over 100 available grids to choose from[2] allowing for faceted plots for US states, counties, and regions

---

[2] https://hafen.github.io/geofacet/articles/geofacet.html#list-available-grids

4.7 Advanced visualization with ggplot2 extensions   89

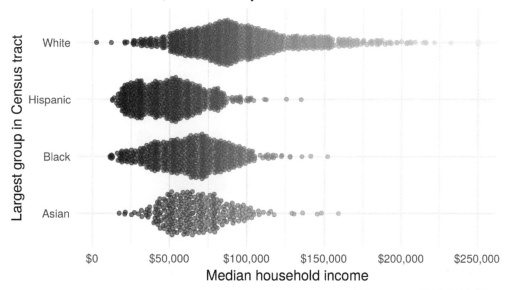

FIGURE 4.20 A beeswarm plot of median household income by most common racial or ethnic group, NYC Census tracts

around the world. The key is to use a column that can map correctly to information in the geofaceted grid that you are using.

In the example below, we replicate the population pyramid code above to generate population pyramids for each state in the United States. However, we also modify the output data so that the population information reflects the proportion of the overall population for each state so that all states are on consistent scales. The population pyramid code is similar, though the axis information is unnecessary as it will be stripped from the final plot.

```
library(geofacet)

us_pyramid_data <- get_estimates(
  geography = "state",
  product = "characteristics",
  breakdown = c("SEX", "AGEGROUP"),
  breakdown_labels = TRUE,
  year = 2019
) %>%
  filter(str_detect(AGEGROUP, "^Age"),
         SEX != "Both sexes") %>%
  group_by(NAME) %>%
  mutate(prop = value / sum(value, na.rm = TRUE)) %>%
  ungroup() %>%
```

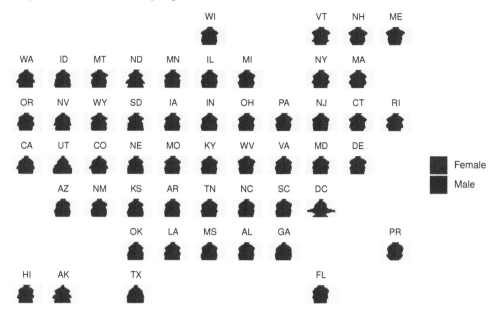

FIGURE 4.21 Geofaceted population pyramids of US states

```
  mutate(prop = ifelse(SEX == "Male", -prop, prop))

ggplot(us_pyramid_data, aes(x = prop, y = AGEGROUP, fill = SEX)) +
  geom_col(width = 1) +
  theme_minimal() +
  scale_fill_manual(values = c("darkred", "navy")) +
  facet_geo(~NAME, grid = "us_state_with_DC_PR_grid2",
            label = "code") +
  theme(axis.text = element_blank(),
        strip.text.x = element_text(size = 8)) +
  labs(x = "",
       y = "",
       title = "Population structure by age and sex",
       fill = "",
       caption = "Data source: US Census Bureau population estimates & tidycensus R
         package")
```

The result in Figure 4.21 is a compelling visualization that expresses general geographic relationships and patterns in the data while showing comparative population pyramids for all states. The unique nature of Washington, DC stands out with its large population in their 20s and 30s; we can also compare Utah, the youngest state in the country, with older states in the Northeast.

If this specific grid arrangement is not to your liking (e.g. Minnesotans may take issue with Wisconsin to the north!), try out some of the other built-in grid options in the package. Also,

4.7 Advanced visualization with ggplot2 extensions

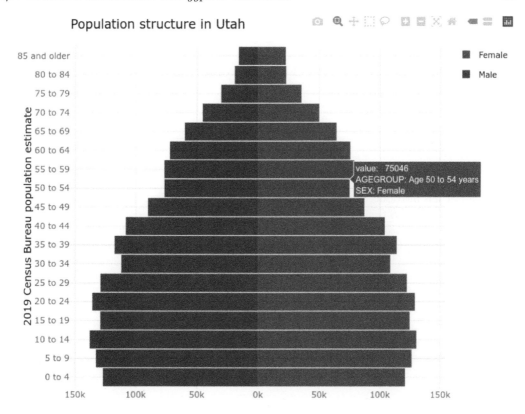

FIGURE 4.22 An interactive population pyramid rendered with ggplotly

you can use the `grid_design()` function in the **geofacet** package to pull up an interactive app in which you can design your own grid for use in geofaceted plots!

4.7.4 Interactive visualization with plotly

The **htmlwidgets** package (Vaidyanathan et al., 2020) provides a bridge between JavaScript libraries for interactive web-based data visualization and the R language. Since 2015, the R community has released hundreds of packages that depend on htmlwidgets and bring interactive data visualization to R. One of the most popular packages for interactive visualization is the **plotly** package (Sievert, 2020), an interface to the Plotly visualization library.

plotly is a well-developed library that allows for many different types of custom visualizations. However, one of the most useful functions in the **plotly** package is arguably the simplest. The `ggplotly()` function can convert an existing **ggplot2** graphic into an interactive web visualization in a single line of code! Let's try it out here with the `utah_pyramid` population pyramid:

```
library(plotly)

ggplotly(utah_pyramid)
```

Try hovering your cursor over the different bars; this reveals a tooltip that shares information about the data. The legend is interactive, as data series can be clicked on and off; viewers can also pan and zoom on the plot using the toolbar that appears in the upper right corner of the visualization. Interactive graphics like this are an excellent way to facilitate additional data exploration, and can be polished further for presentation on the web.

4.8 Learning more about visualization

This chapter has introduced a series of visualization techniques implemented in **ggplot2** that are appropriate for US Census Bureau data. Readers may want to learn more about effective principles for visualization design and communication and apply that to the techniques covered here. While Chapter 6 covers some of these principles in brief, literature that focuses specifically on these topics will be more comprehensive. Munzner (2014) is an in-depth overview of visualization techniques and design principles, and offers a corresponding website with lecture slides[3]. Evergreen (2020) and Knaflic (2015) provide guidelines for effective communication with visualization and both offer excellent design tips for business and general audiences. R users may be interested in Wilke (2019) and Healy (2019), which both offer a comprehensive overview of data visualization best practices along with corresponding R code to reproduce the figures in their books.

4.9 Exercises

- Choose a different variable in the ACS and/or a different location and create a margin of error visualization of your own.
- Modify the population pyramid code to create a different, customized population pyramid. You can choose a different location (state or county), different colors/plot design, or some combination!

---

[3] https://www.cs.ubc.ca/~tmm/vadbook/

5
Census geographic data and applications in R

As discussed in previous chapters of this book, Census and ACS data are associated with *geographies*, which are units at which the data are aggregated. These defined geographies are represented in the US Census Bureau's TIGER/Line database[1], where the acronym **TIGER** stands for *Topologically Integrated Geographic Encoding and Referencing*. This database includes a high-quality series of geographic datasets suitable for both spatial analysis and cartographic visualization . Spatial datasets are made available as *shapefiles*, a common format for encoding geographic data.

The TIGER/Line shapefiles include three general types of data:

- *Legal entities*, which are geographies that have official legal standing in the United States. These include states and counties.
- *Statistical entities*, which are geographies defined by the Census Bureau for purposes of data collection and dissemination. Examples of statistical entities include Census tracts and block groups.
- *Geographic features*, which are geographic datasets that are not linked with aggregate demographic data from the Census Bureau. These datasets include roads and water features.

Traditionally, TIGER/Line shapefiles are downloaded from a web interface as zipped folders, then unzipped for use in a Geographic Information System (GIS) or other software that can work with geographic data. However, the R package **tigris** (Walker, 2016b) allows R users to access these datasets directly from their R sessions without having to go through these steps.

This chapter will cover the core functionality of the **tigris** package for working with Census geographic data in R. In doing so, it will highlight the **sf** package (Pebesma, 2018) for representing spatial data as R objects.

5.1 Basic usage of tigris

The **tigris** R package simplifies the process for R users of obtaining and using Census geographic datasets. Functions in **tigris** *download* a requested Census geographic dataset from the US Census Bureau website, then *load* the dataset into R as a spatial object. Generally speaking, each type of geographic dataset available in the Census Bureau's TIGER/Line database is available with a corresponding function in **tigris**. For example, the states()

---

[1]https://www2.census.gov/geo/pdfs/maps-data/data/tiger/tgrshp2019/TGRSHP2019_TechDoc.pdf

function can be run without arguments to download a boundary file of US states and state equivalents.

```
library(tigris)

st <- states()

## Retrieving data for the year 2020
```

We get a message letting us know that data for the year 2020 are returned; **tigris** typically defaults to the most recent year for which a complete set of Census shapefiles are available, which at the time of this writing is 2020.

Let's take a look at what we got back:

```
class(st)

## [1] "sf"         "data.frame"
```

The returned object is of both class "sf" and "data.frame". We can print out the first 10 rows to inspect the object further:

```
st
```

```
## Simple feature collection with 56 features and 14 fields
## Geometry type: MULTIPOLYGON
## Dimension:     XY
## Bounding box:  xmin: -179.2311 ymin: -14.60181 xmax: 179.8597 ymax: 71.43979
## Geodetic CRS:  NAD83
## First 10 features:
##    REGION DIVISION STATEFP  STATENS GEOID STUSPS           NAME LSAD MTFCC
## 1       3        5      54 01779805    54     WV  West Virginia   00 G4000
## 2       3        5      12 00294478    12     FL        Florida   00 G4000
## 3       2        3      17 01779784    17     IL       Illinois   00 G4000
## 4       2        4      27 00662849    27     MN      Minnesota   00 G4000
## 5       3        5      24 01714934    24     MD       Maryland   00 G4000
## 6       1        1      44 01219835    44     RI   Rhode Island   00 G4000
## 7       4        8      16 01779783    16     ID          Idaho   00 G4000
## 8       1        1      33 01779794    33     NH  New Hampshire   00 G4000
## 9       3        5      37 01027616    37     NC North Carolina   00 G4000
## 10      1        1      50 01779802    50     VT        Vermont   00 G4000
##    FUNCSTAT        ALAND      AWATER    INTPTLAT     INTPTLON
## 1         A  62266296765   489206049 +38.6472854 -080.6183274
## 2         A 138958484319 45975808217 +28.3989775 -082.5143005
## 3         A 143778461053  6216594318 +40.1028754 -089.1526108
## 4         A 206232157570 18949864226 +46.3159573 -094.1996043
## 5         A  25151895765  6979171386 +38.9466584 -076.6744939
## 6         A   2677759219  1323691129 +41.5964850 -071.5264901
## 7         A 214049923496  2391577745 +44.3484222 -114.5588538
## 8         A  23190113978  1025973001 +43.6726907 -071.5843145
## 9         A 125933025759 13456395178 +35.5397100 -079.1308636
```

5.1 Basic usage of tigris

FIGURE 5.1 Default US states data obtained with tigris

```
## 10         A  23873081385  1030243281 +44.0589536 -072.6710173
##                            geometry
## 1  MULTIPOLYGON (((-81.74725 3...
## 2  MULTIPOLYGON (((-86.39964 3...
## 3  MULTIPOLYGON (((-91.18529 4...
## 4  MULTIPOLYGON (((-96.78438 4...
## 5  MULTIPOLYGON (((-77.45881 3...
## 6  MULTIPOLYGON (((-71.7897 41...
## 7  MULTIPOLYGON (((-116.8997 4...
## 8  MULTIPOLYGON (((-72.3299 43...
## 9  MULTIPOLYGON (((-82.41674 3...
## 10 MULTIPOLYGON (((-73.31328 4...
```

The object st, representing all US states and territories, includes a data frame with a series of columns representing characteristics of those states, like a name, postal code, and Census ID (the GEOID column). It also contains a special list-column, geometry, which is made up of a sequence of coordinate of longitude/latitude coordinate pairs that collectively represent the boundary of each state.

The geometry column can be visualized in R with the plot() function, shown in Figure 5.1:

```
plot(st$geometry)
```

Other Census datasets may be available by state or by county within the state. In some cases, this subsetting is optional; in other cases, state and/or county arguments will be required. For example, the counties() function can be used to obtain county boundaries for the entirety of the United States but also can be used with the state argument to return only those counties from a specific state, like New Mexico.

```
nm_counties <- counties("NM")
```

```
plot(nm_counties$geometry)
```

In this case the state postal code "NM" is used to instruct **tigris** to subset the counties dataset for counties in New Mexico. The full name of the state, "New Mexico", would work the same here as well. Obtaining Census shapefiles programmatically requires inputting the Federal Information Processing Standard (FIPS) code; however, **tigris** translates postal

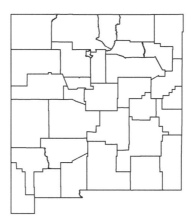

FIGURE 5.2 County boundaries in New Mexico

codes and names of states and counties to their FIPS codes so that R users do not have to look them up.

States and counties are examples of *legal entities* that can be accessed with **tigris**. *Statistical entities* and *geographic features* are similarly accessible if they exist in the TIGER/Line database. For example, a user might request Census tract boundaries for a given county in New Mexico with the corresponding tracts() function.

```
la_tracts <- tracts("NM", "Los Alamos")

plot(la_tracts$geometry)
```

Several geographic features are available in **tigris** as well, including roads and water features, which can be useful for thematic mapping. For example, a user could request area water data for Los Alamos County with the area_water() function.

FIGURE 5.3 Census tract boundaries in Los Alamos County, NM

FIGURE 5.4 Water area in Los Alamos County, NM

```
la_water <- area_water("NM", "Los Alamos")

plot(la_water$geometry)
```

5.1.1 Understanding tigris and simple features

Data returned by the **tigris** package are examples of *vector spatial data*, a spatial data model that represents geographic features as points, lines, and polygons. The vector spatial data model is represented in R with the **sf** package[2], an implementation of simple features in the R language. The **sf** package is an R interface to C libraries that power much of the broader geographic data ecosystem: GDAL[3] for reading & writing spatial data, GEOS[4] for modeling spatial relationships, and PROJ[5] for representing coordinate reference systems. These topics will be outlined in more detail in this chapter and the remainder of this section.

As mentioned earlier, **sf** represents vector spatial data much like a regular R data frame, but with a special column, geometry, that represents the shape of each feature. When a simple features object is printed, the information above the data frame gives some additional geographic context to the coordinates in the geometry column. This includes a *geometry type*, a *bounding box*, and a *coordinate reference system (CRS)* definition. These spatial concepts help define how R represents the data geographically, and will be explored further later in this chapter.

Vector data are typically represented as either *points, lines*, or *polygons*, and **tigris** gives access to all three types.

5.1.1.1 Points

An example point dataset available in the **tigris** package is Census landmarks, which is a point-of-interest dataset that is not comprehensive but is used by the Census Bureau to

---

[2] https://r-spatial.github.io/sf/
[3] https://gdal.org/
[4] https://trac.osgeo.org/geos/
[5] https://proj.org/

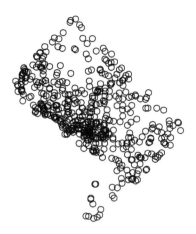

FIGURE 5.5 Census landmarks in Washington, DC

guide field enumerators. Let's acquire landmark point data for the District of Columbia and take a look.

```
dc_landmarks <- landmarks("DC", type = "point")

plot(dc_landmarks$geometry)
```

Points are vector data represented by a single coordinate pair; while they have a location, they do not have length or area and in turn are zero-dimensional. Points are useful for representing geographic phenomena when the physical properties of the features are not of importance to a visualization or analysis. For example, if we are interested in the geographic distribution of Census landmarks in Washington DC, but not in the actual shape or physical area of those specific landmarks, representing landmarks as points makes sense. **sf** represents points with the geometry type POINT.

5.1.1.2 Lines

```
dc_roads <- primary_secondary_roads("DC")

plot(dc_roads$geometry)
```

Lines are one-dimensional representations of geographic features that are used when the length, but not the area, of those features is of primary importance. With respect to the TIGER/Line shapefiles, transportation network features such as roads and railroads are represented as lines. Line features will have at least two linked coordinate pairs, and complex linear representations will have many more. Lines are represented with the geometry type LINESTRING.

5.1 Basic usage of tigris

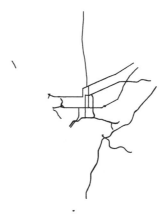

FIGURE 5.6 Primary and secondary roads in Washington, DC

5.1.1.3 Polygons

```
dc_block_groups <- block_groups("DC")

plot(dc_block_groups$geometry)
```

Polygons are enclosed shapes with at least three connected coordinate pairs. With respect to Census geometries, enumeration units like block groups are represented as polygons in the TIGER/Line files. Polygon geometry is useful when an analyst needs to represent the shape or area of geographic features in their project.

These three core geometries (point, line, and polygon) can be encoded in more complex ways in the simple features representation used by sf. For example, the geometry type POLYGON will use one row in a simple features data frame for each discrete shape; the geometry type MULTIPOLYGON, in contrast, can link multiple discrete shapes as part of the same geographic feature. This is important for encoding features that may have detached parts, such as a series of islands that belong to the same county. In this vein, points can be represented as

FIGURE 5.7 Block groups in Washington, DC

`MULTIPOINT` and lines can be represented as `MULTILINESTRING`, respectively, to accommodate similar scenarios.

5.1.2 Data availability in tigris

The above examples have provided a sampling of some of the datasets available in **tigris**; a full enumeration of available datasets and the functions to access them are found in the guide below.

Function	Datasets available	Years available
nation()	cartographic (1:5m; 1:20m)	2013-2021
divisions()	cartographic (1:500k; 1:5m; 1:20m)	2013-2021
regions()	cartographic (1:500k; 1:5m; 1:20m)	2013-2021
states()	TIGER/Line; cartographic (1:500k; 1:5m; 1:20m)	1990, 2000, 2010-2021
counties()	TIGER/Line; cartographic (1:500k; 1:5m; 1:20m)	1990, 2000, 2010-2021
tracts()	TIGER/Line; cartographic (1:500k)	1990, 2000, 2010-2021
block_groups()	TIGER/Line; cartographic (1:500k)	1990, 2000, 2010-2021
blocks()	TIGER/Line	2000, 2010-2021
places()	TIGER/Line; cartographic (1:500k)	2011-2021
pumas()	TIGER/Line; cartographic (1:500k)	2012-2021
school_districts()	TIGER/Line; cartographic	2011-2021
zctas()	TIGER/Line; cartographic (1:500k)	2000, 2010, 2012-2021
congressional_districts()	TIGER/Line; cartographic (1:500k; 1:5m; 1:20m)	2011-2021
state_legislative_districts()	TIGER/Line; cartographic (1:500k)	2011-2021
voting_districts()	TIGER/Line	2012, 2020-2021
area_water()	TIGER/Line	2011-2021
linear_water()	TIGER/Line	2011-2021
coastline()	TIGER/Line()	2013-2021
core_based_statistical_areas()	TIGER/Line; cartographic (1:500k; 1:5m; 1:20m)	2011-2021
combined_statistical_areas()	TIGER/Line; cartographic (1:500k; 1:5m; 1:20m)	2011-2021
metro_divisions()	TIGER/Line	2011-2021
new_england()	TIGER/Line; cartographic (1:500k)	2011-2021
county_subdivisions()	TIGER/Line; cartographic (1:500k)	2010-2021
urban_areas()	TIGER/Line; cartographic (1:500k)	2012-2021
primary_roads()	TIGER/Line	2011-2021
primary_secondary_roads()	TIGER/Line	2011-2021
roads()	TIGER/Line	2011-2021
rails()	TIGER/Line	2011-2021
native_areas()	TIGER/Line; cartographic (1:500k)	2011-2021

Function	Datasets available	Years available
alaska_native_regional_corporations()	TIGER/Line; cartographic (1:500k)	2011-2021
tribal_block_groups()	TIGER/Line	2011-2021
tribal_census_tracts()	TIGER/Line	2011-2021
tribal_subdivisions_national()	TIGER/Line	2011-2021
landmarks()	TIGER/Line	2011-2021
military()	TIGER/Line	2011-2021

Note from the guide that many datasets are available as both **TIGER/Line** and **cartographic boundary** versions, and can be downloaded for multiple years; these distinctions are covered in Section 5.3 below. Years available reflect the time of this writing; tigris supports new yearly datasets as they are released.

5.2 Plotting geographic data

Geographic information science is an inherently visual discipline. For analysts coming to R from a desktop GIS background (e.g. ArcGIS, QGIS), they will be used to having a visual display of their geographic data as central to their interactions with it. This may make the transition to R unfamiliar for geospatial analysts as geographic data will be first and foremost represented as a tabular data frame.

In the previous section, we have used the plot() function to visualize the geometry column of a simple features object obtained with **tigris**. R includes a variety of other options for quick visualization of geographic data that will be useful to geospatial anlaysts.

5.2.1 ggplot2 and `geom_sf()`

As of **ggplot2** version 3.0, the package released support for plotting simple features objects directly with the function geom_sf(). Prior to the release of this functionality, plotting geographic data (and by consequence making maps) was reasonably cumbersome; geom_sf() streamlines the geographic visualization process and makes **ggplot2** a go-to package for visualization of simple features objects.

At a basic level, a couple lines of **ggplot2** code are all that are needed to plot Census shapes obtained with **tigris**. For example, taking a look at our Los Alamos County Census tracts:

```
library(ggplot2)

ggplot(la_tracts) + 
  geom_sf()
```

By default (shown in Figure 5.8), ggplot2 includes its standard grey grid with latitude and longitude values displayed along the axes. For many cartographic applications, an analyst will want to remove this background information. The theme_void() function strips the background grid and axis labels from the plot accordingly, shown in Figure 5.9:

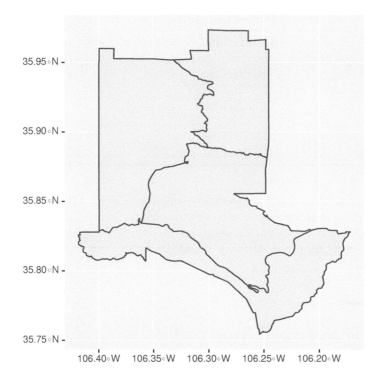

FIGURE 5.8 A default ggplot2 map.

```
ggplot(la_tracts) +
  geom_sf() +
  theme_void()
```

Section 4.6 introduced the concept of *faceted plots* to compare different views, which is also a very useful concept for geographic visualization. Faceted mapping will be addressed directly in the next chapter. For comparative spatial plots, the **patchwork** R package (Pedersen, 2020) works very well for arranging a multi-plot layout. In Figure 5.10 we'll use patchwork to put two **ggplot2** spatial plots – one of Census tracts and one of block groups in Los Alamos County – side-by-side using the + operator.

```
library(patchwork)

la_block_groups <- block_groups("NM", "Los Alamos")

gg1 <- ggplot(la_tracts) +
  geom_sf() +
  theme_void() +
  labs(title = "Census tracts")

gg2 <- ggplot(la_block_groups) +
  geom_sf() +
  theme_void() +
```

5.2 Plotting geographic data

FIGURE 5.9 A ggplot2 map with a blank background

```
  labs(title = "Block groups")

gg1 + gg2
```

Alternatively, **patchwork** allows R users to arrange plots vertically using the / operator.

5.2.2 Interactive viewing with mapview

A major hesitation for geospatial analysts considering a switch from desktop GIS software to R is the strengths of desktop GIS for interactive visual exploration. Both ArcGIS and QGIS allow analysts to quickly load a geographic dataset and interactively pan and zoom to explore geographic trends. Prior to the development of the htmlwidgets framework (discussed in the previous chapter), R had no equivalent capabilities. Frankly, this was the major reason I hesitated to fully transition my work from desktop GIS software to R for spatial analysis.

The **mapview** R package (Appelhans et al., 2020) fills this crucial gap. With a single call to its function mapview(), **mapview** visualizes geographic data on an interactive, zoomable map. Let's try it here with our Census tracts in Los Alamos County.

```
library(mapview)
```

```
mapview(la_tracts)
```

Clicking on a Census tract shape reveals a pop-up with attribute information found in the dataset. Additionally, users can change the underlying basemap to understand the

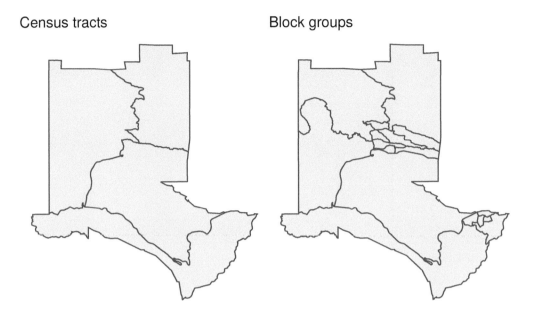

FIGURE 5.10 Comparing Census tracts and block groups

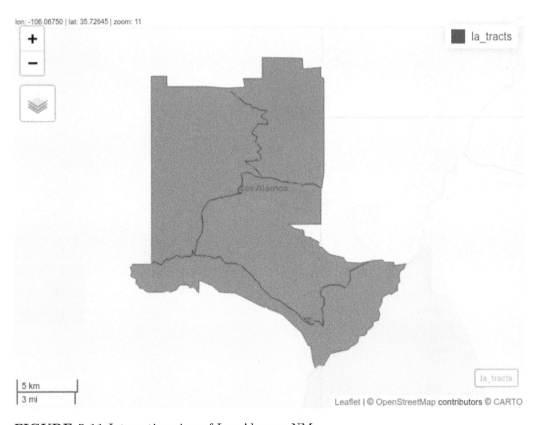

FIGURE 5.11 Interactive view of Los Alamos, NM

geographic context surrounding their data. **mapview** also includes significant functionality for interactive mapping and data display beyond this basic example; some of these features are covered in the next chapter.

5.3 tigris workflows

As covered in the previous sections, tigris is a useful package for getting TIGER/Line shapefiles into your R geospatial projects without having to navigate the Census website. Functions in **tigris** include additional options to allow for customization of output and better integration into geospatial projects. The sections below provide an overview of some of these options.

5.3.1 TIGER/Line and cartographic boundary shapefiles

In addition to the core TIGER/Line shapefiles, after which the **tigris** package is named, the Census Bureau also makes available *cartographic boundary shapefiles*. These files are derived from the TIGER/Line shapefiles but are generalized in the interior and clipped to the shoreline of the United States, making them a better choice in many cases than the TIGER/Line shapefiles for thematic mapping. Most polygon datasets in **tigris** are available as cartographic boundary files, accessible with the argument `cb = TRUE`.

Over the years I have been developing **tigris**, I have gotten multiple inquires akin to "why does this area look so strange when I download it?" Michigan, with its extensive shoreline along the Great Lakes, commonly comes up. The TIGER/Line shapefiles include water area for geographic features, connecting the Upper Peninsula of Michigan with the southern part of the state and giving an unfamiliar representation of Michigan's land area. Using the cartographic boundary alternative resolves this. Let's use **patchwork** to compare the TIGER/Line and cartographic boundary shapefiles for counties in Michigan as an illustration.

```
mi_counties <- counties("MI")
mi_counties_cb <- counties("MI", cb = TRUE)

mi_tiger_gg <- ggplot(mi_counties) + 
  geom_sf() + 
  theme_void() + 
  labs(title = "TIGER/Line")

mi_cb_gg <- ggplot(mi_counties_cb) + 
  geom_sf() + 
  theme_void() + 
  labs(title = "Cartographic boundary")

mi_tiger_gg + mi_cb_gg
```

While the TIGER/Line shapefiles may represent "official" areas of counties – which include water area – they look very unfamiliar to viewers expecting a usual representation of land area in Michigan. As the cartographic boundary file shows islands and a distinct coastline, it will be the better option for most thematic mapping projects. When using the `cb =`

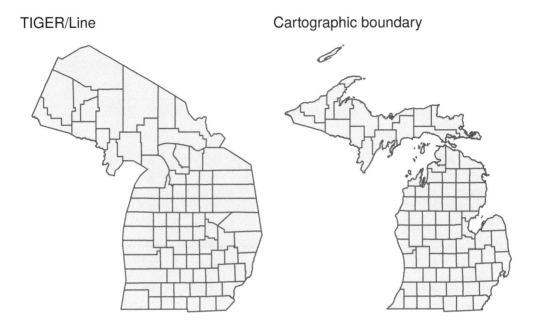

FIGURE 5.12 Comparison of TIGER/Line and cartographic boundary files for Michigan counties

TRUE argument with counties and larger geographies, users can also specify one of three resolutions with the resolution argument: "500k" (the default), "5m", or "20m", with higher values representing more generalized boundaries and smaller file sizes.

5.3.2 Caching tigris data

A common issue that can arise when working with **tigris** is waiting around for large file downloads, especially if the required geographic data is large or if your internet connection is spotty. For example, the Census block shapefile for Texas in 2019 is 441MB in size zipped, which can take a while to obtain without high-speed internet.

tigris offers a solution in the way of *shapefile caching*. By specifying options (tigris_use_cache = TRUE), users can instruct tigris to download shapefiles to a cache directory on their computers for future use rather than a temporary directory as **tigris** does by default. When shapefile caching is turned on, **tigris** will then look first in the cache directory to see if a requested shapefile is already there. If so, it will read it in without re-downloading. If not, it will download the file from the Census website and store it in the cache directory.

The cache directory can be checked with the user_cache_dir() function available in the **rappdirs** package (Ratnakumar et al., 2021). The specific location of your cache directory will depend on your operating system.

```
options(tigris_use_cache = TRUE)

rappdirs::user_cache_dir("tigris")
```

5.3 tigris workflows

```
## [1] "~/.cache/tigris"
```

If desired, users can modify their **tigris** cache directory with the function `tigris_cache_dir()`.

5.3.3 Understanding yearly differences in TIGER/Line files

The US Census Bureau offers a time series of TIGER/Line and cartographic boundary shapefiles dating back to 1990. (For even older geographies, see NHGIS, a topic covered in Chapter 11). While some geographies are reasonably static, such as state boundaries, others change regularly with each decennial US Census, such as Census tracts, block groups, and blocks. An example of these changes is shown below with Census tracts in Tarrant County, Texas displayed, a county that added nearly 1 million people between 1990 and 2020. Given that the US Census Bureau aims to make the population sizes of Census tracts relatively consistent (around 4,000 people), it will subdivide and re-draw tracts in fast-growing areas for each Census to provide better geographic granularity.

We can use some of the tidyverse tools covered earlier in this book to generate a list of tract plots for each year. `purrr::map()` iterates through each year, grabbing a cartographic boundary file of Census tracts for the four decennial Census years then plotting each of them with ggplot2. The `glue()` function in the glue package is used to create a custom title that shows the number of Census tracts in each year.

```
library(tidyverse)
library(patchwork)
library(glue)

yearly_plots <- map(seq(1990, 2020, 10), ~{
  year_tracts <- tracts("TX", "Tarrant", year = .x,
                        cb = TRUE)

  ggplot(year_tracts) +
    geom_sf() +
    theme_void() +
    labs(title = glue("{.x}: {nrow(year_tracts)} tracts"))
})
```

Once the plots are generated, we can use **patchwork** to facet the plots as we did earlier in this chapter. The division operator / places plots on top of one another allowing for the organization of plots in a grid.

```
(yearly_plots[[1]] + yearly_plots[[2]]) /
  (yearly_plots[[3]] + yearly_plots[[4]])
```

Tarrant County added 180 new Census tracts between 1990 and 2020. As the plot shows, many of these tracts are found in fast-growing parts of the county in the northeast and southeast. Notably, these changes in Census tract geography have downstream implications as well for time-series analysis, as covered in Section 3.4.1. Data at the Census tract level in 2010, for example, will be tabulated differently than in 2020 because the tract geographies are different. One common method for adjusting demographic data between disparate zonal configurations is *areal interpolation*, a topic covered in Section 7.4.4.

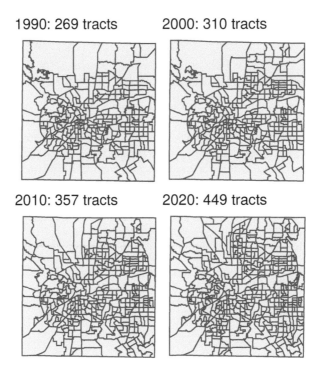

FIGURE 5.13 Tarrant County, TX Census tracts, 1990–2020

The default year of shapefiles in **tigris** is typically updated when cartographic boundary shapefiles for that year become fully available. At the time of this writing, the default year is 2020. This will be changed to 2021 pending full data availability of the cartographic boundary shapefiles for that year later in 2022. If users need the 2021 boundaries in their projects and do not want to type year = 2021 for each dataset, the command options(tigris_year = 2021) can be used. This will direct **tigris** to download the 2021 shapefiles when available without having to specify the year explicitly.

5.3.4 Combining tigris datasets

For years 2019 and later, the US Census Bureau has started releasing national small-area cartographic boundary files, including commonly-requested geographies like block groups, Census tracts, and places. In **tigris**, a user only needs to specify cb = TRUE and leave the state blank to get a national dataset. One line of code in **tigris** is all you need to get 242,303 US block groups for 2020:

```
us_bgs_2020 <- block_groups(cb = TRUE, year = 2020)

nrow(us_bgs_2020)
```

```
## [1] 242303
```

However, this is not an option for years 2018 and earlier, which means that **tigris** users must turn to alternative methods to generate national datasets. Such datasets are straightforward to create with **tidyverse** tools. As covered in several examples thus far in this book, the `purrr::map()` family of functions iterate over a sequence of values and combine function results as directed by a user. For tigris users, the `map_dfr()` function will prove especially useful as it row-binds datasets to create its output. The built-in `state.abb` vector in R gives us postal codes of all 50 US states; if Washington, DC or Puerto Rico are required in your analysis, add these to the vector with `c()` as shown below.

Once the vector of state codes is prepared, a user can iterate over those codes with `map_dfr()` and produce a national block group dataset, shown for 2018 below.

```
state_codes <- c(state.abb, "DC", "PR")

us_bgs_2018 <- map_dfr(
  state_codes,
  ~block_groups(
    state = .x,
    cb = TRUE,
    year = 2018
  )
)

nrow(us_bgs_2018)
```

```
## [1] 220016
```

If you are not using shapefile caching, this process will be slowed by the time it takes to download each block group shapefile from the Census Bureau website. However, if you have a local cache of block group shapefiles as illustrated in this section, loading and combining all 220,016 block groups for use in your analysis should only take a few seconds.

5.4 Coordinate reference systems

For geographic data to appropriately represent locations in mapping and spatial analysis, they must be *referenced* to some model of the Earth's surface correctly. In simpler terms – a data model of the state of Florida should represent where Florida is actually located relative to other locations! This is defined with a *coordinate reference system* (CRS), which specifies not only how data coordinates should be mapped to a model of the Earth's surface but also how measurements should be computed using a given dataset. A more complete discussion of coordinate reference systems is found in (Lovelace et al., 2019); an overview of how to work with coordinate systems in relationship to **tigris** is covered below.

By default, datasets returned by tigris are stored in a *geographic coordinate system*, in which coordinates are represented as longitude and latitude relative to a three-dimensional model of the earth. The `st_crs()` function in the sf package helps us check the CRS of our data; let's do this for counties in Florida.

```
library(sf)
```

```
fl_counties <- counties("FL", cb = TRUE)

st_crs(fl_counties)

## Coordinate Reference System:
##   User input: NAD83
##   wkt:
## GEOGCRS["NAD83",
##     DATUM["North American Datum 1983",
##         ELLIPSOID["GRS 1980",6378137,298.257222101,
##             LENGTHUNIT["metre",1]]],
##     PRIMEM["Greenwich",0,
##         ANGLEUNIT["degree",0.0174532925199433]],
##     CS[ellipsoidal,2],
##         AXIS["latitude",north,
##             ORDER[1],
##             ANGLEUNIT["degree",0.0174532925199433]],
##         AXIS["longitude",east,
##             ORDER[2],
##             ANGLEUNIT["degree",0.0174532925199433]],
##     ID["EPSG",4269]]
```

The function returns a well-known text representation[6] of information about the coordinate reference system. All Census Bureau datasets are stored in the "NAD83" geographic coordinate system, which refers to the North American Datum of 1983. Other relevant information includes the ellipsoid used (GRS 1980, which is a generalized three-dimensional model of the Earth's shape), the prime meridian of the CRS (Greenwich is used here), and the EPSG (European Petroleum Survey Group) ID of **4269**, which is a special code that can be used to represent the CRS in more concise terms.

As of **sf** version 1.0, the package uses the spherical geometry library **s2** to appropriately perform calculations with spatial data stored in geographic coordinate systems. When working with and visualizing geographic data for smaller areas, however, a *projected coordinate reference system* that represents the data in two-dimensions on a planar surface may be preferable. Thousands of projected CRSs exist – each that are appropriate for minimizing data distortion in a specific part of the world. While it can be a challenge to decide on the right projected CRS for your data, the **crsuggest** package (Walker, 2021a) can help narrow down the choices.

5.4.1 Using the crsuggest package

The core function implemented in **crsuggest** is `suggest_crs()`, which returns a tibble of possible choices for a suitable projected CRS for your data. The function works by analyzing the geometry of your input dataset then comparing it to a built-in dataset of CRS extents and choosing the CRSs that minimize the Hausdorff distance[7] between your dataset and those extents.

Let's try this out for the Florida counties dataset.

---

[6]https://www.ogc.org/standards/wkt-crs
[7]https://en.wikipedia.org/wiki/Hausdorff_distance

5.4 Coordinate reference systems

TABLE 5.1 Suggested coordinate reference systems for Florida

crs_code	crs_name	crs_type	crs_gcs	crs_units
6439	NAD83(2011) / Florida GDL Albers	projected	6318	m
3513	NAD83(NSRS2007) / Florida GDL Albers	projected	4759	m
3087	NAD83(HARN) / Florida GDL Albers	projected	4152	m
3086	NAD83 / Florida GDL Albers	projected	4269	m
6443	NAD83(2011) / Florida West (ftUS)	projected	6318	us-ft
6442	NAD83(2011) / Florida West	projected	6318	m
3517	NAD83(NSRS2007) / Florida West (ftUS)	projected	4759	us-ft
3516	NAD83(NSRS2007) / Florida West	projected	4759	m
2882	NAD83(HARN) / Florida West (ftUS)	projected	4152	us-ft
2778	NAD83(HARN) / Florida West	projected	4152	m

```
library(crsuggest)

fl_crs <- suggest_crs(fl_counties)
```

The "best choice" is the CRS "Florida GDL Albers" coordinate reference system, which is available with four different variations on the NAD1983 datum. "Florida GDL" refers to the Florida Geographic Data Library which distributes all of its data in this state-wide equal-area coordinate reference system[8]. Other large states with large or irregular extents like Florida (Texas is one such example) maintain statewide coordinate reference systems like this suitable for statewide mapping and analysis. Let's choose the third entry, "NAD83 (HARN) / Florida GDL Albers", which is recommended on the Florida GDL website. Coordinate reference system transformations in **sf** are implemented in the st_transform() function, used below.

```
fl_projected <- st_transform(fl_counties, crs = 3087)

head(fl_projected)
```

```
## Simple feature collection with 6 features and 12 fields
## Geometry type: MULTIPOLYGON
## Dimension:     XY
## Bounding box:  xmin: 281876.9 ymin: 397330.2 xmax: 669346.5 ymax: 715363.4
## Projected CRS: NAD83(HARN) / Florida GDL Albers
##    STATEFP COUNTYFP COUNTYNS     AFFGEOID  GEOID        NAME
## 11      12      121 00295729 0500000US12121 12121    Suwannee
## 15      12      007 00303634 0500000US12007 12007    Bradford
## 91      12      037 00306911 0500000US12037 12037    Franklin
## 92      12      123 00295728 0500000US12123 12123      Taylor
## 93      12      057 00295757 0500000US12057 12057 Hillsborough
## 94      12      109 00308371 0500000US12109 12109   St. Johns
##          NAMELSAD STUSPS STATE_NAME LSAD      ALAND   AWATER
## 11 Suwannee County     FL    Florida   06 1783430092  9505389
## 15 Bradford County     FL    Florida   06  761362121 16905200
```

---

[8] https://www.fgdl.org/metadataexplorer/fgdlfaq.html#3.1

```
## 91    Franklin County    FL    Florida    06 1411498965 2270440522
## 92      Taylor County    FL    Florida    06 2702224058 1213576997
## 93 Hillsborough County   FL    Florida    06 2646680847  803027226
## 94   St. Johns County    FL    Florida    06 1555661651  572103225
##                            geometry
## 11 MULTIPOLYGON (((471781.3 69...
## 15 MULTIPOLYGON (((552464 6583...
## 91 MULTIPOLYGON (((334863.7 64...
## 92 MULTIPOLYGON (((416379.5 66...
## 93 MULTIPOLYGON (((555139.5 42...
## 94 MULTIPOLYGON (((622427 6727...
```

Note that the coordinates for the bounding box and the feature geometry have changed to much larger numbers; they are expressed in meters rather than the decimal degrees used by the NAD83 geographic coordinate system. Let's take a closer look at our selected CRS:

```
st_crs(fl_projected)
```

```
## Coordinate Reference System:
##   User input: EPSG:3087
##   wkt:
## PROJCRS["NAD83(HARN) / Florida GDL Albers",
##     BASEGEOGCRS["NAD83(HARN)",
##         DATUM["NAD83 (High Accuracy Reference Network)",
##             ELLIPSOID["GRS 1980",6378137,298.257222101,
##                 LENGTHUNIT["metre",1]]],
##         PRIMEM["Greenwich",0,
##             ANGLEUNIT["degree",0.0174532925199433]],
##         ID["EPSG",4152]],
##     CONVERSION["Florida GDL Albers (meters)",
##         METHOD["Albers Equal Area",
##             ID["EPSG",9822]],
##         PARAMETER["Latitude of false origin",24,
##             ANGLEUNIT["degree",0.0174532925199433],
##             ID["EPSG",8821]],
##         PARAMETER["Longitude of false origin",-84,
##             ANGLEUNIT["degree",0.0174532925199433],
##             ID["EPSG",8822]],
##         PARAMETER["Latitude of 1st standard parallel",24,
##             ANGLEUNIT["degree",0.0174532925199433],
##             ID["EPSG",8823]],
##         PARAMETER["Latitude of 2nd standard parallel",31.5,
##             ANGLEUNIT["degree",0.0174532925199433],
##             ID["EPSG",8824]],
##         PARAMETER["Easting at false origin",400000,
##             LENGTHUNIT["metre",1],
##             ID["EPSG",8826]],
##         PARAMETER["Northing at false origin",0,
##             LENGTHUNIT["metre",1],
##             ID["EPSG",8827]]],
##     CS[Cartesian,2],
```

5.5 Coordinate reference systems

```
##         AXIS["easting (X)",east,
##             ORDER[1],
##             LENGTHUNIT["metre",1]],
##         AXIS["northing (Y)",north,
##             ORDER[2],
##             LENGTHUNIT["metre",1]],
##     USAGE[
##         SCOPE["unknown"],
##         AREA["USA - Florida"],
##         BBOX[24.41,-87.63,31.01,-79.97]],
##     ID["EPSG",3087]]
```

There is a lot more information in the CRS's well-known text than for the NAD83 geographic CRS. Information about the base geographic CRS is provided along with parameters for a "false origin." Whereas coordinates for geographic coordinate systems will generally be represented as longitude/latitude relative to the Prime Meridian and Equator, projected coordinate reference systems will be relative to a "false origin" that is specified relative to the area where the CRS is used (noting the USAGE section at the bottom). This "false origin" is located at –84 degrees longitude, 24 degrees latitude (SW of the Florida Keys and north of western Cuba) with a false X value of 400,000 and a false Y value of 0. In turn, X and Y values in the projected data are expressed in meters relative to this origin, which is set so all X and Y values in Florida will be positive numbers. This makes planar geometric calculations like distance, perimeter, and area straightforward.

5.4.2 Plotting with coord_sf()

When visualizing simple feature geometries with ggplot2's geom_sf(), the coord_sf() method allows you to specify a coordinate reference system transformation to be used for the visualization. While coord_sf() will inherit the CRS of the spatial object plotted with geom_sf() by default, it can also modify the displayed CRS of a spatial object without performing a CRS transformation with st_transform(). This is shown in Figure 5.14:

```
options(scipen = 999)

ggplot(fl_counties) +
  geom_sf() +
  coord_sf(crs = 3087)
```

While the data are displayed on the plot in the requested coordinate reference system, the underlying *graticule* (the grid lines and axis tick labels) default to longitude/latitude. To show the coordinates of the projected coordinate reference system, the argument datum can be used which controls the gridlines, shown in Figure 5.15.

```
ggplot(fl_counties) +
  geom_sf() +
  coord_sf(crs = 3087, datum = 3087)
```

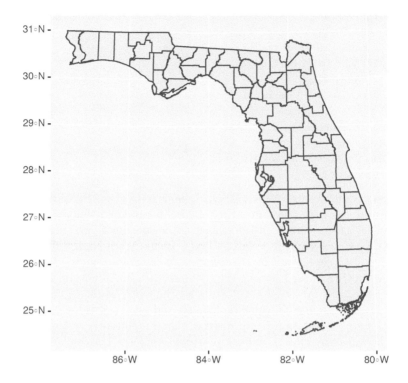

FIGURE 5.14 ggplot2 map with CRS specified

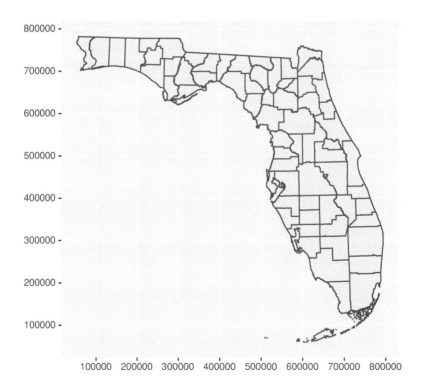

FIGURE 5.15 ggplot2 plot with modified graticule

5.5 Working with geometries

FIGURE 5.16 Default CRS for US states

5.5 Working with geometries

While understanding coordinate reference system transformations is important for spatial data management and geographic visualization, other types of geometric manipulations may be required by the spatial analyst. This can include shifting geometries from their "correct" positions for better cartographic display; changing the geometry type of a spatial object; or "exploding" multipart geometries into individual geometric units. This section gives examples of each of these three scenarios.

5.5.1 Shifting and rescaling geometry for national US mapping

A common problem for national display of the United States is the fragmented nature of US states and territories geographically. The continental United States can be displayed on a map in a relatively straightforward way, and there are a number of projected coordinate reference systems designed for correct display of the continental US. Often, analysts and cartographers then have to make decisions about how to handle Alaska, Hawaii, and Puerto Rico, which cannot be reasonably plotted using default US projections.

For example, let's take a US states shapefile obtained with tigris at low resolution and use ggplot2 to visualize it in the default geographic CRS, NAD 1983:

```
us_states <- states(cb = TRUE, resolution = "20m")

ggplot(us_states) +
  geom_sf() +
  theme_void()
```

The plot in Figure 5.16 does not work well, in part because the Aleutian Islands in far west Alaska cross the 180 degree line of longitude and are plotted on the opposite side of the map. In response, a projected coordinate reference system appropriate for the United States *could* be used, such as the continental US Albers Equal Area projection:

```
ggplot(us_states) +
  geom_sf() +
  coord_sf(crs = 'ESRI:102003') +
  theme_void()
```

FIGURE 5.17 Equal-area CRS for US states

While the representation in Figure 5.17 puts all territories in their appropriate locations, it is clearly not appropriate for Alaska, Hawaii, and Puerto Rico which appear distorted. This coordinate reference system is also not ideal for comparative mapping of states given the large amount of blank space between the states on the map.

tigris offers a solution to this problem with the `shift_geometry()` function. `shift_geometry()` takes an opinionated approach to the shifting and rescaling of Alaska, Hawaii, and Puerto Rico geometries to offer four options for an alternative view of the US. The function works by projecting geometries in Alaska, Hawaii, and Puerto Rico to appropriate coordinate reference systems for those areas, then re-sizing the geometries (if requested) and moving them to an alternative layout in relationship to the rest of the US using the Albers Equal Area CRS.

```
us_states_shifted <- shift_geometry(us_states)

ggplot(us_states_shifted) +
  geom_sf() +
  theme_void()
```

The view in Figure 5.18 uses two default arguments: `preserve_area = FALSE`, which shrinks Alaska and inflates Hawaii and Puerto Rico, and `position = "below"`, which places these areas below the continental United States. Alternatively, we can set `preserve_area = TRUE` and `position = "outside"` (used together below, but they can be mixed and matched) for a different view, shown in Figure 5.19:

FIGURE 5.18 US states with shifted and rescaled geometry

```
us_states_outside <- shift_geometry(us_states,
                                    preserve_area = TRUE,
                                    position = "outside")

ggplot(us_states_outside) +
  geom_sf() +
  theme_void()
```

The areas of Alaska, Hawaii, and Puerto Rico are preserved relative to the continental United States, and the three areas are directionally in their correct positions while still in proximity to the continental US for national display. In addition to spatial objects obtained with tigris, shift_geometry() can shift and rescale other geographic datasets for display in this way. Just make sure you use the same arguments in shift_geometry() for all layers or they will end up misaligned!

5.5.2 Converting polygons to points

As discussed earlier in this chapter, most datasets obtained with tigris are returned with geometry type POLYGON or MULTIPOLYGON, reflecting the fact that Census geometries are generally areal units to which Census data are aggregated. While this makes sense for many applications, there are some instances in which the default geometry type of Census shapes is not necessary.

For example, let's say we are making a simple plot of the largest cities in the state of Texas. The places() function can obtain city geometries, and the states() function gives us the outline of the state of Texas. Two successive calls to geom_sf() create a graphic that displays those cities on top of the state outline.

FIGURE 5.19 US states with shifted geometry and consistent area

```
tx_places <- places("TX", cb = TRUE) %>%
  filter(NAME %in% c("Dallas", "Fort Worth", "Houston",
                     "Austin", "San Antonio", "El Paso")) %>%
  st_transform(6580)

tx_outline <- states(cb = TRUE) %>%
  filter(NAME == "Texas") %>%
  st_transform(6580)

ggplot() +
  geom_sf(data = tx_outline) +
  geom_sf(data = tx_places, fill = "red", color = NA) +
  theme_void()
```

The issue with the graphic in Figure 5.20 is that city geographies are actually quite irregular and disjoint in practice. All six cities spread across large areas, have holes, and even in some cases include portions that are detached from the main part of the city. This information is important for local planning purposes, but are unnecessary for a state-wide map.

An alternative representation is possible by converting the city polygons to points where each point represents the *centroid* of each polygon, placed at their geometric centers. In sf, this conversion is implemented with the function st_centroid(). As shown in Figure 5.21, we use st_centroid() to convert the polygons to central points, and plot those points over the outline of Texas.

```
tx_centroids <- st_centroid(tx_places)
```

5.5 Working with geometries

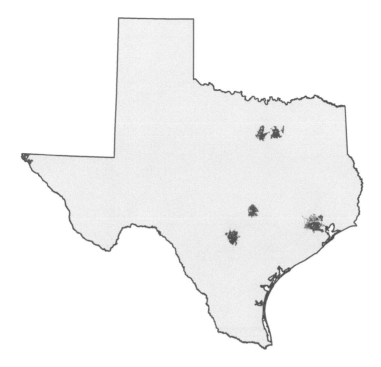

FIGURE 5.20 Large cities in Texas

```
ggplot() +
  geom_sf(data = tx_outline) +
  geom_sf(data = tx_centroids, color = "red", size = 3) +
  theme_void()
```

The cities are displayed as circles rather than as irregular polygons, which makes more sense for visualization of the cities' locations at this scale.

5.5.3 Exploding multipolygon geometries to single parts

Generally speaking, areal Census features are returned with the geometry type MULTIPOLYGON. This makes sense as many Census shapes – including several states – include disconnected areas such as islands that belong to the same Census area. As this is particularly significant in Florida, let's return to the Florida counties dataset used earlier in this chapter, and take the example of Lee County on Florida's western coast.

```
lee <- fl_projected %>%
  filter(NAME == "Lee")
```

```
mapview(lee)
```

The Lee County polygon shown in Figure 5.22 has four distinct parts as displayed on the map: the mainland area that contains the cities of Cape Coral and Fort Myers, and three

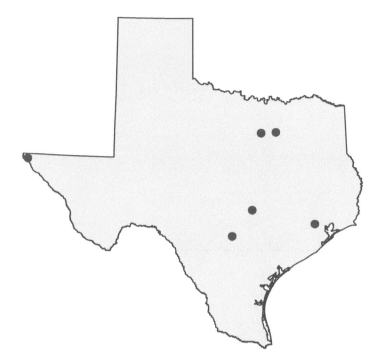

FIGURE 5.21 Large cities in Texas represented as points

disjoint island areas. Other islands can be further extracted using techniques covered in Chapter 7. Understandably, these four areas are interpreted by R as a single feature:

```
lee
```

```
## Simple feature collection with 1 feature and 12 fields
## Geometry type: MULTIPOLYGON
## Dimension:    XY
## Bounding box: xmin: 571477.3 ymin: 258768.2 xmax: 642726 ymax: 310584
## Projected CRS: NAD83(HARN) / Florida GDL Albers
##   STATEFP COUNTYFP COUNTYNS      AFFGEOID  GEOID NAME    NAMELSAD STUSPS
## 1      12      071 00295758 0500000US12071  12071  Lee  Lee County     FL
##   STATE_NAME LSAD      ALAND    AWATER                       geometry
## 1    Florida   06 2022803068 1900583561 MULTIPOLYGON (((580415.7 30...
```

Specific parts of the multipolygon Lee County object can be extracted by *exploding* the multipart geometry into single parts. This is accomplished with sf's function st_cast(), which can convert spatial objects from one geometry type to another. In this example, we will "cast" Lee County as a POLYGON object which will create a separate row for each non-contiguous area. For analysts coming from a desktop GIS background, this will perform a similar operation to "Multipart to Singlepart" geoprocessing tools.

```
lee_singlepart <- st_cast(lee, "POLYGON")
```

5.5 Working with geometries

FIGURE 5.22 Lee County, Florida

```
lee_singlepart
```

```
## Simple feature collection with 4 features and 12 fields
## Geometry type: POLYGON
## Dimension:     XY
## Bounding box:  xmin: 571477.3 ymin: 258768.2 xmax: 642726 ymax: 310584
## Projected CRS: NAD83(HARN) / Florida GDL Albers
##     STATEFP COUNTYFP COUNTYNS      AFFGEOID GEOID NAME    NAMELSAD STUSPS
## 1        12      071 00295758 0500000US12071 12071  Lee Lee County     FL
## 1.1      12      071 00295758 0500000US12071 12071  Lee Lee County     FL
## 1.2      12      071 00295758 0500000US12071 12071  Lee Lee County     FL
## 1.3      12      071 00295758 0500000US12071 12071  Lee Lee County     FL
##     STATE_NAME LSAD      ALAND     AWATER                       geometry
## 1      Florida   06 2022803068 1900583561 POLYGON ((580415.7 300219.6...
## 1.1    Florida   06 2022803068 1900583561 POLYGON ((576540.8 289935.7...
## 1.2    Florida   06 2022803068 1900583561 POLYGON ((572595.8 298881, ...
## 1.3    Florida   06 2022803068 1900583561 POLYGON ((571477.3 310583.5...
```

The resulting spatial object now has four rows. Using row indexing, we can extract any of these rows as an individual object, such as the area representing Sanibel Island.

FIGURE 5.23 Sanibel Island, Florida

```
sanibel <- lee_singlepart[2,]

mapview(sanibel)
```

5.6 Exercises

- Give **tigris** a try for yourselves! Go through the available geographies in **tigris** and fetch data for a state and/or county of your choosing.

- Using your data, try some of the plotting options covered in this chapter. Plot the data with plot(), geom_sf(), and with mapview().

- Use suggest_crs() from the **crsuggest** package to identify an appropriate projected coordinate reference system for your data. Then, use st_transform() from the sf package to apply a CRS transformation to your data.

6

Mapping Census data with R

Data from the United States Census Bureau are commonly visualized using maps, given that Census and ACS data are aggregated to enumeration units. This chapter will cover the process of map-making using the **tidycensus** R package. Notably, **tidycensus** enables R users to download *simple feature geometry* for common geographies, linking demographic information with their geographic locations in a dataset. In turn, this data model facilitates the creation of both static and interactive demographic maps.

In this chapter, readers will learn how to use the `geometry` parameter in tidycensus functions to download geographic data along with demographic data from the US Census Bureau. The chapter will then cover how to make static maps of Census demographic data using the popular **ggplot2** and **tmap** visualization packages. The closing parts of the chapter will then turn to interactive mapping, with a focus on the **mapview** and **Leaflet** R packages for interactive cartographic visualization.

6.1 Using geometry in tidycensus

As covered in the previous chapter, Census geographies are available from the **tigris** R package as simple features objects, using the data model from the **sf** R package. tidycensus wraps several common geographic data functions in the **tigris** package to allow R users to return simple feature geometry pre-linked to downloaded demographic data with a single function call. The key argument to accomplish this is `geometry = TRUE`, which is available in the core data download functions in tidycensus, `get_acs()`, `get_decennial()`, and `get_estimates()`.

Traditionally, getting "spatial" Census data requires a tedious multi-step process that can involve several software platforms. These steps include:

- Fetching shapefiles from the Census website;
- Downloading a CSV of data, then cleaning and formatting it;
- Loading geometries and data into your desktop GIS of choice;
- Aligning key fields in your desktop GIS and joining your data.

A major motivation for developing **tidycensus** was my frustration with having to go through this process over and over again before making a simple map of Census data. `geometry = TRUE` combines the automated data download functionality of tidycensus and tigris to allow R users to bypass this process entirely. The following example illustrates the use of the `geometry = TRUE` argument, fetching information on median household income for Census tracts in the District of Columbia. As discussed in the previous chapter, the option tigris_use_cache = TRUE is used to cache the downloaded geographic data on the user's computer.

```
library(tidycensus)
options(tigris_use_cache = TRUE)

dc_income <- get_acs(
  geography = "tract",
  variables = "B19013_001",
  state = "DC",
  year = 2020,
  geometry = TRUE
)

dc_income
```

```
## Simple feature collection with 206 features and 5 fields
## Geometry type: POLYGON
## Dimension:     XY
## Bounding box:  xmin: -77.11976 ymin: 38.79165 xmax: -76.9094 ymax: 38.99511
## Geodetic CRS:  NAD83
## First 10 features:
##          GEOID                                                   NAME
## 1  11001005003 Census Tract 50.03, District of Columbia, District of Columbia
## 2  11001002503 Census Tract 25.03, District of Columbia, District of Columbia
## 3  11001006801 Census Tract 68.01, District of Columbia, District of Columbia
## 4  11001009802 Census Tract 98.02, District of Columbia, District of Columbia
## 5  11001008904 Census Tract 89.04, District of Columbia, District of Columbia
## 6  11001003301 Census Tract 33.01, District of Columbia, District of Columbia
## 7  11001000704  Census Tract 7.04, District of Columbia, District of Columbia
## 8  11001007601 Census Tract 76.01, District of Columbia, District of Columbia
## 9  11001009604 Census Tract 96.04, District of Columbia, District of Columbia
## 10 11001009803 Census Tract 98.03, District of Columbia, District of Columbia
##       variable estimate   moe                       geometry
## 1  B19013_001   116250 15922 POLYGON ((-77.03195 38.9096...
## 2  B19013_001    80682 37845 POLYGON ((-77.02971 38.9376...
## 3  B19013_001   109863 24472 POLYGON ((-76.98364 38.8898...
## 4  B19013_001    54432 50966 POLYGON ((-77.00057 38.8309...
## 5  B19013_001    47694  8992 POLYGON ((-76.98341 38.9001...
## 6  B19013_001   160536 52157 POLYGON ((-77.01493 38.9206...
## 7  B19013_001    86733 23440 POLYGON ((-77.07957 38.9334...
## 8  B19013_001    40330  5556 POLYGON ((-76.9901 38.87135...
## 9  B19013_001    72872 13143 POLYGON ((-76.9618 38.89612...
## 10 B19013_001    29000 24869 POLYGON ((-77.00774 38.8356...
```

As shown in the example call, the structure of the object returned by tidycensus resembles the object we've become familiar with to this point in the book. For example, median household income data are found in the estimate column with associated margins of error in the moe column, along with a variable ID, GEOID, and Census tract name. However, there are some notable differences. The geometry column contains polygon feature geometry for each Census tract, allowing for a linking of the estimates and margins of error with

6.2 Using geometry in tidycensus

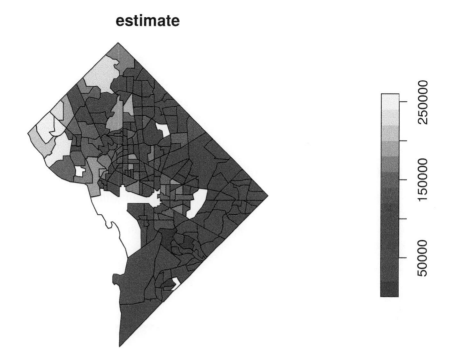

FIGURE 6.1 Base R plot of median household income by tract in DC

their corresponding locations in Washington, DC. Beyond that, the object is associated with coordinate system information – using the NAD 1983 geographic coordinate system in which Census geographic datasets are stored by default.

6.1.1 Basic mapping of sf objects with `plot()`

Such geographic information can be difficult to understand without visualization. As the returned object is a simple features object, both geometry and attributes can be visualized with `plot()`. Key here is specifying the name of the column to be plotted inside of brackets, which in this case is `"estimate"`.

```
plot(dc_income["estimate"])
```

The `plot()` function returns a simple map (Figure 6.1) showing income variation in Washington, DC. Wealthier areas, as represented with warmer colors, tend to be located in the northwestern part of the District. NA values are represented on the map in white. If desired, the map can be modified further with base plotting functions.

The remainder of this chapter, however, will focus on map-making with additional data visualization packages in R. This includes the popular **ggplot2** package for visualization, which supports direct visualization of simple features objects; the **tmap** package for thematic mapping, and the **leaflet** package for interactive map-making which calls the Leaflet JavaScript framework directly from R.

6.2 Map-making with ggplot2 and geom_sf

As illustrated in Section 5.2, geom_sf() in **ggplot2** can be used for quick plotting of sf objects using familiar **ggplot2** syntax. geom_sf() goes far beyond simple cartographic display. The full power of **ggplot2** is available to create highly customized maps and geographic data visualizations.

6.2.1 Choropleth mapping

One of the most common ways to visualize statistical information on a map is with *choropleth mapping*. Choropleth maps use shading to represent how underlying data values vary by feature in a spatial dataset. The income plot of Washington, DC shown earlier in this chapter is an example of a choropleth map.

In the example below, **tidycensus** is used to obtain linked ACS and spatial data on median age by state for the 50 US states plus the District of Columbia and Puerto Rico. For national maps, it is often preferable to generate insets of Alaska, Hawaii, and Puerto Rico so that they can all be viewed comparatively with the continental United States. We'll use the shift_geometry() function in **tigris** to shift and rescale these areas for national mapping.

```
library(tidycensus)
library(tidyverse)
library(tigris)

us_median_age <- get_acs(
  geography = "state",
  variables = "B01002_001",
  year = 2019,
  survey = "acs1",
  geometry = TRUE,
  resolution = "20m"
) %>%
  shift_geometry()

plot(us_median_age$geometry)
```

The state polygons can be styled using **ggplot2** conventions and the geom_sf() function. With two lines of **ggplot2** code, a basic map of median age by state can be created with **ggplot2** defaults.

```
ggplot(data = us_median_age, aes(fill = estimate)) +
  geom_sf()
```

The geom_sf() function in the above example interprets the geometry of the sf object (in this case, polygon) and visualizes the result as a filled choropleth map. In this case, the ACS estimate of median age is mapped to the default blue dark-to-light color ramp in **ggplot2**,

6.2 Map-making with ggplot2 and geom_sf

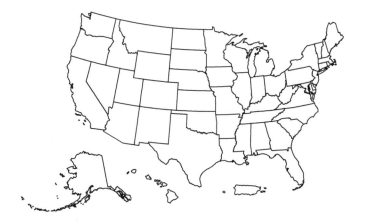

FIGURE 6.2 Plot of shifted and rescaled US geometry

highlighting the youngest states (such as Utah) with darker blues and the oldest states (such as Maine) with lighter blues.

6.2.2 Customizing ggplot2 maps

In many cases, map-makers using **ggplot2** will want to customize this graphic further. For example, a designer may want to modify the color palette and reverse it so that darker colors represent older areas. The map would also benefit from some additional information describing its content and data sources. These modifications can be specified in the same way a user would update a regular ggplot2 graphic. The `scale_fill_distiller()` function

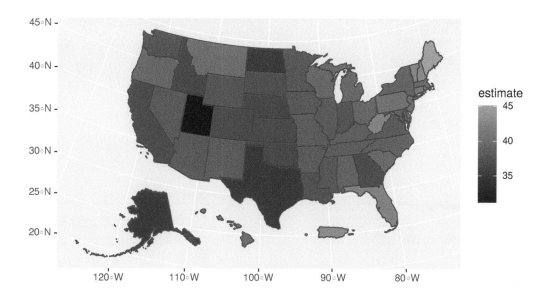

FIGURE 6.3 US choropleth map with ggplot2 defaults

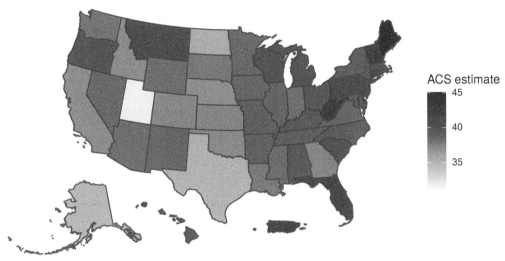

FIGURE 6.4 Styled choropleth of US median age with ggplot2

allows users to specify a ColorBrewer[1] palette to use for the map, which includes a wide range of sequential, diverging, and qualitative color palettes (Brewer et al., 2003). The labs() function can then be used to add a title, caption, and better legend label to the plot. Finally, **ggplot2** cartographers will often want to use the theme_void() function to remove the background and gridlines from the map. The result is found in Figure 6.4.

```
ggplot(data = us_median_age, aes(fill = estimate)) +
  geom_sf() +
  scale_fill_distiller(palette = "RdPu",
                       direction = 1) +
  labs(title = "  Median Age by State, 2019",
       caption = "Data source: 2019 1-year ACS, US Census Bureau",
       fill = "ACS estimate") +
  theme_void()
```

6.3 Map-making with tmap

For **ggplot2** users, geom_sf() offers a familiar interface for mapping data obtained from the US Census Bureau. However, **ggplot2** is far from the only option for cartographic visualization in R. The **tmap** package (Tennekes, 2018) is an excellent alternative for mapping in R that includes a wide range of functionality for custom cartography. The section

---

[1] https://colorbrewer2.org/#type=sequential&scheme=BuGn&n=3

6.3 Map-making with tmap

TABLE 6.1 Race and ethnicity in Hennepin County, MN

GEOID	NAME	variable	value	summary_value	geometry	percent
27053026707	Census Tract 267.07, Hennepin County, Minnesota	Hispanic	175	5188	MULTIPOLYGON (((-93.49271 4...	3.3731689
27053026707	Census Tract 267.07, Hennepin County, Minnesota	White	4215	5188	MULTIPOLYGON (((-93.49271 4...	81.2451812
27053026707	Census Tract 267.07, Hennepin County, Minnesota	Black	258	5188	MULTIPOLYGON (((-93.49271 4...	4.9730146
27053026707	Census Tract 267.07, Hennepin County, Minnesota	Native	34	5188	MULTIPOLYGON (((-93.49271 4...	0.6553585
27053026707	Census Tract 267.07, Hennepin County, Minnesota	Asian	178	5188	MULTIPOLYGON (((-93.49271 4...	3.4309946
27053021602	Census Tract 216.02, Hennepin County, Minnesota	Hispanic	224	5984	MULTIPOLYGON (((-93.38036 4...	3.7433155
27053021602	Census Tract 216.02, Hennepin County, Minnesota	White	4581	5984	MULTIPOLYGON (((-93.38036 4...	76.5541444
27053021602	Census Tract 216.02, Hennepin County, Minnesota	Black	658	5984	MULTIPOLYGON (((-93.38036 4...	10.9959893
27053021602	Census Tract 216.02, Hennepin County, Minnesota	Native	21	5984	MULTIPOLYGON (((-93.38036 4...	0.3509358
27053021602	Census Tract 216.02, Hennepin County, Minnesota	Asian	188	5984	MULTIPOLYGON (((-93.38036 4...	3.1417112

that follows is an overview of several cartographic techniques implemented with **tmap** for visualizing US Census data. A full treatment of best practices in cartographic design is beyond the scope of this section; recommended resources for learning more include Peterson (2020) and Brewer (2016).

To begin, we will obtain race and ethnicity data from the 2020 decennial US Census using the get_decennial() function. We'll be looking at data on non-Hispanic white, non-Hispanic Black, Asian, and Hispanic populations for Census tracts in Hennepin County, Minnesota.

```
hennepin_race <- get_decennial(
  geography = "tract",
  state = "MN",
  county = "Hennepin",
  variables = c(
    Hispanic = "P2_002N",
    White = "P2_005N",
    Black = "P2_006N",
    Native = "P2_007N",
    Asian = "P2_008N"
  ),
  summary_var = "P2_001N",
  year = 2020,
  geometry = TRUE
) %>%
  mutate(percent = 100 * (value / summary_value))
```

In Figure 6.5, we returned ACS data in tidycensus's regular "tidy" or long format, which will be useful in a moment for comparative map-making, and completed some basic data wrangling tasks learned in Chapter 3 to calculate group percentages. To get started mapping this data, we'll extract a single group from the dataset to illustrate how **tmap** works.

6.3.1 Choropleth maps with tmap

tmap's map-making syntax will be somewhat familiar to users of **ggplot2**, as it uses the concept of *layers* to specify modifications to the map. The map object is initialized with tm_shape(), which then allows us to view the Census tracts with tm_polygons(). We'll first filter our long-form spatial dataset to get a unique set of tract polygons, then visualize them.

```
library(tmap)
hennepin_black <- filter(hennepin_race,
```

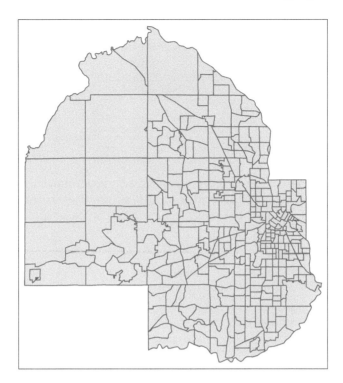

FIGURE 6.5 Basic polygon plot with tmap

```
                    variable == "Black")

tm_shape(hennepin_black) +
  tm_polygons()
```

In Figure 6.5, we get a default view of Census tracts in Hennepin County, Minnesota. Alternatively, the `tm_fill()` function can be used to produce choropleth maps, as in the ggplot2 examples above.

```
tm_shape(hennepin_black) +
  tm_polygons(col = "percent")
```

You'll notice that **tmap** uses a classed color scheme rather than the continuous palette used by **ggplot2**, by default. This involves the identification of "classes" in the distribution of data values and mapping a color from a color palette to data values that belong to each class. The default classification scheme used by `tm_fill()` is `"pretty"`, which identifies clean-looking intervals in the data based on the data range. In this example, data classes change every 20 percent. However, this approach will always be sensitive to the distribution of data values. Let's take a look at our data distribution to understand why:

```
hist(hennepin_black$percent)
```

As the histogram in Figure 6.7 illustrates, most Census tracts in Hennepin County have Black populations below 20 percent. In turn, variation within this bucket is not visible on the

6.3 Map-making with tmap

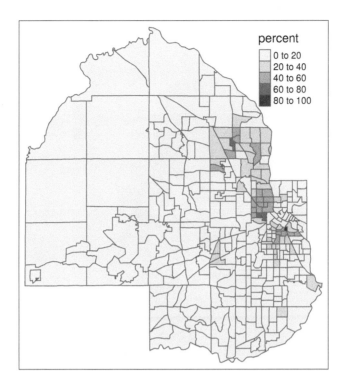

FIGURE 6.6 Basic choropleth with tmap

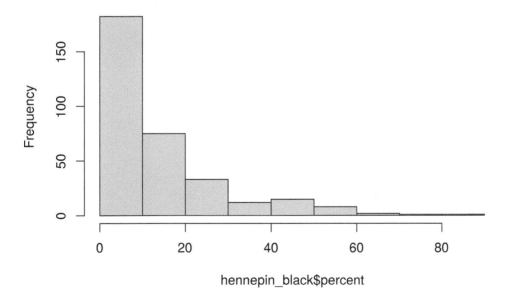

FIGURE 6.7 Base R histogram of percent Black by Census tract

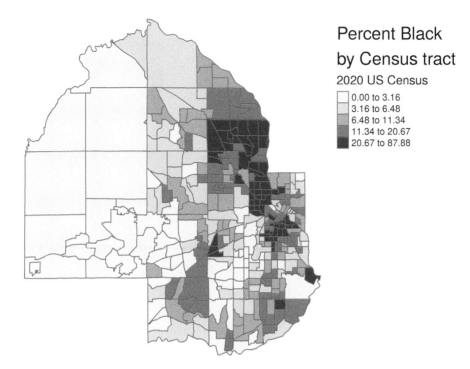

FIGURE 6.8 A tmap choropleth map with options

map given that most tracts fall into one class. The style argument in tm_fill() supports a number of other methods for classification, including quantile breaks ("quantile"), equal intervals ("equal"), and Jenks natural breaks ("jenks"). Let's switch to quantiles below, where each class will contain the same number of Census tracts. We can also change the color palette and add some contextual text as we did with ggplot2.

```
tm_shape(hennepin_black) +
  tm_polygons(col = "percent",
          style = "quantile",
          n = 5,
          palette = "Purples",
          title = "2020 US Census") +
  tm_layout(title = "Percent Black\nby Census tract",
          frame = FALSE,
          legend.outside = TRUE)
```

Switching from the default classification scheme to quantiles reveals additional neighborhood-level heterogeneity in Hennepin County's Black population in suburban areas. However, it does mask some heterogeneity in Minneapolis as the top class now includes values ranging from 21 percent to 88 percent. A "compromise" solution commonly used in GIS cartography applications is the Jenks natural-breaks method, which uses an algorithm to identify meaningful breaks in the data for bin boundaries (Jenks, 1967). To assist with understanding how the different classification methods work, the legend.hist argument in

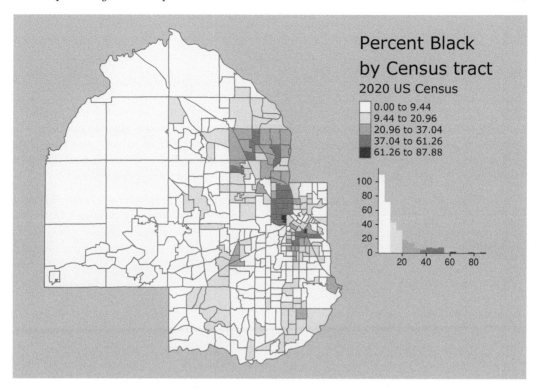

FIGURE 6.9 A styled tmap choropleth map with a histogram

`tm_polygons()` can be set to TRUE, adding a histogram to the map with bars colored by the values used on the map.

```
tm_shape(hennepin_black) +
  tm_polygons(col = "percent",
          style = "jenks",
          n = 5,
          palette = "Purples",
          title = "2020 US Census",
          legend.hist = TRUE) +
  tm_layout(title = "Percent Black\nby Census tract",
          frame = FALSE,
          legend.outside = TRUE,
          bg.color = "grey70",
          legend.hist.width = 5,
          fontfamily = "Verdana")
```

The `tm_layout()` function is used to customize the styling of the map and histogram, and has many more options beyond those shown that can be viewed in the function's documentation.

6.3.2 Adding reference elements to a map

The choropleth map as illustrated in the previous example represents the data as a statistical graphic, with both a map and histogram showing the underlying data distribution.

Cartographers coming to R from a desktop GIS background will be accustomed to adding a variety of reference elements to their map layouts to provide additional geographical context to the map. These elements may include a basemap, north arrow, and scale bar, all of which can be accommodated by **tmap**.

The quickest way to get basemaps for use with **tmap** is the **rosm** package Dunnington (2019), which helps users access freely-available basemaps from a variety of different providers. In some cases, users will want to design their own basemaps for use in their R projects. The example shown here uses the **mapboxapi** R package Walker (2021c), which gives users with a Mapbox account access to pre-designed Mapbox basemaps as well as custom-designed basemaps in Mapbox Studio[2].

To use Mapbox basemaps with **mapboxapi** and **tmap**, you'll first need a Mapbox account and access token. Mapbox accounts are free; register at the Mapbox website[3] then find your access token. In R, this token can be set with the `mb_access_token()` function in **mapboxapi**.

```
library(mapboxapi)

# Replace with your token below
mb_access_token("pk.ey92lksd...")
```

Once set, basemap tiles for an input spatial dataset can be fetched with the `get_static_tiles()` function in **mapboxapi**, which interacts with the Mapbox Static Tiles API[4]. Mapbox Studio users can design a custom basemap style and use the custom style ID along with their username to fetch tiles from that style for mapping; any Mapbox user can also use the default Mapbox styles[5] by supplying `username = "mapbox"` and the appropriate style ID. The level of detail of the underlying basemap can be adjusted with the `zoom` argument.

```
# If you don't have a Mapbox style to use, replace style_id with "light-v9"
# and username with "mapbox".  If you do, replace those arguments with your
# style ID and user name.
hennepin_tiles <- get_static_tiles(
  location = hennepin_black,
  zoom = 10,
  style_id = "ckedp72zt059t19nssixpgapb",
  username = "kwalkertcu"
)
```

In most cases, users should choose a muted, monochrome basemap when designing a map with a choropleth overlay to avoid confusing blending of colors.

These basemap tiles are layered into the familiar **tmap** workflow with the `tm_rgb()` function. To show the underlying basemap, users should modify the transparency of the Census tract polygons with the `alpha` argument. Additional **tmap** functions then add ancillary map elements. `tm_scale_bar()` adds a scale bar; `tm_compass()` adds a north arrow; and

---

[2]https://studio.mapbox.com/
[3]https://account.mapbox.com/
[4]https://docs.mapbox.com/api/maps/static-tiles/
[5]https://docs.mapbox.com/api/maps/styles/#mapbox-styles

6.3 Map-making with tmap

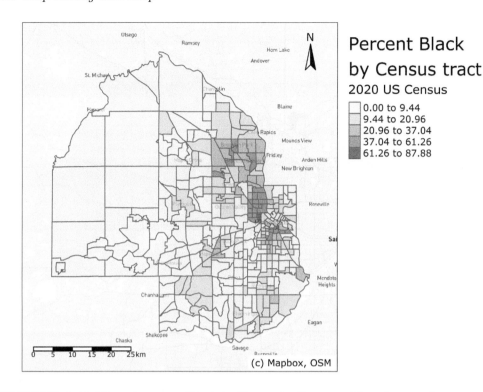

FIGURE 6.10 Map of percent Black in Hennepin County with reference elements

`tm_credits()` helps cartographers give credit for the basemap, which is required when using Mapbox and OpenStreetMap tiles.

```
tm_shape(hennepin_tiles) +
  tm_rgb() +
  tm_shape(hennepin_black) +
  tm_polygons(col = "percent",
          style = "jenks",
          n = 5,
          palette = "Purples",
          title = "2020 US Census",
          alpha = 0.7) +
  tm_layout(title = "Percent Black\nby Census tract",
          legend.outside = TRUE,
          fontfamily = "Verdana") +
  tm_scale_bar(position = c("left", "bottom")) +
  tm_compass(position = c("right", "top")) +
  tm_credits("(c) Mapbox, OSM      ",
          bg.color = "white",
          position = c("RIGHT", "BOTTOM"))
```

Depending on the shape of your Census data, the `position` arguments in `tm_scale_bar()`, `tm_compass()`, and `tm_credits()` can be modified to organize ancillary map elements

FIGURE 6.11 Sequential "Purples" color palette

FIGURE 6.12 Sequential "viridis" color palette

appropriately. When capitalized (as used in tm_credits()), the element will be positioned tighter to the map frame.

6.3.3 Choosing a color palette

The examples shown in this chapter thus far have used a variety of *color palettes* to display statistical variation on choropleth maps. Software packages like sf, ggplot2, and tmap will have color palettes built in as "defaults"; I've shown the default palettes for all three, then changed the palettes used in **ggplot2** and in **tmap**. So how do you go about choosing an appropriate color palette? There are a variety of considerations to take into account.

First, it is important to understand the type of data you are working with. If your data are *quantitative* – that is, expressed with numbers, which you'll commonly be working with using Census data – you'll want a color palette that can show the statistical variation in your data correctly. In the demographic examples shown above, decennial Census data range from a low value to a high value. This type of information is effectively represented with a *sequential* color palette. **Sequential** color palettes use either a single hue or related hues then modify the color lightness or intensity to generate a sequence of colors. An example single-hue palette is the "Purples" ColorBrewer palette used in the map above.

With this palette, lighter colors should generally be used to represent lower data values, and darker values should represent higher values, suggesting a greater density/concentration of that attribute. In other color palettes, however, the more intense colors may be the lighter colors and should be used accordingly to represent data values. This is the case with the popular viridis color palettes, implemented in R with the **viridis** package (Garnier et al., 2021) and shown in Figure 6.12.

Diverging color palettes are best used when the cartographer wants to highlight extreme values on either end of the data distribution and represent neutral values in the middle. The example shown in Figure 6.13 is the ColorBrewer "RdBu" palette.

For Census data mapping, diverging palettes are well-suited to maps that visualize change over time. A map of population change using a diverging palette would highlight extreme population loss and extreme population gain with intense colors on either end of the palette, and represent minimal population change with a muted, neutral color in the middle.

6.3 Map-making with tmap

FIGURE 6.13 Diverging "RdBu" color palette

FIGURE 6.14 Categorical "Set1" color palette

Qualitative palettes are appropriate for categorical data, as they represent data values with unique, unordered hues. A good example is the "Set1" color palette shown in Figure 6.14.

While maps of Census data as returned by **tidycensus** will generally use sequential or diverging color palettes (given the quantitative nature of Census data), derived data products may require qualitative palettes. Illustrative examples in this book include categorical dot-density maps (addressed later in this chapter) and visualizations of geodemographic clusters, explored in Section 8.5.1.

Choosing an appropriate color palette for your maps can be a challenge. Fortunately, the ColorBrewer and viridis palettes are appropriate for a wide range of cartographic use cases and have built-in support in **ggplot2** and **tmap**. An excellent tool for helping decide on a color palette is **tmap**'s Palette Explorer app, accessible with the command `tmaptools::palette_explorer()`. Run this command in your R console to launch an interactive app that helps you explore different color scenarios using ColorBrewer and viridis palettes. Notably, the app includes a color blindness simulator to help you choose color palettes that are color blindness friendly.

6.3.4 Alternative map types with tmap

Choropleth maps are a core part of the Census data analyst's toolkit, but they are not ideal for every application. In particular, choropleth maps are best suited for visualizing rates, percentages, or statistical values that are normalized for the population of areal units. They are not ideal when the analyst wants to compare counts (or estimated counts) themselves, however. Choropleth maps of count data may ultimately reflect the underlying size of a baseline population; additionally, given that the counts are compared visually relative to the irregular shape of the polygons, choropleth maps can make comparisons difficult.

6.3.4.1 Graduated symbols

An alternative commonly used to visualize count data is the **graduated symbol map**. Graduated symbol maps use shapes referenced to geographic units that are sized relative to a data attribute. The example below uses **tmap**'s `tm_bubbles()` function to create a

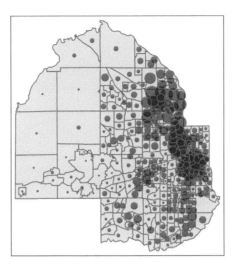

FIGURE 6.15 Graduated symbols with tmap

graduated symbol map of the Black population in Hennepin County, mapping the `estimate` column.

```
tm_shape(hennepin_black) +
  tm_polygons() +
  tm_bubbles(size = "value", alpha = 0.5,
             col = "navy",
             title.size = "Non-Hispanic Black - 2020 US Census") +
  tm_layout(legend.outside = TRUE,
            legend.outside.position = "bottom")
```

The visual comparisons on the map are made between the circles, not the polygons themselves, reflecting differences between population sizes.

6.3.4.2 Faceted maps

Given that the long-form race & ethnicity dataset returned by tidycensus includes information on five groups, a cartographer may want to visualize those groups comparatively. A single choropleth map cannot effectively visualize five demographic attributes simultaneously, and creating five separate maps can be tedious. A solution to this is using *faceting*, a concept introduced in Chapter 4.

Faceted maps in **tmap** are created with the `tm_facets()` function. The `by` argument specifies the column to be used to identify unique groups in the data. The remaining code is familiar **tmap** code; in this example, `tm_fill()` is preferred to `tm_polygons()` to hide the Census

6.3 Map-making with tmap

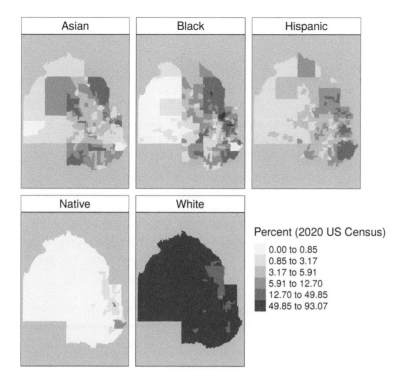

FIGURE 6.16 Faceted map with tmap

tract borders given the smaller sizes of the maps. The legend is also moved with the `legend.position` argument in `tm_layout()` to fill the empty space in the faceted map.

```
tm_shape(hennepin_race) +
  tm_facets(by = "variable", scale.factor = 4) +
  tm_fill(col = "percent",
          style = "quantile",
          n = 6,
          palette = "Blues",
          title = "Percent (2020 US Census)",) +
  tm_layout(bg.color = "grey",
            legend.position = c(-0.7, 0.15),
            panel.label.bg.color = "white")
```

The faceted maps do a good job of showing variations in each group in comparative context. However, the common legend and classification scheme used means that within-group variation is suppressed relative to the need to show consistent comparisons between groups. In turn, the "White" subplot shows little variation among Census tracts in Hennepin County given the large size of that group in the area. One additional disadvantage of separate maps by group is that they do not show neighborhood heterogeneity and diversity as well as they could. A popular alternative for visualizing within-unit heterogeneity is the dot-density map, covered below.

6.3.4.3 Dot-density maps

Dot-density maps scatter dots within areal units relative to the size of a data attribute. This cartographic method is intended to show attribute density through the dot distributions; for a Census map, in areas where the dots are dense, more people live there, whereas sparsely positioned dots reflect sparsity of population. Dot-density maps can also incorporate categories in the data to visualize densities of different subgroups simultaneously.

The `as_dot_density()` function in **tidycensus** helps users get Census data ready for dot-density visualization. For an input dataset, the function requires specifying a `value` column that represents the data attribute to be visualized, and a `values_per_dot` value that determines how many data values each dot should represent. The `group` column then partitions dots by group and shuffles their visual ordering on the map so that no one group occludes other groups.

In this example, we specify `value = "estimate"` to visualize the data from the 2020 US Census; `values_per_dot = 100` for a data to dots ratio of 100 people per dot; and `group = "variables"` to partition dots by racial / ethnic group on the map.

```
hennepin_dots <- hennepin_race %>%
  as_dot_density(
    value = "value",
    values_per_dot = 100,
    group = "variable"
  )
```

The map itself is created with the `tm_dots()` function, which in this example is combined with a background map using `tm_polygons()` to show the relative racial and ethnic heterogeneity of Census tracts in Hennepin County.

```
background_tracts <- filter(hennepin_race, variable == "White")

tm_shape(background_tracts) +
  tm_polygons(col = "white",
              border.col = "grey") +
  tm_shape(hennepin_dots) +
  tm_dots(col = "variable",
          palette = "Set1",
          size = 0.005,
          title = "1 dot = 100 people") +
  tm_layout(legend.outside = TRUE,
            title = "Race/ethnicity,\n2020 US Census")
```

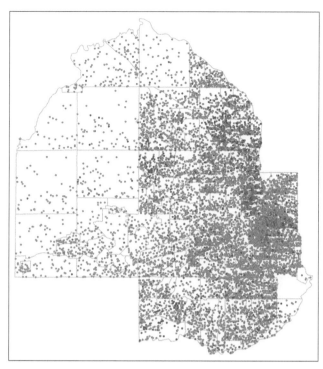

Issues with dot-density maps can include overplotting of dots, which can make legibility a problem; experiment with different dot sizes and dots to data ratios to improve this. Additionally, the use of Census tract polygons to generate the dots can cause visual issues. As dots are placed randomly within Census tract polygons, they in many cases will be placed in locations where no people live (such as lakes in Hennepin County). Dot distributions will also follow tract boundaries, which can create an artificial impression of abrupt changes in population distributions along polygon boundaries (as seen on the example map). A solution is the *dasymetric dot-density map* (Walker, 2018), which first removes areas from polygons which are known to be uninhabited then runs the dot-generation algorithm over those modified areas. `as_dot_density()` includes an argument, `erase_water`, that will automatically remove water areas from the input shapes before generating dots, avoiding dot placement in large bodies of water. This technique uses the `erase_water()` function in the **tigris** package, which is covered in more detail in Section 7.5.

6.4 Cartographic workflows with non-Census data

In many instances, an analyst may possess data that is available at a Census geography but is not available through the ACS or decennial Census. This means that the `geometry = TRUE` functionality in **tidycensus**, which automatically enriches data with geographic information, is not possible. In these cases, Census shapes obtained with **tigris** can be joined to tabular data and then visualized.

This section covers two such workflows. The first reproduces the popular red/blue election map common in presidential election cycles. The second focuses on mapping zip code tabulation areas, or ZCTAs, a geography that represents the spatial location of zip codes (postal codes) in the United States.

6.4.1 National election mapping with tigris shapes

While enumeration units like Census tracts and block groups will generally be used to map Census data, Census shapes representing legal entities are useful for a variety of cartographic purposes. A popular example is the political map, which shows the winner or poll results from an election in a region. We'll use data from the Cook Political Report[6] to generate a basic red state/blue state map of the 2020 US Presidential election results. This dataset was downloaded on June 5, 2021 and is available at `"data/us_vote_2020.csv"` in the book GitHub repository.

```
library(tidyverse)
library(tigris)

# Data source: https://cookpolitical.com/2020-national-popular-vote-tracker
vote2020 <- read_csv("data/us_vote_2020.csv")

names(vote2020)
```

```
##  [1] "state"          "called"           "final"           "dem_votes"
##  [5] "rep_votes"      "other_votes"      "dem_percent"     "rep_percent"
##  [9] "other_percent"  "dem_this_margin"  "margin_shift"    "vote_change"
## [13] "stateid"        "EV"               "X"               "Y"
## [17] "State_num"      "Center_X"         "Center_Y"        "...20"
## [21] "2016 Margin"    "Total 2016 Votes"
```

The data include a wide variety of columns that can be visualized on a map. As discussed in the previous chapter, a comparative map of the United States can use the `shift_geometry()` function in the **tigris** package to shift and rescale Alaska and Hawaii. State geometries are available in **tigris** with the `states()` function, which should be used with the arguments `cb = TRUE` and `resolution = "20m"` to appropriately generalize the state geometries for national mapping.

To create the map, the geometry data obtained with tigris must be joined to the election data from the Cook Political Report. This is accomplished with the `left_join()` function from **dplyr**. dplyr's `*_join()` family of functions are supported by simple features objects, and work in this context analogous to the common "Join" operations in desktop GIS software. The join functions work by matching values in one or more "key fields" between two tables and merging data from those two tables into a single output table. The most common join functions you'll use for spatial data are `left_join()`, which retains all rows from the first dataset and fills non-matching rows with `NA` values, and `inner_join()`, which drops non-matching rows in the output dataset.

Let's try this out by obtaining low-resolution state geometry with **tigris**, shifting and rescaling with `shift_geometry()`, then merging the political data to those shapes, matching the `NAME` column in `us_states` to the `state` column in `vote2020`.

---

[6]https://cookpolitical.com/2020-national-popular-vote-tracker

6.4 Cartographic workflows with non-Census data

```
us_states <- states(cb = TRUE, resolution = "20m") %>%
  filter(NAME != "Puerto Rico") %>%
  shift_geometry()

us_states_joined <- us_states %>%
  left_join(vote2020, by = c("NAME" = "state"))
```

Before proceeding we'll want to do some quality checks. In `left_join()`, values must match exactly between `NAME` and `state` to merge correctly – and this is not always guaranteed when using data from different sources. Let's check to see if we have any problems:

```
table(is.na(us_states_joined$state))
```

```
## 
## FALSE 
##    51
```

We've matched all the 50 states plus the District of Columbia correctly. In turn, the joined dataset has retained the shifted and rescaled geometry of the US states and now includes the election information from the tabular dataset, which can be used for mapping. To achieve this structure, specifying the directionality of the join was critical. For spatial information to be retained in a join, the spatial object *must* be on the left-hand side of the join. In our pipeline, we specified the `us_states` object first and used `left_join()` to join the election information *to* the states object. If we had done this in reverse, we would have lost the spatial class information necessary to make the map.

For a basic red state/blue state map using **ggplot2** and `geom_sf()`, a manual color palette can be supplied to the `scale_fill_manual()` function, filling state polygons based on the `called` column which represents the party for whom the state was called.

```
ggplot(us_states_joined, aes(fill = called)) +
  geom_sf(color = "white", lwd = 0.2) +
  scale_fill_manual(values = c("blue", "red")) +
  theme_void() +
  labs(fill = "Party",
       title = " 2020 US presidential election results by state",
       caption = paste0("Note: Nebraska and Maine split electoral ",
                        "college votes by congressional district"))
```

6.4.2 Understanding and working with ZCTAs

The most granular geography at which many agencies release data is at the zip code level. This is not an ideal geography for visualization, given that zip codes represent collections of US Postal Service routes (or sometimes even a single building, or Post Office box) that are not guaranteed to form coherent geographies. The US Census Bureau allows for an approximation of zip code mapping with Zip Code Tabulation Areas, or ZCTAs. ZCTAs are shapes built from Census blocks in which the most common zip code for addresses in each block determines how blocks are allocated to corresponding ZCTAs. While ZCTAs are not recommended for spatial analysis due to these irregularities, they can be useful for visualizing data distributions when no other granular geographies are available.

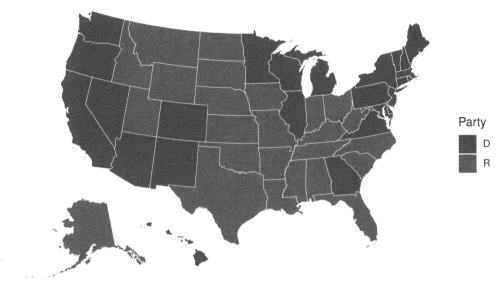

FIGURE 6.17 Map of the 2020 US presidential election results with ggplot2

An example of this is the Internal Revenue Service's Statistics of Income (SOI) data[7], which includes a wide range of indicators derived from tax returns. The most detailed geography available is the zip code level in this dataset, meaning that within-county visualizations require using ZCTAs. Let's read in the data for 2018 from the IRS website:

```
irs_data <- read_csv("https://www.irs.gov/pub/irs-soi/18zpallnoagi.csv")

ncol(irs_data)
```

```
## [1] 153
```

The dataset contains 153 columns which are identified in the linked codebook[8]. Geographies are identified by the ZIPCODE column, which shows aggregated data by state (ZIPCODE == "000000") and by zip code. We might be interested in understanding the geography of self-employment income within a given region. We'll retain the variables N09400, which represents the number of tax returns with self-employment tax, and N1, which represents the total number of returns.

```
self_employment <- irs_data %>%
  select(ZIPCODE, self_emp = N09400, total = N1)
```

---

[7]https://www.irs.gov/statistics/soi-tax-stats-individual-income-tax-statistics-2018-zip-code-data-soi
[8]https://www.irs.gov/pub/irs-soi/18zpdoc.docx

6.4 Cartographic workflows with non-Census data

From here, we'll need to identify a region of zip codes for analysis. In **tigris**, the `zctas()` function allows us to fetch a Zip Code Tabulation Areas shapefile. Given that some ZCTA geography is irregular and sometimes stretches across multiple states, a shapefile for the entire United States must first be downloaded. It is recommended that shapefile caching with `options(tigris_use_cache = TRUE)` be used with ZCTAs to avoid long data download times.

In the next chapter, you'll learn how to use spatial overlay to extract geographic data within a specific region. That said, the `starts_with` parameter in `zctas()` allows users to filter down ZCTAs based on a vector of prefixes, which can identify an area without using a spatial process. For example, we can get ZCTA data near Boston, MA by using the appropriate prefixes.

```
library(mapview)
library(tigris)
options(tigris_use_cache = TRUE)

boston_zctas <- zctas(
  cb = TRUE,
  starts_with = c("021", "022", "024"),
  year = 2018
)
```

Next, we can use `mapview()` to inspect the results:

```
mapview(boston_zctas)
```

The ZCTA prefixes `021`, `022`, and `024` cover much of the Boston metropolitan area; "holes" in the region represent areas like Boston Common which are not covered by ZCTAs. Let's take a quick look at its attributes:

```
names(boston_zctas)
```

```
## [1] "ZCTA5CE10"  "AFFGEOID10" "GEOID10"    "ALAND10"    "AWATER10"
## [6] "geometry"
```

Either the `ZCTA4CE10` column or the `GEOID10` column can be matched to the appropriate zip code information in the IRS dataset for visualization. The code below joins the IRS data to the spatial dataset and computes a new column representing the percentage of returns with self-employment income.

```
boston_se_data <- boston_zctas %>%
  left_join(self_employment, by = c("GEOID10" = "ZIPCODE")) %>%
  mutate(pct_self_emp = 100 * (self_emp / total)) %>%
  select(GEOID10, self_emp, pct_self_emp)
```

There are a variety of ways to visualize this information. One such method is a choropleth map, which you've learned about earlier this chapter:

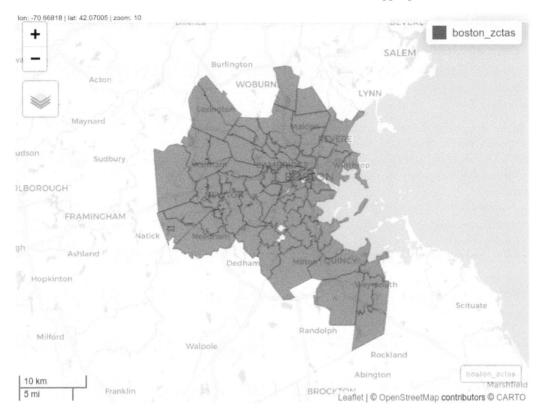

FIGURE 6.18 ZCTAs in the Boston, MA area

TABLE 6.2 Self-employment percentages by ZCTA in the Boston, MA area

GEOID10	self_emp	pct_self_emp	geometry
02461	860	24.15730	MULTIPOLYGON (((-71.22275 4...
02141	930	11.90781	MULTIPOLYGON (((-71.09475 4...
02139	2820	14.58872	MULTIPOLYGON (((-71.1166 42...
02180	1680	13.45076	MULTIPOLYGON (((-71.11976 4...
02457	NA	NA	MULTIPOLYGON (((-71.27642 4...

```
tm_shape(boston_se_data, projection = 26918) +
  tm_fill(col = "pct_self_emp",
          palette = "Purples",
          title = "% self-employed,\n2018 IRS SOI data")
```

The choropleth map in Figure 6.19 shows that self-employment filings are more common in suburban Boston ZCTAs than nearer to the urban core, generally speaking. However, we might also be interested in understanding where most self-employment income filings are located rather than their share relative to the total number of returns filed. This requires visualizing the self_emp column directly. As discussed earlier in this chapter, a graduated symbol map with tm_bubbles() is preferable to a choropleth map for this purpose.

6.5 Interactive mapping

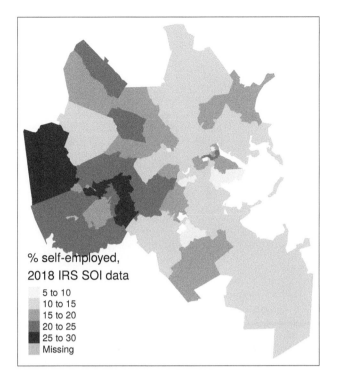

FIGURE 6.19 Simple choropleth of self-employment in Boston

```
tm_shape(boston_se_data) +
  tm_polygons() +
  tm_bubbles(size = "self_emp",
             alpha = 0.5,
             col = "navy",
             title.size = "Self-employed filers,\n2018 IRS SOI data")
```

6.5 Interactive mapping

The examples addressed in this chapter thus far have all focused on *static maps*, where the output is fixed after rendering the map. Modern web technologies – and the integration of those technologies into R with the htmlwidgets package, as discussed in Section 4.7.4 – make the creation of interactive, explorable Census data maps straightforward.

6.5.1 Interactive mapping with Leaflet

With over 31,000 GitHub stars as of July 2021, the Leaflet JavaScript library (Agafonkin, 2020) is one of the most popular frameworks worldwide for interactive mapping. The RStudio team brought the Leaflet to R with the **leaflet** R package (Cheng et al., 2021), which now powers several approaches to interactive mapping in R. The following examples cover how

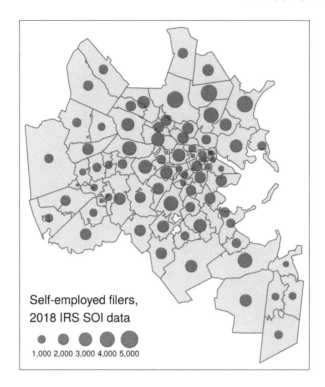

FIGURE 6.20 Graduated symbol map of self-employment by ZCTA in Boston

to visualize Census data on an interactive Leaflet map using approaches from **mapview**, **tmap**, and the core **leaflet** package.

Let's start by getting some illustrative data on the percentage of the population aged 25 and up with a bachelor's degree or higher from the 2016-2020 ACS. We'll look at this information by Census tract in Dallas County, Texas.

```
dallas_bachelors <- get_acs(
  geography = "tract",
  variables = "DP02_0068P",
  year = 2020,
  state = "TX",
  county = "Dallas",
  geometry = TRUE
)
```

In Chapter 5, you learned how to quickly visualize geographic data obtained with tigris on an interactive map by using the mapview() function in the **mapview** package. The mapview() function also includes a parameter zcol that takes a column in the dataset as an argument, and visualizes that column with an interactive choropleth map, as shown in Figure 6.21.

```
library(mapview)
mapview(dallas_bachelors, zcol = "estimate")
```

6.5 Interactive mapping

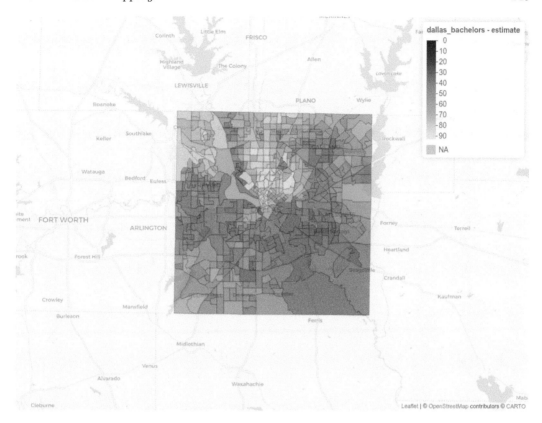

FIGURE 6.21 Interactive mapview choropleth

Conversion of **tmap** maps to interactive Leaflet maps is also straightforward with the command `tmap_mode("view")`. After entering this command, all subsequent tmap maps in your R session will be rendered as interactive Leaflet maps using the same tmap syntax you'd use to make static maps.

```
library(tmap)
tmap_mode("view")

tm_shape(dallas_bachelors) +
  tm_fill(col = "estimate", palette = "magma",
          alpha = 0.5)
```

To switch back to static plotting mode, run the command `tmap_mode("plot")`.

For more fine-grained control over your Leaflet maps, the core **leaflet** package can be used. Below, we'll reproduce the **mapview/tmap** examples using the leaflet package's native syntax. First, a color palette will be defined using the `colorNumeric()` function. This function itself creates a function we're calling `pal()`, which translates data values to color values for a given color palette. Our chosen color palette in this example is the viridis magma palette.

```
library(leaflet)
```

FIGURE 6.22 Interactive map with tmap in view mode

```
pal <- colorNumeric(
  palette = "magma",
  domain = dallas_bachelors$estimate
)

pal(c(10, 20, 30, 40, 50))
```

```
## [1] "#170F3C" "#420F75" "#6E1E81" "#9A2D80" "#C73D73"
```

The map itself is built with a magrittr pipeline and the following steps:

- The leaflet() function initializes the map. A data object can be specified here or in a function that comes later in the pipeline.
- addProviderTiles() helps you add a basemap to the map that will be shown beneath your data as a reference. Several providers are built-in to the Leaflet package, including the popular Stamen[9] reference maps. If you are only interested in a basic basemap, the addTiles() function returns the standard OpenStreetMap[10] basemap. Use the built-in providers object to try out different basemaps; it is good practice for choropleth mapping to use a greyscale or muted basemap.

---

[9]https://stamen.com/
[10]https://www.openstreetmap.org/#map=5/38.007/-95.844

6.5 Interactive mapping

- `addPolygons()` adds the tract polygons to the map and styles them. In the code below, we are using a series of options to specify the input data; to color the polygons relative to the defined color palette; and to adjust the smoothing between polygon borders, the opacity, and the line weight. The `label` argument will create a hover tooltip on the map for additional information about the polygons.
- `addLegend()` then creates a legend for the map, providing critical information about how the colors on the map relate to the data values.

```
leaflet() %>%
  addProviderTiles(providers$Stamen.TonerLite) %>%
  addPolygons(data = dallas_bachelors,
              color = ~pal(estimate),
              weight = 0.5,
              smoothFactor = 0.2,
              fillOpacity = 0.5,
              label = ~estimate) %>%
  addLegend(
    position = "bottomright",
    pal = pal,
    values = dallas_bachelors$estimate,
    title = "% with bachelor's<br/>degree"
  )
```

This example only scratches the surface of what the **leaflet** R package can accomplish; I'd encourage you to review the documentation for more examples[11].

6.5.2 Alternative approaches to interactive mapping

Like most interactive mapping platforms, Leaflet uses *tiled mapping*[12] in the Web Mercator coordinate reference system. Web Mercator works well for tiled web maps that need to fit within rectangular computer screens, and preserves angles at large scales (zoomed-in areas) which is useful for local navigation. However, it grossly distorts the area of geographic features near the poles, making it inappropriate for small-scale thematic mapping of the world or world regions (Battersby et al., 2014).

Let's illustrate this by mapping median home value by state from the 1-year ACS using **leaflet**. We'll first acquire the data with geometry using tidycensus, setting the output resolution to "20m" to get low-resolution boundaries to speed up our interactive mapping.

```
us_value <- get_acs(
  geography = "state",
  variables = "B25077_001",
  year = 2019,
  survey = "acs1",
  geometry = TRUE,
  resolution = "20m"
)
```

---
[11] https://rstudio.github.io/leaflet/
[12] https://wiki.openstreetmap.org/wiki/Slippy_Map

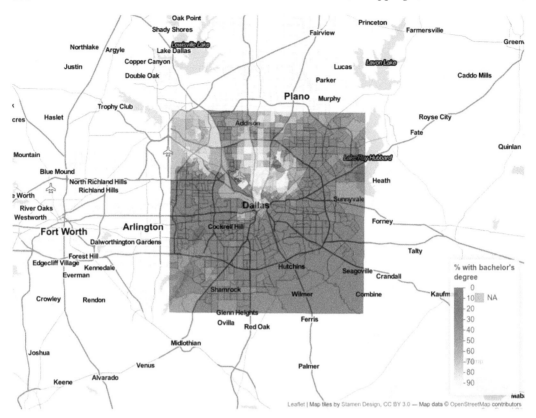

FIGURE 6.23 Interactive leaflet map

The acquired ACS data for the US can be mapped using the same techniques as with the educational attainment map for Dallas County.

```
library(leaflet)

us_pal <- colorNumeric(
  palette = "plasma",
  domain = us_value$estimate
)

leaflet() %>%
  addProviderTiles(providers$Stamen.TonerLite) %>%
  addPolygons(data = us_value,
              color = ~us_pal(estimate),
              weight = 0.5,
              smoothFactor = 0.2,
              fillOpacity = 0.5,
              label = ~estimate) %>%
  addLegend(
    position = "bottomright",
    pal = us_pal,
    values = us_value$estimate,
```

6.5 Interactive mapping

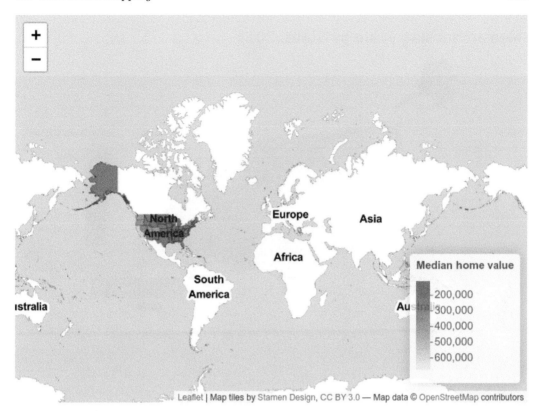

FIGURE 6.24 Interactive US map using Web Mercator

```
    title = "Median home value"
)
```

The disadvantages of Web Mercator – as well as this general approach to mapping the United States – are on full display. Alaska's area is grossly distorted relative to the rest of the United States. It is also difficult on this map to compare Alaska and Hawaii to the continental United States – which is particularly important in this example as Hawaii's median home value is the highest in the entire country. The solution proposed elsewhere in this book is to use `tigris::shift_geometry()` which adopts appropriate coordinate reference systems for Alaska, Hawaii, and the continental US and arranges them in a better comparative fashion on the map. However, this approach risks losing the interactivity of a Leaflet map.

A compromise solution can involve other R packages that allow for interactive mapping. An excellent option is the **ggiraph** package (Gohel and Skintzos, 2021), which like the plotly package can convert ggplot2 graphics into interactive plots. In the example shown in Figure 6.25, interactivity is added to a ggplot2 plot with **ggiraph**, allowing for panning and zooming with a hover tooltip on a shifted and rescaled map of the US.

```
library(ggiraph)
library(scales)

us_value_shifted <- us_value %>%
```

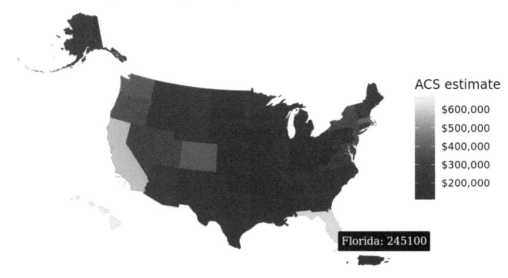

FIGURE 6.25 Interactive US map with ggiraph

```
  shift_geometry(position = "outside") %>%
  mutate(tooltip = paste(NAME, estimate, sep = ": "))

gg <- ggplot(us_value_shifted, aes(fill = estimate)) +
  geom_sf_interactive(aes(tooltip = tooltip, data_id = NAME),
                      size = 0.1) +
  scale_fill_viridis_c(option = "plasma", labels = label_dollar()) +
  labs(title = "Median housing value by State, 2019",
       caption = "Data source: 2019 1-year ACS, US Census Bureau",
       fill = "ACS estimate") +
  theme_void()

girafe(ggobj = gg) %>%
  girafe_options(opts_hover(css = "fill:cyan;"),
                 opts_zoom(max = 10))
```

6.6 Advanced examples

The examples discussed in this chapter thus far likely cover a large proportion of cartographic use cases for Census data analysts. However, R allows cartographers to go beyond these core map types. This final section of the chapter introduces some options for more advanced visualization using data from **tidycensus**.

6.6 Advanced examples

TABLE 6.3 Top origins for migrants to Travis County, TX

GEOID1	GEOID2	FULL1_NAME	FULL2_NAME	variable	estimate	moe	centroid1
48453	48491	Travis County, Texas	Williamson County, Texas	MOVEDIN	10147	1198	POINT (-97.78195 30.33469)
48453	48201	Travis County, Texas	Harris County, Texas	MOVEDIN	5746	742	POINT (-97.78195 30.33469)
48453	48209	Travis County, Texas	Hays County, Texas	MOVEDIN	4240	839	POINT (-97.78195 30.33469)
48453	48029	Travis County, Texas	Bexar County, Texas	MOVEDIN	3758	631	POINT (-97.78195 30.33469)
48453	48113	Travis County, Texas	Dallas County, Texas	MOVEDIN	3005	657	POINT (-97.78195 30.33469)
48453	48439	Travis County, Texas	Tarrant County, Texas	MOVEDIN	2053	527	POINT (-97.78195 30.33469)
48453	06037	Travis County, Texas	Los Angeles County, California	MOVEDIN	1770	426	POINT (-97.78195 30.33469)
48453	48021	Travis County, Texas	Bastrop County, Texas	MOVEDIN	1423	334	POINT (-97.78195 30.33469)
48453	48085	Travis County, Texas	Collin County, Texas	MOVEDIN	1172	514	POINT (-97.78195 30.33469)
48453	48141	Travis County, Texas	El Paso County, Texas	MOVEDIN	1108	445	POINT (-97.78195 30.33469)

6.6.1 Mapping migration flows

In 2021, **tidycensus** co-author Matt Herman added support for the ACS Migration Flows API in the package, covered briefly in Section 2.5. One notable feature of this new functionality, available in the `get_flows()` function, is built-in support for flow mapping with the argument `geometry = TRUE`. Geometry operates differently for migration flows data given that the geography of interest is not a single location for a given row but rather the connection between those locations. In turn, for `get_flows()`, `geometry = TRUE` returns two POINT geometry columns: one for the location itself, and one for the location to which it is linked.

Let's take a look for one of the most popular recent migration destinations in the United States: Travis County Texas, home to Austin.

```
travis_inflow <- get_flows(
  geography = "county",
  state = "TX",
  county = "Travis",
  geometry = TRUE
) %>%
  filter(variable == "MOVEDIN") %>%
  na.omit() %>%
  arrange(desc(estimate))
```

The dataset is filtered to focus on in-migration, represented by the MOVEDIN variable, and drops migrations from outside the United States with `na.omit()` (as these areas do not have a GEOID value).

The **mapdeck** R package (Cooley, 2020) offers excellent support for interactive flow mapping with minimal code. **mapdeck** is a wrapper of deck.gl[13], a tremendous visualization library originally developed at Uber that offers 3D mapping capabilities built on top of Mapbox's GL JS library[14]. Users will need to sign up for a Mapbox account and get a Mapbox access token to use **mapdeck**; see the mapdeck documentation for more information[15].

Once set, flow maps can be created by initializing a mapdeck map with `mapdeck()` then using the `add_arc()` function, which can link either X/Y coordinate columns or POINT geometry columns, as shown below. In this example, we are using the top 30 origins for migrants to Travis County, and generating a new `weight` column that makes the proportional arc widths less bulky.

---

[13]https://deck.gl/
[14]https://www.mapbox.com/mapbox-gljs
[15]https://symbolixau.github.io/mapdeck/articles/mapdeck.html#the-basics

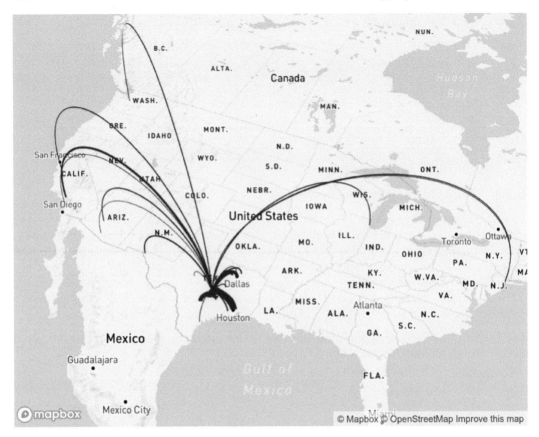

FIGURE 6.26 Flow map of in-migration to Travis County, TX with mapdeck

```
library(mapdeck)

token <- "YOUR TOKEN HERE"

travis_inflow %>%
  slice_max(estimate, n = 30) %>%
  mutate(weight = estimate / 500) %>%
  mapdeck(token = token) %>%
  add_arc(origin = "centroid2",
          destination = "centroid1",
          stroke_width = "weight",
          update_view = FALSE)
```

6.6.2 Linking maps and charts

This chapter and Chapter 4 are linked in many ways. The visualization principles discussed in each chapter apply to each other; the key difference is that this chapter focuses on *geographic visualization* whereas Chapter 4 does not. In many cases, you'll want to take advantage of the strengths of both geographic and non-geographic visualization. Maps are excellent at

6.6 Advanced examples

showing patterns and trends over space, and offer a familiar reference to viewers; charts are better at showing rankings and ordering of data values between places.

The example below illustrates how to combine two chart types discussed in this book: a choropleth map and a margin of error plot. Margins of error are notoriously difficult to display on maps; possible options include superimposing patterns on choropleth maps to highlight areas with high levels of uncertainty (Wong and Sun, 2013) or using bivariate choropleth maps to simultaneously visualize estimates and MOEs (Lucchesi and Wikle, 2017).

R's visualization tools offer an alternative approach: interactive linking of a choropleth map with a chart that clearly visualizes the uncertainty around estimates. This approach involves generating a map and chart with ggplot2, then combining the plots with patchwork and rendering them as interactive, linked graphics with ggiraph. The key aesthetic to be used here is `data_id`, which if set in the code for both plots will highlight corresponding data points on both plots on user hover.

Example code to generate such a linked visualization is below.

```r
library(tidycensus)
library(ggiraph)
library(tidyverse)
library(patchwork)
library(scales)

vt_income <- get_acs(
  geography = "county",
  variables = "B19013_001",
  state = "VT",
  year = 2020,
  geometry = TRUE
) %>%
  mutate(NAME = str_remove(NAME, " County, Vermont"))

vt_map <- ggplot(vt_income, aes(fill = estimate)) +
  geom_sf_interactive(aes(data_id = GEOID)) +
  scale_fill_distiller(palette = "Greens",
                       direction = 1,
                       guide = "none") +
  theme_void()

vt_plot <- ggplot(vt_income, aes(x = estimate, y = reorder(NAME, estimate),
                                 fill = estimate)) +
  geom_errorbar(aes(xmin = estimate - moe, xmax = estimate + moe)) +
  geom_point_interactive(color = "black", size = 4, shape = 21,
                         aes(data_id = GEOID)) +
  scale_fill_distiller(palette = "Greens", direction = 1,
                       labels = label_dollar()) +
  scale_x_continuous(labels = label_dollar()) +
  labs(title = "Household income by county in Vermont",
       subtitle = "2016-2020 American Community Survey",
```

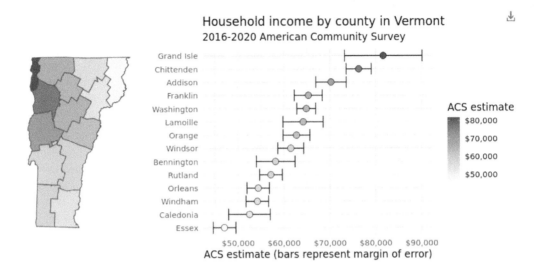

FIGURE 6.27 Linked map and chart with ggiraph

```
    y = "",
    x = "ACS estimate (bars represent margin of error)",
    fill = "ACS estimate") +
  theme_minimal(base_size = 14)

girafe(ggobj = vt_map + vt_plot, width_svg = 10, height_svg = 5) %>%
  girafe_options(opts_hover(css = "fill:cyan;"))
```

This example largely re-purposes visualization code readers have learned in other examples in this book. Try hovering your cursor over any county in Vermont on the map – or any data point on the chart – and notice what happens on the other plot. The corresponding county or data point will also be highlighted, allowing for a linked representation of geographic location and margin of error visualization.

6.6.3 Reactive mapping with Shiny

Advanced Census cartographers may want to take these examples a step further and build them into full-fledged data dashboards and web-based visualization applications. Fortunately, R users don't have to do this from scratch. Shiny (Chang et al., 2021) is a tremendously popular and powerful framework for the development of interactive web applications in R that can execute R code in response to user inputs. A full treatment of Shiny is beyond the scope of this book; however, Shiny is a "must-learn" for R data analysts.

An example Shiny visualization app that extends the race/ethnicity example from this chapter is shown below. It includes a drop-down menu that allows users to select a racial or ethnic group in the Twin Cities and visualizes the distribution of that group on an interactive choropleth map that uses Leaflet and the viridis color palette.

6.6 Advanced examples

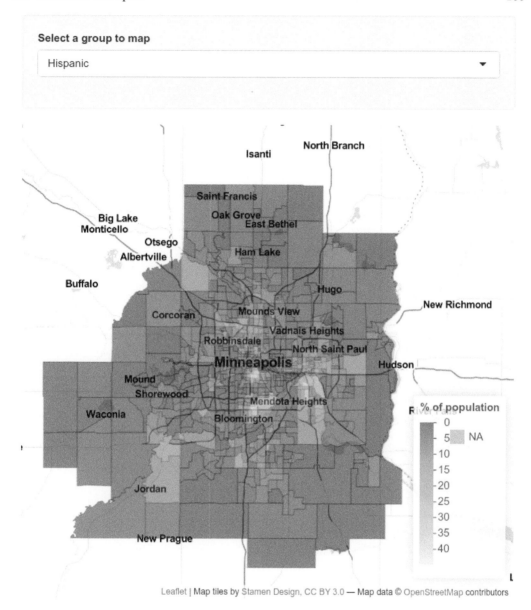

FIGURE 6.28 Interactive mapping app with Shiny

The code used to create the app is found below; copy-paste this code into your own R script, set your Census API key, and run it! To learn more, I encourage you to review Hadley Wickham's *Mastering Shiny* book[16] and the Leaflet package's documentation on Shiny integration[17].

```
# app.R
library(tidycensus)
```

---
[16]https://mastering-shiny.org/index.html
[17]https://rstudio.github.io/leaflet/shiny.html

```r
library(shiny)
library(leaflet)
library(tidyverse)

census_api_key("YOUR KEY HERE")

twin_cities_race <- get_acs(
  geography = "tract",
  variables = c(
    hispanic = "DP05_0071P",
    white = "DP05_0077P",
    black = "DP05_0078P",
    native = "DP05_0079P",
    asian = "DP05_0080P",
    year = 2019
  ),
  state = "MN",
  county = c("Hennepin", "Ramsey", "Anoka", "Washington",
             "Dakota", "Carver", "Scott"),
  geometry = TRUE
)

groups <- c("Hispanic" = "hispanic",
            "White" = "white",
            "Black" = "black",
            "Native American" = "native",
            "Asian" = "asian")

ui <- fluidPage(
  sidebarLayout(
    sidebarPanel(
      selectInput(
        inputId = "group",
        label = "Select a group to map",
        choices = groups
      )
    ),
    mainPanel(
      leafletOutput("map", height = "600")
    )
  )
)

server <- function(input, output) {

  # Reactive function that filters for the selected group in the drop-down menu
  group_to_map <- reactive({
    filter(twin_cities_race, variable == input$group)
  })
```

```r
  # Initialize the map object, centered on the Minneapolis-St. Paul area
  output$map <- renderLeaflet({

    leaflet(options = leafletOptions(zoomControl = FALSE)) %>%
      addProviderTiles(providers$Stamen.TonerLite) %>%
      setView(lng = -93.21,
              lat = 44.98,
              zoom = 8.5)

  })

  observeEvent(input$group, {

    pal <- colorNumeric("viridis", group_to_map()$estimate)

    leafletProxy("map") %>%
      clearShapes() %>%
      clearControls() %>%
      addPolygons(data = group_to_map(),
                  color = ~pal(estimate),
                  weight = 0.5,
                  fillOpacity = 0.5,
                  smoothFactor = 0.2,
                  label = ~estimate) %>%
    addLegend(
      position = "bottomright",
      pal = pal,
      values = group_to_map()$estimate,
      title = "% of population"
    )
  })

}

shinyApp(ui = ui, server = server)
```

6.7 Working with software outside of R for cartographic projects

The examples shown in this chapter display all maps *within* R. RStudio users running the code in this chapter, for example, will display static plots in the Plots pane, or interactive maps in the interactive Viewer pane. In many cases, R users will want to export out their maps for display on a website or in a report. In other cases, R users might be interested in using R and tidycensus as a "data pipeline" that can generate appropriate Census data for mapping in an external software package like Tableau, ArcGIS, or QGIS. This section covers those use-cases.

6.7.1 Exporting maps from R

Cartographers exporting maps made with **ggplot2** will likely want to use the `ggsave()` command. The map export process with `ggsave()` is similar to the process described in Section 4.2.3. If `theme_void()` is used for the map, however, the cartographer may want to supply a color to the `bg` parameter, as it would otherwise default to `"transparent"` for that theme.

The `tmap_save()` command is the best option for exporting maps made with **tmap**. `tmap_save()` requires that a map be stored as an object first; this example will re-use a map of Hennepin County from earlier in the chapter and assign it to a variable named `hennepin_map`.

```
hennepin_map <- tm_shape(hennepin_black) +
  tm_polygons(col = "percent",
          style = "jenks",
          n = 5,
          palette = "Purples",
          title = "ACS estimate",
          legend.hist = TRUE) +
  tm_layout(title = "Percent Black\nby Census tract",
          frame = FALSE,
          legend.outside = TRUE,
          bg.color = "grey70",
          legend.hist.width = 5,
          fontfamily = "Verdana")
```

That map can be saved using similar options to `ggsave()`. `tmap_save()` allows for specification of width, height, units, and dpi. If small values are passed to `width` and `height`, **tmap** will assume that the units are inches; if large values are passed (greater than 50), **tmap** will assume that the units represent pixels.

```
tmap_save(
  tm = hennepin_map,
  filename = "~/images/hennepin_black_map.png",
  height = 5.5,
  width = 8,
  dpi = 300
)
```

Interactive maps designed with **leaflet** can be written to HTML documents using the `saveWidget()` function from the **htmlwidgets** package. A Leaflet map should first be assigned to a variable, which is then passed to `saveWidget()` along with a specified name and path for the output HTML file. Interactive maps created with `mapview()` are written to HTML files the same way, though the object to be saved will need to be accessed from the `map` slot with the notation `@map`, as shown below.

```
library(htmlwidgets)

dallas_map <- mapview(dallas_bachelors, zcol = "estimate")
```

```
saveWidget(dallas_map@map, "dallas_mapview_map.html", selfcontained = TRUE)
```

The argument `selfcontained = TRUE` is an important one to consider when writing interactive maps to HTML files. If `TRUE` as shown in the example, `saveWidget()` will bundle any necessary assets (e.g. CSS, JavaScript) as a base64-encoded string in the HTML file. This makes the HTML file more portable but can lead to large file sizes. The alternative, `selfcontained = FALSE`, places these assets into an accompanying directory when the HTML file is written out. For interactive maps generated with **tmap**, `tmap_save()` can also be used to write out HTML files in this way.

6.7.2 Interoperability with other visualization software

Although R packages have rich capabilities for designing both static and interactive maps as well as map-based dashboards, some analysts will want to turn to other specialized tools for data visualization. Such tools might include drag-and-drop dashboard builders like Tableau, or dedicated GIS software like ArcGIS or QGIS that allow for more manual control over cartographic layouts and outputs.

This workflow will often involve the use of R, and packages like **tidycensus**, for data acquisition and wrangling, then the use of an external visualization program for data visualization and cartography. In this workflow, an R object produced with **tidycensus** can be written to an external spatial file with the `st_write()` function in the **sf** package. The code below illustrates how to write Census data from R to a *shapefile*, a common vector spatial data format readable by desktop GIS software and Tableau.

```
library(tidycensus)
library(sf)
options(tigris_use_cache = TRUE)

dc_income <- get_acs(
  geography = "tract",
  variables = "B19013_001",
  state = "DC",
  year = 2020,
  geometry = TRUE
)

st_write(dc_income, "dc_income.shp")
```

The output file `dc_income.shp` will be written to the user's current working directory. Other spatial data formats like GeoPackage (`.gpkg`) and GeoJSON (`.geojson`) are available by specifying the appropriate file extension.

QGIS cartographers can also take advantage of functionality within the software to run R (and consequently **tidycensus** functions) directly within the platform. In QGIS, this is enabled with the Processing R Provider plugin[18]. QGIS users should install the plugin from the **Plugins** drop-down menu, then click **Processing > Toolbox** to access QGIS's suite of tools. Clicking the R icon then **Create New R Script...** will open the R script editor.

---

[18] https://north-road.github.io/qgis-processing-r/

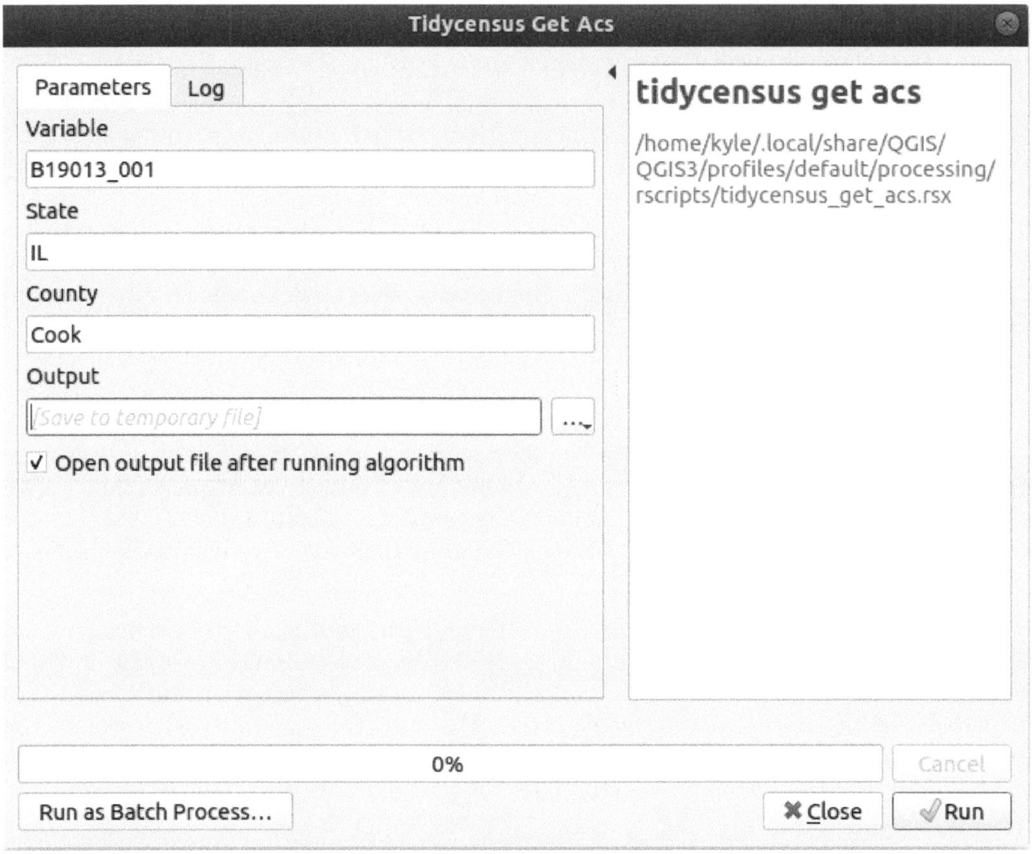

FIGURE 6.29 Example tidycensus tool in QGIS

The example script below will be translated by QGIS into a GIS tool that uses the version of **tidycensus** installed on a user's system to add a demographic layer at the Census tract level from the ACS to a QGIS project.

```
##Variable=string
##State=string
##County=string
##Output=output vector
library(tidycensus)

Output = get_acs(
    geography = "tract",
    variables = Variable,
    state = State,
    county = County,
    geometry = TRUE
)
```

Tool parameters are defined at the beginning of the script with ## notation. Variable, State, and County will all accept strings (text) as input, and the result of get_acs() in the tool will

6.8 Exercises

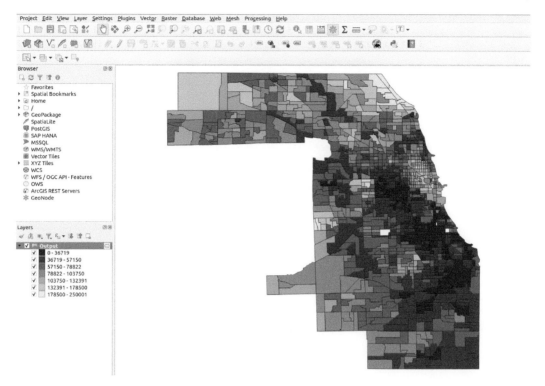

FIGURE 6.30 Styled layer from tidycensus in QGIS

be written to Output, which is added to the QGIS project as a vector layer. Once finished, the script should be saved with an appropriate name and the file extension .rsx in the suggested output directory, then located from the R section of the Processing Toolbox and opened. The GIS tool will look something like the image below.

Fill in the text boxes with a desired ACS variable, state, and county, then click **Run**. The QGIS tool will call the user's R installation to execute the tool with the specified inputs. If everything runs correctly, a layer will be added to the user's QGIS project ready for mapping with QGIS's suite of cartographic tools.

The example shown displays Census tracts in Cook County, Illinois obtained from **tidycensus** in a QGIS project; these tracts have been styled as a choropleth in QGIS by median household income (the requested variable) after the tool added the layer to the project.

6.8 Exercises

Using one of the mapping frameworks introduced in this chapter (either **ggplot2**, **tmap**, or **leaflet**) complete the following tasks:

- If you are just getting started with **tidycensus**/the tidyverse, make a race/ethnicity map by adapting the code provided in this section but for a different county.

- Next, find a different variable to map with `tidycensus::load_variables()`. Review the discussion of cartographic choices in this chapter and visualize your data appropriately.

7

Spatial analysis with US Census data

A very common use-case of spatial data from the US Census Bureau is *spatial analysis*. Spatial analysis refers to the performance of analytic tasks that explicitly incorporate the spatial properties of a dataset. Principles in spatial analysis are closely related to the field of *geographic information science*, which incorporates both theoretical perspectives and methodological insights with regards to the use of geographic data.

Traditionally, geographic information science has been performed in a *geographic information system*, or "GIS", which refers to an integrated software platform for the management, processing, analysis, and visualization of geographic data. As evidenced in this book already, R packages exist for handling these tasks, allowing R to function as a capable substitute for desktop GIS software like ArcGIS or QGIS.

Traditionally, spatial analytic tasks in R have been handled by the **sp** package and allied packages such as **rgeos**. In recent years, however, the **sf** package has emerged as the next-generation alternative to **sp** for spatial data analysis in R. In addition to the simpler representation of vector spatial data in R, as discussed in previous chapters, **sf** also includes significant functionality for spatial data analysis that integrates seamlessly with tidyverse tools.

This chapter covers how to perform common spatial analysis tasks with Census data using the **sf** package. As with previous chapters, the examples will focus on data acquired with the **tidycensus** and **tigris** packages, and will cover common workflows of use to practitioners who work with US Census Bureau data.

7.1 Spatial overlay

Spatial data analysis allows practitioners to consider how geographic datasets interrelate in geographic space. This analytic functionality facilitates answering a wide range of research questions that would otherwise prove difficult without reference to a dataset's geographic properties.

One common use-case employed by the geospatial analyst is *spatial overlay*. Key to the concept of spatial overlay is the representation of geographic datasets as *layers* in a GIS. This representation is exemplified by Figure 7.1 (credit to Rafael Pereira for the implementation[1]).

In this representation, different components of the landscape that interact in the real world are abstracted out into different layers, represented by different geometries. For example, Census tracts might be represented as polygons; customers as points; and roads as linestrings. Separating out these components has significant utility for the geospatial analyst, however. By

---

[1] https://www.urbandemographics.org/post/figures-map-layers-r/

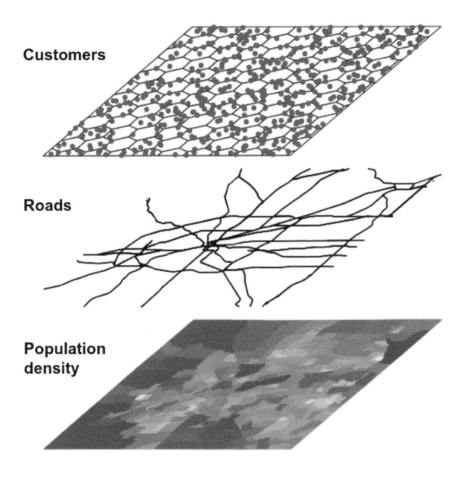

FIGURE 7.1 Conceptual view of GIS layers

using spatial analytic tools, a researcher could answer questions like "How many customers live within a given Census tract?" or "Which roads intersect a given Census tract?".

7.1.1 Note: aligning coordinate reference systems

Section 5.4 covered *coordinate reference systems* in R, their importance to spatial data, and how to select appropriate projected coordinate reference systems using the **crsuggest** package. In any workflow using spatial overlay, including all of the methods discussed in this chapter, it is essential that all layers share the same CRS for overlay methods to work.

Spatial datasets obtained with **tigris** or **tidycensus** will by default share the same geographic CRS, NAD 1983. For geographic coordinate reference systems, the **sf** package uses the **s2** spherical geometry library (Dunnington et al., 2021) to compute three-dimensional overlay rather than assuming planar geometries for geographic coordinates. This represents a significant technical advancement; however, I have found that it can be much slower to compute spatial overlay operations in this way than if the same workflow were using a projected coordinate reference system.

7.1 Spatial overlay

In turn, a recommended spatial analysis data preparation workflow is as follows:

1. Download the datasets you plan to use in your spatial analysis;
2. Use `suggest_crs()` in the **crsuggest** package to identify an appropriate projected CRS for your layers;
3. Transform your data to the projected CRS using `st_transform()`;
4. Compute the spatial overlay operation.

To avoid redundancy, step 2 is implied in the examples in this chapter and an appropriate projected coordinate reference system has been pre-selected for all sections.

7.1.2 Identifying geometries within a metropolitan area

One example of the utility of spatial overlay for the Census data analyst is the use of overlay techniques to find out which geographies lie within a given metropolitan area. Core-based statistical areas, also known as metropolitan or micropolitan areas, are common geographies defined by the US Census Bureau for regional analysis. Core-based statistical areas are defined as agglomerations of counties that are oriented around a central core or cores, and have a significant degree of population interaction as measured through commuting patterns. Metropolitan areas are those core-based statistical areas that have a population exceeding 50,000.

A Census data analyst in the United States will often need to know which Census geographies, such as Census tracts or block groups, fall within a given metropolitan area. However, these geographies are only organized by state and county, and don't have metropolitan area identification included by default. Given that Census spatial datasets are designed to align with one another, spatial overlay can be used to identify geographic features that fall within a given metropolitan area and extract those features.

Let's use the example of the Kansas City metropolitan area, which includes Census tracts in both Kansas and Missouri. We'll first use **tigris** to acquire 2020 Census tracts for the two states that comprise the Kansas City region as well as the boundary of the Kansas City metropolitan area.

```
library(tigris)
library(tidyverse)
library(sf)
options(tigris_use_cache = TRUE)

# CRS used: NAD83(2011) Kansas Regional Coordinate System
# Zone 11 (for Kansas City)
ks_mo_tracts <- map_dfr(c("KS", "MO"), ~{
  tracts(.x, cb = TRUE, year = 2020)
}) %>%
  st_transform(8528)

kc_metro <- core_based_statistical_areas(cb = TRUE, year = 2020) %>%
  filter(str_detect(NAME, "Kansas City")) %>%
  st_transform(8528)

ggplot() +
```

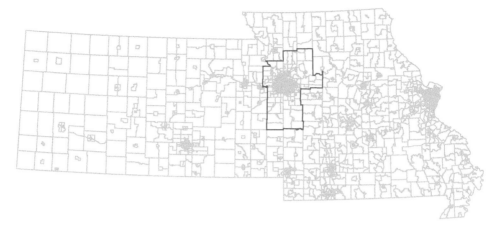

FIGURE 7.2 The Kansas City CBSA relative to Kansas and Missouri

```
  geom_sf(data = ks_mo_tracts, fill = "white", color = "grey") +
  geom_sf(data = kc_metro, fill = NA, color = "red") +
  theme_void()
```

We can see visually in Figure 7.2 which Census tracts are *within* the Kansas City metropolitan area, and which lay outside. This spatial relationship represented in the image can be expressed through code using *spatial subsetting*, enabled by functionality in the **sf** package.

7.1.3 Spatial subsets and spatial predicates

Spatial subsetting uses the extent of one spatial dataset to extract features from another spatial dataset based on co-location, defined by a *spatial predicate*. Spatial subsets can be expressed through base R indexing notation:

```
kc_tracts <- ks_mo_tracts[kc_metro, ]

ggplot() +
  geom_sf(data = kc_tracts, fill = "white", color = "grey") +
  geom_sf(data = kc_metro, fill = NA, color = "red") +
  theme_void()
```

The spatial subsetting operation returns all the Census tracts that *intersect* the extent of the Kansas City metropolitan area, using the default spatial predicate, st_intersects(). This gives us back tracts that fall within the metro area's boundary and those that cross or touch the boundary. For many analysts, however, this will be insufficient as they will want to tabulate statistics exclusively for tracts that fall *within* the metropolitan area's boundaries. In this case, a different spatial predicate can be used with the op argument.

Generally, Census analysts will want to use the st_within() spatial predicate to return tracts within a given metropolitan area. As long as objects within the core Census hierarchy are obtained for the same year from **tigris**, the st_within() spatial predicate will cleanly return geographies that fall within the larger geography when requested. The example

7.2 Spatial joins

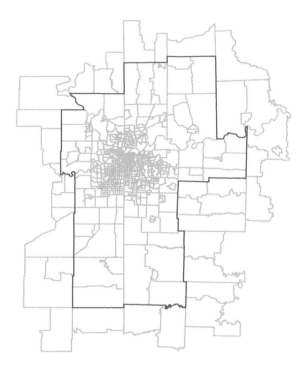

FIGURE 7.3 Census tracts that intersect the Kansas City CBSA

below illustrates the same process using the st_filter() function in **sf**, which allows spatial subsetting to be used cleanly within a tidyverse-style pipeline. The key difference between these two approaches to spatial subsetting is the argument name for the spatial predicate (op vs. .predicate).

```
kc_tracts_within <- ks_mo_tracts %>%
  st_filter(kc_metro, .predicate = st_within)

# Equivalent syntax:
# kc_metro2 <- kc_tracts[kc_metro, op = st_within]

ggplot() +
  geom_sf(data = kc_tracts_within, fill = "white", color = "grey") +
  geom_sf(data = kc_metro, fill = NA, color = "red") +
  theme_void()
```

7.2 Spatial joins

Spatial joins extend the aforementioned concepts in spatial overlay by transferring attributes between spatial layers. Conceptually, spatial joins can be thought of like the table joins covered in Section 6.4.1 where the equivalent of a "key field" used to match rows is a

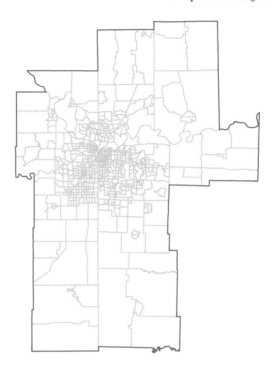

FIGURE 7.4 Census tracts that are within the Kansas City CBSA

spatial relationship defined by a spatial predicate. Spatial joins in R are implemented in sf's st_join() function. This section covers two common use cases for spatial joins with Census data. The first topic is the *point-in-polygon spatial join*, where a table of coordinates is matched to Census polygons to determine demographic characteristics around those locations. The second topic covers *polygon-in-polygon spatial joins*, where smaller Census shapes are matched to larger shapes.

7.2.1 Point-in-polygon spatial joins

Analysts are commonly tasked with matching point-level data to Census shapes in order to study demographic differences. For example, a marketing analyst may have a dataset of customers and needs to understand the characteristics of those customers' neighborhoods in order to target products efficiently. Similarly, a health data analyst may need to match neighborhood demographic data to patient information to understand inequalities in patient outcomes. This scenario is explored in this section.

Let's consider a hypothetical task where a health data analyst in Gainesville, Florida needs to determine the percentage of residents age 65 and up who lack health insurance in patients' neighborhoods. The analyst has a dataset of patients with patient ID along with longitude and latitude information.

```
library(tidyverse)
library(sf)
library(tidycensus)
library(mapview)
```

7.2 Spatial joins

TABLE 7.1 Hypothetical dataset of patients in Gainesville, Florida

patient_id	longitude	latitude
1	-82.30813	29.64593
2	-82.31197	29.65519
3	-82.36175	29.62176
4	-82.37438	29.65358
5	-82.38177	29.67720
6	-82.25946	29.67492
7	-82.36744	29.71099
8	-82.40403	29.71159
9	-82.43289	29.64823
10	-82.46184	29.62404

```
gainesville_patients <- tibble(
  patient_id = 1:10,
  longitude = c(-82.308131, -82.311972, -82.361748, -82.374377,
                -82.38177, -82.259461, -82.367436, -82.404031,
                -82.43289, -82.461844),
  latitude = c(29.645933, 29.655195, 29.621759, 29.653576,
               29.677201, 29.674923, 29.71099, 29.711587,
               29.648227, 29.624037)
)
```

Whereas the spatial overlay example in the previous section used spatial datasets from **tigris** that already include geographic information, this dataset needs to be converted to a simple features object. The st_as_sf() function in the **sf** package can take an R data frame or tibble with longitude and latitude columns like this and create a dataset of geometry type POINT. By convention, the coordinate reference system used for longitude / latitude data is WGS 1984, represented with the EPSG code 4326. We'll need to specify this CRS in st_as_sf() so that **sf** can locate the points correctly before we transform to an appropriate projected coordinate reference system with st_transform().

```
# CRS: NAD83(2011) / Florida North
gainesville_sf <- gainesville_patients %>%
  st_as_sf(coords = c("longitude", "latitude"),
           crs = 4326) %>%
  st_transform(6440)
```

Once prepared as a spatial dataset, the patient information can be mapped.

```
mapview(
  gainesville_sf,
  col.regions = "red",
  legend = FALSE
)
```

FIGURE 7.5 Map of hypothetical patient locations in Gainesville, Florida

As the patient data are now formatted as a simple features object, the next step is to acquire data on health insurance from the American Community Survey. A pre-computed percentage from the ACS Data Profile is available at the Census tract level, which will be used in the example below. Users who require a more granular geography can construct this information from the ACS Detailed Tables at the block group level using table B27001 and techniques learned in Section 3.3.2. As Gainesville is contained within Alachua County, Florida, we can obtain data from the 2015-2019 5-year ACS accordingly.

```
alachua_insurance <- get_acs(
  geography = "tract",
  variables = "DP03_0096P",
  state = "FL",
  county = "Alachua",
  year = 2019,
  geometry = TRUE
) %>%
  select(GEOID, pct_insured = estimate,
         pct_insured_moe = moe) %>%
  st_transform(6440)
```

After obtaining the spatial & demographic data with `get_acs()` and the `geometry = TRUE` argument, two additional commands help pre-process the data for the spatial join. The

7.2 Spatial joins

FIGURE 7.6 Layered interactive view of patients and Census tracts in Gainesville

call to `select()` retains three non-geometry columns in the simple features object: `GEOID`, which is the Census tract ID, and the ACS estimate and MOE renamed to `pct_insured` and `pct_insured_moe`, respectively. This formats the information that will be appended to the patient data in the spatial join. The `st_transform()` command then aligns the coordinate reference system of the Census tracts with the CRS used by the patient dataset.

Before computing the spatial join, the spatial relationships between patient points and Census tract demographics can be visualized interactively with `mapview()`, layering two interactive views with the + operator and shown in Figure 7.6.

```
mapview(
  alachua_insurance,
  zcol = "pct_insured",
  layer.name = "% with health<br/>insurance"
) +
  mapview(
    gainesville_sf,
    col.regions = "red",
    legend = FALSE
  )
```

The interrelationships between patient points and tract neighborhoods can be explored on the map. These relationships can be formalized with a *spatial join*, implemented with the

TABLE 7.2 Patients dataset after spatial join to Census tracts

patient_id	geometry	GEOID	pct_insured	pct_insured_moe
1	POINT (812216.5 73640.54)	12001000700	81.6	7.0
2	POINT (811825 74659.85)	12001000500	91.0	5.1
3	POINT (807076.2 70862.84)	12001001515	85.2	6.2
4	POINT (805787.6 74366.12)	12001001603	88.3	5.1
5	POINT (805023.3 76971.06)	12001001100	96.2	2.7
6	POINT (816865 76944.93)	12001001902	86.0	5.9
7	POINT (806340.4 80741.63)	12001001803	92.3	4.0
8	POINT (802798.8 80742.12)	12001001813	97.9	1.4
9	POINT (800134.1 73669.13)	12001002207	95.7	2.4
10	POINT (797379.1 70937.74)	12001002205	96.5	1.6

st_join() function in the **sf** package. st_join() returns a new simple features object that inherits geometry and attributes from a first dataset x with attributes from a second dataset y appended. Rows in x are matched to rows in y based on a spatial relationship defined by a spatial predicate, which defaults in st_join() to st_intersects(). For point-in-polygon spatial joins, this default will be sufficient in most cases unless a point falls directly on the boundary between polygons (which is not true in this example).

```
patients_joined <- st_join(
  gainesville_sf,
  alachua_insurance
)
```

The output dataset includes the patient ID and the original POINT feature geometry, but also now includes GEOID information from the Census tract dataset along with neighborhood demographic information from the ACS. This workflow can be used for analyses of neighborhood characteristics in a wide variety of applications and to generate data suitable for hierarchical modeling.

An issue to avoid when interpreting the results of point-in-polygon spatial joins is the *ecological fallacy*, where individual-level characteristics are inferred from that of the neighborhood. While neighborhood demographics are useful for inferring the characteristics of the environment in which an observation is located, they do not necessarily provide information about the demographics of the observation itself – particularly important when the observations represent people.

7.2.2 Spatial joins and group-wise spatial analysis

Spatial data operations can also be embedded in workflows where analysts are interested in understanding how characteristics vary by group. For example, while demographic data for metropolitan areas can be readily acquired using **tidycensus** functions, we might also be interested in learning about how demographic characteristics of *neighborhoods within metropolitan areas* vary across the United States. The example below illustrates this with some important new concepts for spatial data analysts. It involves a *polygon-on-polygon spatial join* in which attention to the spatial predicate used will be very important. Additionally, as all polygons involved are acquired with tidycensus and get_acs(), the section will show how st_join() handles column names that are duplicated between datasets.

7.2 Spatial joins

TABLE 7.3 Large CBSAs in Texas

GEOID	NAME	variable	estimate	moe	geometry
19100	Dallas-Fort Worth-Arlington, TX Metro Area	B01003_001	7573136	NA	MULTIPOLYGON (((1681247 760...
26420	Houston-The Woodlands-Sugar Land, TX Metro Area	B01003_001	7066140	NA	MULTIPOLYGON (((2009903 730...
41700	San Antonio-New Braunfels, TX Metro Area	B01003_001	2550960	NA	MULTIPOLYGON (((1538306 729...
12420	Austin-Round Rock-Georgetown, TX Metro Area	B01003_001	2227083	NA	MULTIPOLYGON (((1664195 732...

7.2.2.1 Spatial join data setup

Let's say that we are interested in analyzing the distributions of neighborhoods (defined here as Census tracts) by Hispanic population for the four largest metropolitan areas in Texas. We'll use the variable B01003_001 from the 2019 1-year ACS to acquire population data by core-based statistical area (CBSA) along with simple feature geometry which will eventually be used for the spatial join.

```
library(tidycensus)
library(tidyverse)
library(sf)

# CRS: NAD83(2011) / Texas Centric Albers Equal Area
tx_cbsa <- get_acs(
  geography = "cbsa",
  variables = "B01003_001",
  year = 2019,
  survey = "acs1",
  geometry = TRUE
) %>%
  filter(str_detect(NAME, "TX")) %>%
  slice_max(estimate, n = 4) %>%
  st_transform(6579)
```

The filtering steps used merit some additional explanation. The expression filter(str_detect(NAME, "TX")) first subsets the core-based statistical area data for only those metropolitan or micropolitan areas in (or partially in) Texas. Given that string matching in str_detect() is case-sensitive, using "TX" as the search string will match rows correctly. slice_max(), introduced in Section 4.1, then retains the four rows with the largest population values, found in the estimate column. Finally, the spatial dataset is transformed to an appropriate projected coordinate reference system for the state of Texas.

Given that all four of these metropolitan areas are completely contained within the state of Texas, we can obtain data on percent Hispanic by tract from the ACS Data Profile for 2015-2019.

```
pct_hispanic <- get_acs(
  geography = "tract",
  variables = "DP05_0071P",
  state = "TX",
  year = 2019,
  geometry = TRUE
) %>%
  st_transform(6579)
```

TABLE 7.4 Percent Hispanic by Census tract in Texas

GEOID	NAME	variable	estimate	moe	geometry
48113019204	Census Tract 192.04, Dallas County, Texas	DP05_0071P	58.1	5.7	MULTIPOLYGON (((1801563 765...
48377950200	Census Tract 9502, Presidio County, Texas	DP05_0071P	95.6	3.7	MULTIPOLYGON (((1060841 729...
48029190601	Census Tract 1906.01, Bexar County, Texas	DP05_0071P	86.2	6.7	MULTIPOLYGON (((1642736 726...
48355002301	Census Tract 23.01, Nueces County, Texas	DP05_0071P	80.6	3.9	MULTIPOLYGON (((1755212 707...
48441012300	Census Tract 123, Taylor County, Texas	DP05_0071P	26.4	6.3	MULTIPOLYGON (((1522575 758...
48051970500	Census Tract 9705, Burleson County, Texas	DP05_0071P	20.4	6.1	MULTIPOLYGON (((1812744 736...
48441010400	Census Tract 104, Taylor County, Texas	DP05_0071P	64.4	7.7	MULTIPOLYGON (((1522727 759...
48201311900	Census Tract 3119, Harris County, Texas	DP05_0071P	84.9	5.8	MULTIPOLYGON (((1950592 729...
48113015500	Census Tract 155, Dallas County, Texas	DP05_0071P	56.6	7.5	MULTIPOLYGON (((1778712 763...
48217960500	Census Tract 9605, Hill County, Texas	DP05_0071P	9.3	4.8	MULTIPOLYGON (((1741591 753...

TABLE 7.5 Census tracts after spatial join operation

	GEOID_tracts	NAME_tracts	variable_tracts	estimate_tracts	moe_tracts	GEOID_metro	NAME_metro	variable_metro
1	48113019204	Census Tract 192.04, Dallas County, Texas	DP05_0071P	58.1	5.7	19100	Dallas-Fort Worth-Arlington, TX Metro Area	B01003_001
3	48029190601	Census Tract 1906.01, Bexar County, Texas	DP05_0071P	86.2	6.7	41700	San Antonio-New Braunfels, TX Metro Area	B01003_001
8	48201311900	Census Tract 3119, Harris County, Texas	DP05_0071P	84.9	5.8	26420	Houston-The Woodlands-Sugar Land, TX Metro Area	B01003_001
9	48113015500	Census Tract 155, Dallas County, Texas	DP05_0071P	56.6	7.5	19100	Dallas-Fort Worth-Arlington, TX Metro Area	B01003_001
11	48439102000	Census Tract 1020, Tarrant County, Texas	DP05_0071P	15.7	6.4	19100	Dallas-Fort Worth-Arlington, TX Metro Area	B01003_001
13	48201450200	Census Tract 4502, Harris County, Texas	DP05_0071P	10.7	3.9	26420	Houston-The Woodlands-Sugar Land, TX Metro Area	B01003_001
14	48201450400	Census Tract 4504, Harris County, Texas	DP05_0071P	26.8	13.1	26420	Houston-The Woodlands-Sugar Land, TX Metro Area	B01003_001
22	48157670300	Census Tract 6703, Fort Bend County, Texas	DP05_0071P	27.0	5.8	26420	Houston-The Woodlands-Sugar Land, TX Metro Area	B01003_001
25	48201554502	Census Tract 5545.02, Harris County, Texas	DP05_0071P	13.4	3.0	26420	Houston-The Woodlands-Sugar Land, TX Metro Area	B01003_001
27	48121020503	Census Tract 205.03, Denton County, Texas	DP05_0071P	35.3	8.7	19100	Dallas-Fort Worth-Arlington, TX Metro Area	B01003_001

The returned dataset covers Census tracts in the entirety of the state of Texas; however, we only need to retain those tracts that fall within our four metropolitan areas of interest. We can accomplish this with a spatial join using st_join().

7.2.2.2 Computing and visualizing the spatial join

We know that in st_join(), we request that a given spatial dataset x, for which geometry will be retained, gains attributes from a second spatial dataset y based on their spatial relationship. This spatial relationship, as in the above examples, will be defined by a spatial predicate passed to the join parameter. The argument suffix defines the suffixes to be used for columns that share the same names, which will be important given that both datasets came from **tidycensus**. The argument left = FALSE requests an inner spatial join, returning only those tracts that fall within the four metropolitan areas.

```
hispanic_by_metro <- st_join(
  pct_hispanic,
  tx_cbsa,
  join = st_within,
  suffix = c("_tracts", "_metro"),
  left = FALSE
)
```

The output dataset has been reduced from 5,265 Census tracts to 3,189 as a result of the inner spatial join. Notably, the output dataset now includes information for each Census tract about the metropolitan area that it falls within. This enables group-wise data visualization and analysis across metro areas such as a faceted plot, shown in Figure 7.7:

```
hispanic_by_metro %>%
  mutate(NAME_metro = str_replace(NAME_metro, ", TX Metro Area", "")) %>%
  ggplot() +
  geom_density(aes(x = estimate_tracts), color = "navy", fill = "navy",
```

7.2 Spatial joins

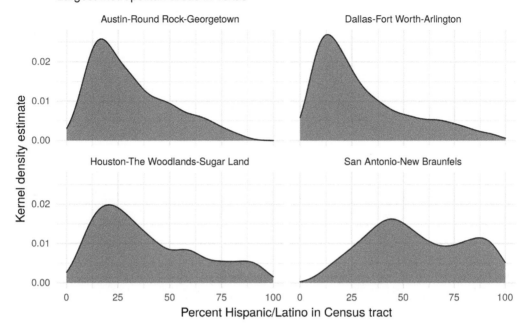

FIGURE 7.7 Faceted density plot of tract Hispanic populations by CBSA in Texas

```
                alpha = 0.4) +
  theme_minimal() +
  facet_wrap(~NAME_metro) +
  labs(title = "Distribution of Hispanic/Latino population by Census tract",
       subtitle = "Largest metropolitan areas in Texas",
       y = "Kernel density estimate",
       x = "Percent Hispanic/Latino in Census tract")
```

Output from a spatial join operation can also be "rolled up" to a larger geography through group-wise data analysis. For example, let's say we want to know the median value of the four distributions visualized in the plot above. As explained in Section 3.3, we can accomplish this by grouping our dataset by metro area then summarizing using the `median()` function.

```
median_by_metro <- hispanic_by_metro %>%
  group_by(NAME_metro) %>%
  summarize(median_hispanic = median(estimate_tracts, na.rm = TRUE))
```

The grouping column (`NAME_metro`) and the output of `summarize()` (`median_hispanic`) are returned as expected. However, the `group_by() %>% summarize()` operations also return the dataset as a simple features object with geometry, but in this case with only 4 rows. Let's take a look at the output geometry (visualized in Figure 7.8):

TABLE 7.6 Summarized median Hispanic population by metro

NAME_metro	median_hispanic
Austin-Round Rock-Georgetown, TX Metro Area	25.9
Dallas-Fort Worth-Arlington, TX Metro Area	22.6
Houston-The Woodlands-Sugar Land, TX Metro Area	32.4
San Antonio-New Braunfels, TX Metro Area	53.5

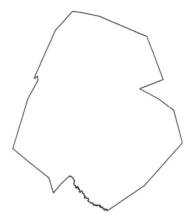

FIGURE 7.8 Dissolved geometry of Census tracts identified within the Austin CBSA

```
plot(median_by_metro[1,]$geometry)
```

The returned geometry represents the extent of the given metropolitan area (in the above example, Austin-Round Rock). The analytic process we carried out not only summarized the data by group, it also summarized the geometry by group. The typical name for this geometric process in geographic information systems is a *dissolve* operation, where geometries are identified by group and combined to return a single larger geometry. In this case, the Census tracts are dissolved by metropolitan area, returning metropolitan area geometries. This type of process is extremely useful when creating custom geographies (e.g. sales territories) from Census geometry building blocks that may belong to the same group.

7.3 Small area time-series analysis

Previous chapters of this book covered techniques and methods for analyzing demographic change over time in the US Census. Section 3.4 introduced the ACS Comparison Profile along with how to use iteration to get multiple years of data from the ACS Detailed Tables; Section 4.4 then illustrated how to visualize time-series ACS data. These techniques, however, are typically only appropriate for larger geographies like counties that rarely change shape over time. In contrast, smaller geographies like Census tracts and block groups are re-drawn by the US Census Bureau with every decennial US Census, making time-series analysis for smaller areas difficult.

7.3 Small area time-series analysis

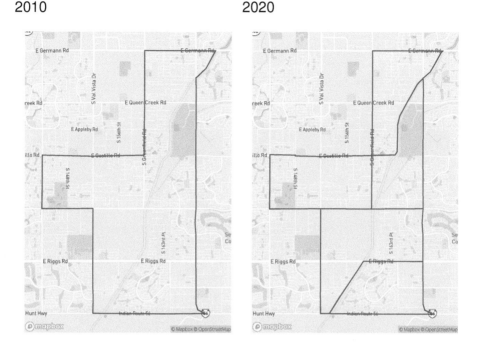

FIGURE 7.9 Comparison of Census tracts in Gilbert, AZ from the 2010 and 2020 Census

For example, we can compare Census tract boundaries for a fast-growing area of Gilbert, Arizona (southeast of Phoenix) for 2010 and 2020.

As discussed in Section 5.3.3, the US Census Bureau tries to keep Census tract sizes relatively consistent at around 4,000 people. If a tract grows too large between Census years, the Census Bureau will subdivide it into multiple Census tracts when re-drawing tracts for the next decennial Census. In this example from Arizona, the tract shown was divided into five tracts in 2020.

While this partitioning of Census tracts makes practical sense and allows for more granular demographic analysis in 2020, it also makes time-series comparisons difficult. This is particularly important with the release of the 2016-2020 ACS, which is the first ACS dataset to use 2020 Census boundaries. A common method for resolving this issue in geographic information science is *areal interpolation*. Areal interpolation refers to the allocation of data from one set of zones to a second overlapping set of zones that may or may not perfectly align spatially. In cases of mis-alignment, some type of weighting scheme needs to be specified to determine how to allocate partial data in areas of overlap. Two such approaches for interpolation are outlined here: *area-weighted interpolation* and *population-weighted interpolation*.

To get started, let's obtain some comparison data for Maricopa County, AZ on the number of people working from home in the 2011-2015 ACS (which uses 2010 boundaries) and the 2016-2020 ACS (which uses 2020 boundaries). We will use both interpolation methods to allocate 2011-2015 data to 2020 Census tracts.

```
library(tidycensus)
library(tidyverse)
```

```
library(tigris)
library(sf)
options(tigris_use_cache = TRUE)

# CRS: NAD 83 / Arizona Central
wfh_15 <- get_acs(
  geography = "tract",
  variables = "B08006_017",
  year = 2015,
  state = "AZ",
  county = "Maricopa",
  geometry = TRUE
) %>%
  select(estimate) %>%
  st_transform(26949)

wfh_20 <- get_acs(
  geography = "tract",
  variables = "B08006_017",
  year = 2020,
  state = "AZ",
  county = "Maricopa",
  geometry = TRUE
) %>%
  st_transform(26949)
```

7.3.1 Area-weighted areal interpolation

Area-weighted areal interpolation is implemented in **sf** with the st_interpolate_aw() function. This method uses the area of overlap of geometries as the interpolation weights. From a technical standpoint, an intersection is computed between the origin geometries and the destination geometries. Weights are then computed as the proportion of the overall origin area comprised by the intersection. Area weights used to estimate data at 2020 geographies for the Census tract in Gilbert are illustrated in Figure 7.10.

Those weights are applied to target variables (in this case, the information on workers from home) in accordance with the value of the extensive argument. If extensive = TRUE, as used below, weighted sums will be computed. Alternatively, if extensive = FALSE, the function returns weighted means.

```
wfh_interpolate_aw <- st_interpolate_aw(
  wfh_15,
  wfh_20,
  extensive = TRUE
) %>%
  mutate(GEOID = wfh_20$GEOID)
```

7.3 Small area time-series analysis

Illustration of area weights

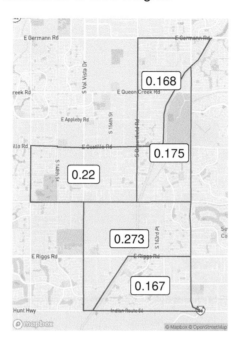

FIGURE 7.10 Illustration of area weights

7.3.2 Population-weighted areal interpolation

When a user computes area-weighted areal interpolation with `st_interpolate_aw()`, the function prints the following warning: `st_interpolate_aw assumes attributes are constant or uniform over areas of x`. This assumption that proportionally larger *areas* also have proportionally more *people* is often incorrect with respect to the geography of human settlements, and can be a source of error when using this method. An alternative method, *population-weighted areal interpolation*, can represent an improvement. As opposed to using area-based weights, population-weighted techniques estimate the populations of the intersections between origin and destination from a third dataset, then use those values for interpolation weights.

This method is implemented in **tidycensus** with the `interpolate_pw()` function. This function is specified in a similar way to `st_interpolate_aw()`, but also requires a third dataset to be used as weights, and optionally a weight column to determine the relative influence of each feature in the weights dataset. For many purposes, **tidycensus** users will want to use Census blocks as the weights dataset, though users can bring alternative datasets as well. 2020 Census blocks acquired with the **tigris** package have the added benefit of `POP20` and `HU20` columns in the dataset that represent population and housing unit counts, respectively, either one of which could be used to weight each block.

```
maricopa_blocks <- blocks(
  state = "AZ",
  county = "Maricopa",
  year = 2020
```

Illustration of block points   Illustration of population weights

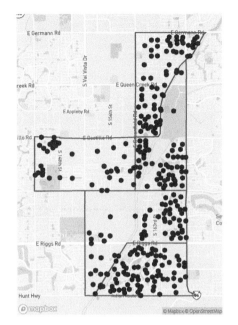 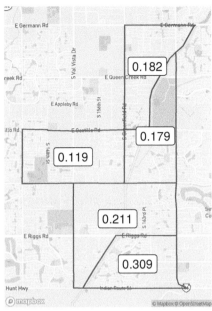

FIGURE 7.11 Illustration of block points and population weights

```
)

wfh_interpolate_pw <- interpolate_pw(
  wfh_15,
  wfh_20,
  to_id = "GEOID",
  extensive = TRUE,
  weights = maricopa_blocks,
  weight_column = "POP20",
  crs = 26949
)
```

interpolate_pw() as implemented here uses a *weighted block point* approach to interpolation, where the input Census blocks are first converted to points using the st_point_on_surface() function from the **sf** package, then joined to the origin/destination intersections to produce population weights. An illustration of this process is found below; the map on the left-hand side shows the block weights as points, and the map on the right-hand side shows the population weights used for each 2020 Census tract.

The population-based weights differ significantly from the area-based weights for the Census tract in Gilbert. Notably, the southern-most Census tract in the example only had an area weight of 0.167, whereas the population weighting revealed that over 30 percent of the origin tract's population is actually located there. This leads to substantive differences in the results of the area- and population-weighted approaches, as illustrated in Figure 7.12.

7.3 Small area time-series analysis

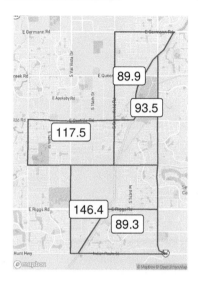
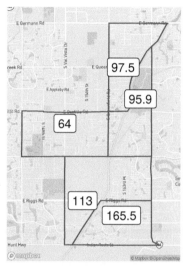

FIGURE 7.12 Comparison of area-weighted and population-weighted interpolation results

The area-weighted method under-estimates the population in geographically smaller tracts, and over-estimates in larger ones; in contrast, the population-weighted method takes the underlying population distribution into account.

7.3.3 Making small-area comparisons

As these methods have interpolated 2011-2015 ACS estimates to 2020 Census tracts, 2011-2015 and 2016-2020 ACS data can now be compared at consistent geographies. To do this, we will join the population-weighted interpolated 2011-2015 data to the original 2016-2020 data using `left_join()` (as covered in Section 6.4), taking care to drop the geometry of the dataset on the right-hand side of the join and to specify a `suffix` argument to distinguish the two ACS estimates. We then calculate change over time from these estimates and map the result.

```
library(mapboxapi)

wfh_shift <- wfh_20 %>%
  left_join(st_drop_geometry(wfh_interpolate_pw),
            by = "GEOID",
            suffix = c("_2020", "_2015")) %>%
  mutate(wfh_shift = estimate_2020 - estimate_2015)

maricopa_basemap <- layer_static_mapbox(
  location = wfh_shift,
```

FIGURE 7.13 Map of shift in workers from home, Maricopa County Arizona

```
  style_id = "dark-v9",
  username = "mapbox"
)

ggplot() + 
  maricopa_basemap + 
  geom_sf(data = wfh_shift, aes(fill = wfh_shift), color = NA, 
          alpha = 0.8) + 
  scale_fill_distiller(palette = "PuOr", direction = -1) + 
  labs(fill = "Shift, 2011-2015 to\n2016-2020 ACS", 
       title = "Change in work-from-home population", 
       subtitle = "Maricopa County, Arizona") + 
  theme_void()
```

Notable increases in tract-level working from home are found in locations like Gilbert, Scottsdale, and Tempe on the eastern side of the metropolitan area. That said, these results may simply be a function of overall population growth in those tracts, which means that a follow-up analysis should examine change in the share of the population working from home. This would require interpolating a total workforce denominator column and calculating a percentage. Fortunately, both interpolation methods introduced in this section will interpolate all numeric columns in an input dataset, so wide-form data or data with a summary variable acquired by **tidycensus** will work well for this purpose.

An advantage of using either area-weighted or population-weighted areal interpolation as covered in this section is that they can be implemented entirely with data available in tidycensus and tigris. Some users may be interested in alternative weights using datasets not included in these packages, like land use/land cover data, or may want to use more sophisticated regression-based approaches. While they are not covered here, Schroeder and Van Riper (2013) provides a good overview of these methods.

As discussed in Section 3.5.1, derived margins of error (even for sums) require special methods. Given the complexity of the interpolation methods covered here, direct interpolation of margin of error columns will not take these methods into account. Analysts should interpret such columns with caution.

7.4 Distance and proximity analysis

A common use case for spatially-referenced demographic data is the analysis of *accessibility*. This might include studying the relative accessibility of different demographic groups to resources within a given region, or analyzing the characteristics of potential customers who live within a given distance of a store. Conceptually, there are a variety of ways to measure accessibility. The most straightforward method, computationally, is using straight-line (Euclidean) distances over geographic data in a projected coordinate system. A more computationally complex – but potentially more accurate – method involves the use of transportation networks to model accessibility, where proximity is measured not based on distance from a given location but instead based on travel times for a given transit mode, such as walking, cycling, or driving. This section will illustrate both types of approaches. Let's consider the topic of accessibility to Level I and Level II trauma hospitals by Census tract in the state of Iowa. 2019 Census tract boundaries are acquired from **tigris**, and we use st_read() to read in a shapefile of hospital locations acquired from the US Department of Homeland Security.

```
library(tigris)
library(sf)
library(tidyverse)
options(tigris_use_cache = TRUE)

# CRS: NAD83 / Iowa North
ia_tracts <- tracts("IA", cb = TRUE, year = 2019) %>%
  st_transform(26975)

hospital_url <- paste0("https://opendata.arcgis.com/api/v3/datasets/",
                       "6ac5e325468c4cb9b905f1728d6fbf0f_0/downloads/data",
                       "?format=geojson&spatialRefId=4326")

trauma <- st_read(hospital_url) %>%
  filter(str_detect(TRAUMA, "LEVEL I\\b|LEVEL II\\b|RTH|RTC")) %>%
  st_transform(26975) %>%
  distinct(ID, .keep_all = TRUE)

names(trauma)
```

```
##  [1] "OBJECTID"   "ID"          "NAME"        "ADDRESS"     "CITY"
##  [6] "STATE"      "ZIP"         "ZIP4"        "TELEPHONE"   "TYPE"
## [11] "STATUS"     "POPULATION"  "COUNTY"      "COUNTYFIPS"  "COUNTRY"
## [16] "LATITUDE"   "LONGITUDE"   "NAICS_CODE"  "NAICS_DESC"  "SOURCE"
## [21] "SOURCEDATE" "VAL_METHOD"  "VAL_DATE"    "WEBSITE"     "STATE_ID"
## [26] "ALT_NAME"   "ST_FIPS"     "OWNER"       "TTL_STAFF"   "BEDS"
## [31] "TRAUMA"     "HELIPAD"     "GlobalID"    "geometry"
```

7.4.1 Calculating distances

To determine accessibility of Iowa Census tracts to Level I or II trauma centers, we need to identify not only those hospitals that are located in Iowa, but also those in other states near to the Iowa border, such as in Omaha, Nebraska and Rock Island, Illinois. We can accomplish this by applying a distance threshold in st_filter(). In this example, we use the spatial predicate st_is_within_distance, and set a 100km distance threshold with the dist = 100000 argument (specified in meters, the base measurement unit of our coordinate system used).

```
ia_trauma <- trauma %>%
  st_filter(ia_tracts,
            .predicate = st_is_within_distance,
            dist = 100000)

ggplot() +
  geom_sf(data = ia_tracts, color = "NA", fill = "grey50") +
  geom_sf(data = ia_trauma, color = "red") +
  theme_void()
```

As illustrated in Figure 7.14, the st_filter() operation has retained Level I and II trauma centers *within* the state of Iowa, but also within the 100km threshold beyond the state's borders.

With the Census tract and hospital data in hand, we can calculate distances from Census tracts to trauma centers by using the st_distance() function in the **sf** package. st_distance(x, y) by default returns the dense matrix of distances computed from the geometries in x to the geometries in y. In this example, we will calculate the distances from the *centroids* of Iowa Census tracts (reflecting the center points of each tract geometry) to each trauma center.

```
dist <- ia_tracts %>%
  st_centroid() %>%
  st_distance(ia_trauma)

dist[1:5, 1:5]
```

```
## Units: [m]
##             [,1]       [,2]      [,3]      [,4]      [,5]
## [1,]    279570.18  279188.81  385140.7  383863.7  257745.5
## [2,]    298851.01  298409.46  400428.5  399022.9  276955.8
## [3,]    350121.53  347800.57  353428.8  350263.3  404616.0
```

FIGURE 7.14 Level I or II trauma centers within 100 km of Iowa

```
## [4,] 361742.20 360450.59 421691.6 419364.9 369415.7
## [5,]  66762.19  63479.67 143552.1 143935.5 194001.0
```

A glimpse at the matrix shows distances (in meters) between the first five Census tracts in the dataset and the first five hospitals. When considering *accessibility*, we may be interested in the distance to the *nearest* hospital to each Census tract. The code below extracts the minimum distance from the matrix for each row, converts to a vector, and divides each value by 1000 to convert values to kilometers. A quick histogram visualizes the distribution of minimum distances.

```
min_dist <- dist %>%
  apply(1, min) %>%
  as.vector() %>%
  magrittr::divide_by(1000)

hist(min_dist)
```

The code that extracts minimum distances from the distance matrix includes some notation that may be unfamiliar to readers.

- The apply() function from base R is used to iterate over rows of the matrix. Matrices are a data structure not handled by the map_*() family of functions in the tidyverse, so base R methods must be used. In the example pipeline, the apply() function inherits the dist matrix object as its first argument. The second argument, 1, refers to the margin of the matrix that apply() will iterate over; 1 references rows (which we want), whereas 2 would

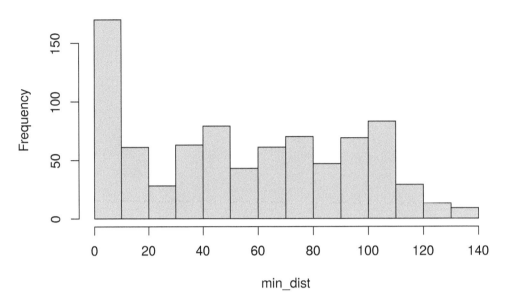

FIGURE 7.15 Base R histogram of minimum distances to trauma centers

be used for columns. `min` then is the function to be applied to each row, giving us the minimum distance to a hospital for each Census tract.

- The `divide_by()` function in the **magrittr** package is a convenience arithmetic function to be used in analytic pipelines as R's arithmetic operators (e.g. `/` for division) won't work in this way. In this example, it divides all the values by 1000 to convert meters to kilometers.

While many tracts are within 10 km of a trauma center, around 16 percent of Iowa Census tracts in 2019 are beyond 100 km from a Level I or II trauma center, suggesting significant accessibility issues for these areas.

7.4.2 Calculating travel times

An alternative way to model accessibility to hospitals is through *travel times* rather than distance, as the way that people experience access to locations is through time expended given a transportation network. While network-based accessibility may be a more accurate representation of people's lived experiences, it is more computationally complex and requires additional tools. To perform spatial network analyses, R users will either need to obtain network data (like roadways) and use appropriate tools that can model the network; set up a routing engine that R can connect to; or connect to a hosted routing engine via a web API. In this example, we'll use the **mapboxapi** R package (Walker, 2021c) to perform network analysis using Mapbox's travel-time Matrix API[2].

The function `mb_matrix()` in **mapboxapi** works much like `st_distance()` in that it only requires arguments for origins and destinations, and will return the dense matrix of travel

---

[2]https://docs.mapbox.com/api/navigation/matrix/

7.4 Distance and proximity analysis

times by default. In turn, much of the computational complexity of routing is abstracted away by the function. However, as routes will be computed across the state of Iowa and API usage is subject to rate-limitations, the function can take several minutes to compute for larger matrices like this one.

If you are using **mapboxapi** for the first time, visit mapbox.com[3], register for an account, and obtain an access token. The function `mb_access_token()` installs this token in your .Renviron for future use.

```
library(mapboxapi)
# mb_access_token("pk.eybcasq...", install = TRUE)

times <- mb_matrix(ia_tracts, ia_trauma)
```

```
times[1:5, 1:5]
```

```
##              [,1]      [,2]      [,3]      [,4]      [,5]
## [1,]     211.43833 212.50667 278.5733  284.9717 212.1050
## [2,]     218.15167 214.06000 280.1267  286.5250 226.7733
## [3,]     274.84500 270.75333 291.8367  290.7117 292.2767
## [4,]     274.58333 270.49167 291.5750  290.4500 292.0150
## [5,]      56.80333  52.71167 122.3017  128.7000 161.2617
```

A glimpse at the travel-time matrix shows a similar format to the distance matrix, but with travel times in minutes used instead of meters. As with the distance-based example, we can determine the minimum travel time from each tract to a Level I or Level II trauma center. In this instance, we will visualize the result on a map.

```
min_time <- apply(times, 1, min)

ia_tracts$time <- min_time

ggplot(ia_tracts, aes(fill = time)) +
  geom_sf(color = NA) +
  scale_fill_viridis_c(option = "magma") +
  theme_void() +
  labs(fill = "Time (minutes)",
       title = "Travel time to nearest Level I or Level II trauma hospital",
       subtitle = "Census tracts in Iowa",
       caption = "Data sources: US Census Bureau, US DHS, Mapbox")
```

The map illustrates considerable accessibility gaps to trauma centers across the state. Whereas urban residents typically live within 20 minutes of a trauma center, travel times in rural Iowa can exceed two hours.

An advantage to using a package like **mapboxapi** for routing and travel times is that users can connect directly to a hosted routing engine using an API. Due to rate limitations, however, web APIs are likely inadequate for more advanced users who need to compute travel times at scale. There are several R packages that can connect to user-hosted routing

---

[3]https://mapbox.com

Travel time to nearest Level I or Level II trauma hospital
Census tracts in Iowa

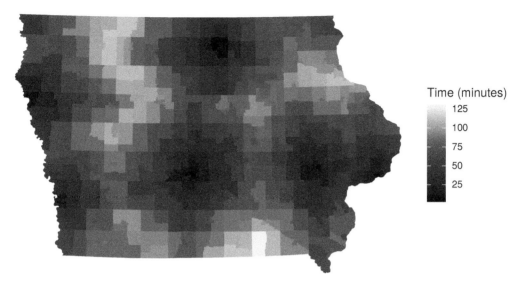

Data sources: US Census Bureau, US DHS, Mapbox

FIGURE 7.16 Map of travel-times to trauma centers by Census tract in Iowa

engines which may be better-suited to such tasks. These packages include **osrm**[4] for the Open Source Routing Machine; **opentripplanner**[5] for OpenTripPlanner; and **r5r**[6] for R5.

7.4.3 Catchment areas with buffers and isochrones

The above example considers a broader accessibility analysis across the state of Iowa. In many cases, however, you'll want to analyze accessibility in a more local way. A common use case might involve a study of the demographic characteristics of a hospital *catchment area*, defined as the area around a hospital from which patients will likely come.

As with the matrix-based accessibility approach outlined above, catchment area-based proximity can be modeled with either Euclidean distances or network travel times as well. Let's consider the example of Iowa Methodist Medical Center in Des Moines[7], one of two Level I trauma centers in the state of Iowa.

The example below illustrates the distance-based approach using a *buffer*, implemented with the st_buffer() function in **sf**. A buffer is a common GIS operation that represents the area within a given distance of a location. The code below creates a 5km buffer around Iowa Methodist Medical Center by using the argument dist = 5000.

---

[4]https://github.com/rCarto/osrm
[5]https://docs.ropensci.org/opentripplanner/
[6]https://ipeagit.github.io/r5r/
[7]https://www.unitypoint.org/desmoines/iowa-methodist-medical-center.aspx

7.4 Distance and proximity analysis

```
iowa_methodist <- filter(ia_trauma, NAME == "IOWA METHODIST MEDICAL CENTER")

buf5km <- st_buffer(iowa_methodist, dist = 5000)
```

An alternative option is to create network-based *isochrones*, which are polygons that represent the accessible area around a given location within a given travel time for a given travel mode. Isochrones are implemented in the **mapboxapi** package with the `mb_isochrone()` function. Mapbox isochrones default to typical driving conditions around a location; this can be adjusted with the `depart_at` parameter for historical traffic and the argument profile = "driving-traffic". The example below draws a 10-minute driving isochrone around Iowa Methodist for a Tuesday during evening rush hour.

```
iso10min <- mb_isochrone(
  iowa_methodist,
  time = 10, profile = "driving-traffic",
  depart_at = "2022-04-05T17:00"
)
```

We can visualize the comparative extents of these two methods in Des Moines. Run the code on your own computer to get a synced interactive map showing the two methods. The `makeAwesomeIcon()` function in **leaflet** creates a custom icon appropriate for a medical facility; many other icons are available for common points of interest[8].

```
library(leaflet)
library(leafsync)

hospital_icon <- makeAwesomeIcon(icon = "ios-medical",
                                 markerColor = "red",
                                 library = "ion")

# The Leaflet package requires data be in CRS 4326
map1 <- leaflet() %>%
  addTiles() %>%
  addPolygons(data = st_transform(buf5km, 4326)) %>%
  addAwesomeMarkers(data = st_transform(iowa_methodist, 4326),
                    icon = hospital_icon)

map2 <- leaflet() %>%
  addTiles() %>%
  addPolygons(data = iso10min) %>%
  addAwesomeMarkers(data = st_transform(iowa_methodist, 4326),
                    icon = hospital_icon)

sync(map1, map2)
```

The comparative maps in Figure 7.17 illustrate the differences between the two methods quite clearly. Many areas of equal distance to the hospital do not have the same level of

---

[8] http://rstudio.github.io/leaflet/markers.html#awesome-icons

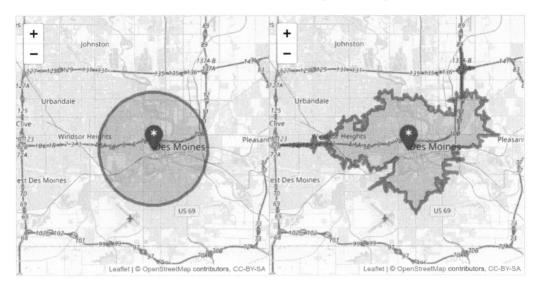

FIGURE 7.17 Synced map showing buffer and isochrone-based catchment areas in Des Moines

access; this is particularly true of areas to the south of the Raccoon/Des Moines River. Conversely, due to the location of highways, there are some areas outside the 5km buffer area that can reach the hospital within 10 minutes.

7.4.4 Computing demographic estimates for zones with areal interpolation

Common to both methods, however, is a mis-alignment between their geometries and those of any Census geographies we may use to infer catchment area demographics. As opposed to the spatial overlay analysis matching Census tracts to metropolitan areas earlier in this chapter, Census tracts or block groups on the edge of the catchment area will only be partially included in the catchment. Areal interpolation methods like those introduced in Section 7.3 can be used here to estimate the demographics of both the buffer zone and isochrone.

Let's produce interpolated estimates of the percentage of population in poverty for both catchment area definitions. This will require obtaining block group-level poverty information from the ACS for Polk County, Iowa, which encompasses both the buffer and the isochrone. The variables requested from the ACS include the number of family households with incomes below the poverty line along with total number of family households to serve as a denominator.

```
polk_poverty <- get_acs(
  geography = "block group",
  variables = c(poverty_denom = "B17010_001",
                poverty_num = "B17010_002"),
  state = "IA",
  county = "Polk",
  geometry = TRUE,
  output = "wide",
  year = 2020
```

7.5 Distance and proximity analysis

TABLE 7.7 Comparison of buffer and isochrone catchment areas

Method	Families in poverty	Total families	Percent
5km buffer	2961.475	21175.79	14.0
10min isochrone	2983.521	21688.47	13.8

```
) %>%
  select(poverty_denomE, poverty_numE) %>%
  st_transform(26975)
```

We can then use population-weighted areal interpolation with `interpolate_pw()` function in **tidycensus** to estimate family poverty in both the buffer zone and the isochrone. Block weights for Polk County are obtained with **tigris**, and both the numerator and denominator columns are interpolated.

```
library(glue)

polk_blocks <- blocks(
  state = "IA",
  county = "Polk",
  year = 2020
)

buffer_pov <- interpolate_pw(
  from = polk_poverty,
  to = buf5km,
  extensive = TRUE,
  weights = polk_blocks,
  weight_column = "POP20",
  crs = 26975
) %>%
  mutate(pct_poverty = 100 * (poverty_numE / poverty_denomE))

iso_pov <- interpolate_pw(
  from = polk_poverty,
  to = iso10min,
  extensive = TRUE,
  weights = polk_blocks,
  weight_column = "POP20",
  crs = 26975
) %>%
  mutate(pct_poverty = 100 * (poverty_numE / poverty_denomE))
```

The two methods return slightly different results in Table 7.7, illustrating how the definition of catchment area impacts downstream analyses.

7.5 Better cartography with spatial overlay

As discussed in Section 6.1, one of the major benefits of working with the **tidycensus** package to get Census data in R is its ability to retrieve pre-joined feature geometry for Census geographies with the argument `geometry = TRUE`. tidycensus uses the **tigris** package to fetch these geometries, which default to the Census Bureau's cartographic boundary shapefiles[9]. Cartographic boundary shapefiles are preferred to the core TIGER/Line shapefiles[10] in **tidycensus** as their smaller size speeds up processing and because they are pre-clipped to the US coastline.

However, there may be circumstances in which your mapping requires more detail. A good example of this would be maps of New York City, in which even the cartographic boundary shapefiles include water area. For example, take this example of median household income by Census tract in Manhattan (New York County), NY:

```
library(tidycensus)
library(tidyverse)
options(tigris_use_cache = TRUE)

ny <- get_acs(
  geography = "tract",
  variables = "B19013_001",
  state = "NY",
  county = "New York",
  year = 2020,
  geometry = TRUE
)

ggplot(ny) +
  geom_sf(aes(fill = estimate)) +
  scale_fill_viridis_c(labels = scales::label_dollar()) +
  theme_void() +
  labs(fill = "Median household\nincome")
```

As illustrated in Figure 7.18, the boundaries of Manhattan include water boundaries – stretching into the Hudson and East Rivers. In turn, a more accurate representation of Manhattan's land area might be desired. To accomplish this, a **tidycensus** user can use the core TIGER/Line shapefiles instead, then erase water area from Manhattan's geometry.

7.5.1 "Erasing" areas from Census polygons

tidycensus allows users to get TIGER/Line instead of cartographic boundary shapefiles with the keyword argument `cb = FALSE`. This argument will be familiar to users of the **tigris** package, as it is used by **tigris** to distinguish between cartographic boundary and TIGER/Line shapefiles in the package.

---

[9] https://www.census.gov/geo/maps-data/data/tiger-cart-boundary.html
[10] https://www.census.gov/geo/maps-data/data/tiger-line.html

7.5 Better cartography with spatial overlay

FIGURE 7.18 Map of Manhattan with default CB geometries

```
ny2 <- get_acs(
  geography = "tract",
  variables = "B19013_001",
  state = "NY",
  county = "New York",
  geometry = TRUE,
  year = 2020,
  cb = FALSE
) %>%
  st_transform(6538)
```

Next, the `erase_water()` function in the **tigris** package will be used to remove water area from the Census tracts. `erase_water()` works by auto-detecting US counties that surround an input dataset, obtaining an area water shapefile from the Census Bureau for those counties, then computing an *erase* operation to remove those water areas from the input dataset. Using TIGER/Line geometries with `cb = FALSE` is recommended as they will align with the input water areas and minimize the creation of *sliver polygons*, which are small polygons that can be created from the overlay of inconsistent spatial datasets.

```
ny_erase <- erase_water(ny2)
```

Although it is not used here, `erase_water()` has an optional argument, `area_threshold`, that defines the area percentile threshold at which water areas are kept for the erase operation. The default of 0.75, used here, erases water areas with a size percentile of 75 percent and up

FIGURE 7.19 Map of Manhattan with water areas erased

(so, the top 25 percent). A lower area threshold can produce more accurate shapes, but can slow down the operation.

After erasing water area from Manhattan's Census tracts with erase_water(), we can map the result:

```
ggplot(ny_erase) +
  geom_sf(aes(fill = estimate)) +
  scale_fill_viridis_c(labels = scales::label_dollar()) +
  theme_void() +
  labs(fill = "Median household\nincome")
```

The map appears in Figure 7.19 with a more familiar representation of the extent of Manhattan.

7.6 Spatial neighborhoods and spatial weights matrices

The spatial capabilities of **tidycensus** also allow for exploratory spatial data analysis (ESDA) within R. ESDA refers to the use of datasets' spatial properties in addition to their attributes to explore patterns and relationships. This may involve exploration of spatial patterns in datasets or identification of spatial clustering of a given demographic attribute.

7.6 Spatial neighborhoods and spatial weights matrices

To illustrate how an analyst can apply ESDA to Census data, let's acquire a dataset on median age by Census tract in the Dallas-Fort Worth, TX metropolitan area. Census tracts in the metro area will be identified using methods introduced earlier in this chapter.

```
library(tidycensus)
library(tidyverse)
library(tigris)
library(sf)
library(spdep)
options(tigris_use_cache = TRUE)

# CRS: NAD83 / Texas North Central
dfw <- core_based_statistical_areas(cb = TRUE, year = 2020) %>%
  filter(str_detect(NAME, "Dallas")) %>%
  st_transform(32138)

dfw_tracts <- get_acs(
  geography = "tract",
  variables = "B01002_001",
  state = "TX",
  year = 2020,
  geometry = TRUE
) %>%
  st_transform(32138) %>%
  st_filter(dfw, .predicate = st_within) %>%
  na.omit()

ggplot(dfw_tracts) +
  geom_sf(aes(fill = estimate), color = NA) +
  scale_fill_viridis_c() +
  theme_void()
```

7.6.1 Understanding spatial neighborhoods

Exploratory spatial data analysis relies on the concept of a *neighborhood*, which is a representation of how a given geographic feature (e.g. a given point, line, or polygon) interrelates with other features nearby. The workhorse package for exploratory spatial data analysis in R is **spdep**, which includes a wide range of tools for exploring and modeling spatial data. As part of this framework, **spdep** supports a variety of neighborhood definitions. These definitions include:

- *Proximity-based neighbors*, where neighboring features are identified based on some measure of distance. Neighbors might be defined as those that fall within a given distance threshold (e.g. all features within 2km of a given feature) or as k-nearest neighbors (e.g. the nearest eight features to a given feature).
- *Graph-based neighbors*, where neighbors are defined through network relationships (e.g. along a street network).
- *Contiguity-based neighbors*, used when geographic features are polygons. Options for contiguity-based spatial relationships include *queen's case neighbors*, where all polygons that share at least one vertex are considered neighbors; and *rook's case neighbors*, where polygons must share at least one line segment to be considered neighbors.

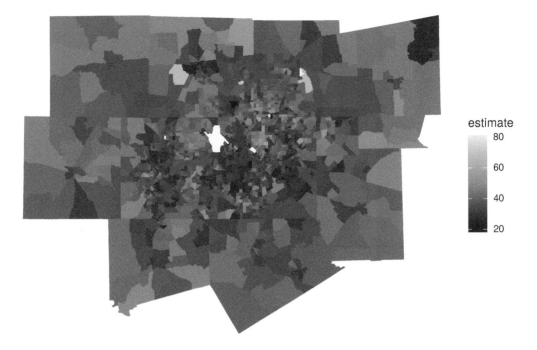

In this example, we'll choose a queen's case contiguity-based neighborhood definition for our Census tracts. We implement this with the function poly2nb(), which can take an **sf** object as an argument and produce a neighbors list object. We use the argument queen = TRUE to request queen's case neighbors explicitly (though this is the function default).

```
neighbors <- poly2nb(dfw_tracts, queen = TRUE)

summary(neighbors)
```

```
## Neighbour list object:
## Number of regions: 1699
## Number of nonzero links: 10930
## Percentage nonzero weights: 0.378646
## Average number of links: 6.433196
## Link number distribution:
##
##   2   3   4   5   6   7   8   9  10  11  12  13  14  15  17
##   8  50 173 307 395 343 220 109  46  28  11   5   2   1   1
## 8 least connected regions:
## 33 620 697 753 1014 1358 1579 1642 with 2 links
## 1 most connected region:
## 1635 with 17 links
```

On average, the Census tracts in the Dallas-Fort Worth metropolitan area have 6.43 neighbors. The minimum number of neighbors in the dataset is 2 (there are eight such tracts), and the maximum number of neighbors is 17 (the tract at row index 1635). An important caveat to keep in mind here is that tracts with few neighbors may actually have more neighbors than listed here given that we have restricted the tract dataset to those tracts within the

7.6 Spatial neighborhoods and spatial weights matrices

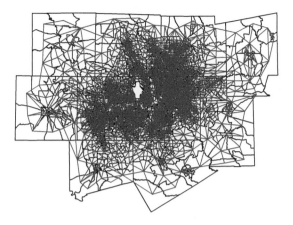

FIGURE 7.20 Visualization of queens-case neighborhood relationships

Dallas-Fort Worth metropolitan area. In turn, our analysis will be influenced by *edge effects* as neighborhoods on the edge of the metropolitan area are artificially restricted.

Neighborhood relationships can be visualized using plotting functionality in **spdep**, with blue lines connecting each polygon with its neighbors.

```
dfw_coords <- dfw_tracts %>%
  st_centroid() %>%
  st_coordinates()

plot(dfw_tracts$geometry)
plot(neighbors,
     coords = dfw_coords,
     add = TRUE,
     col = "blue",
     points = FALSE)
```

Additionally, row indices for the neighbors of a given feature can be readily extracted from the neighbors list object.

```
# Get the row indices of the neighbors of the Census tract at row index 1
neighbors[[1]]
```

```
## [1]   45  585  674 1152 1580
```

7.6.2 Generating the spatial weights matrix

To perform exploratory spatial data analysis, we can convert the neighbors list object into *spatial weights*. Spatial weights define how metrics associated with a feature's neighbors should be weighted. Weight generation is implemented in the nb2listw() function, to which we pass the neighbors object and specify a style of weights. The default, style = "W", produces a row-standardized weights object where the weights for all neighbors of a given feature sum to 1. This is the option you would choose when analyzing neighborhood means.

An alternative option, `style = "B"`, produces binary weights where neighbors are given the weight of 1 and non-neighbors take the weight of 0. This style of weights is useful for producing neighborhood sums.

In the example below, we create row-standardized spatial weights for the Dallas-Fort Worth Census tracts and check their values for the feature at row index 1.

```
weights <- nb2listw(neighbors, style = "W")

weights$weights[[1]]
```

```
## [1] 0.2 0.2 0.2 0.2 0.2
```

Given that the Census tract at row index 1 has five neighbors, each neighbor is assigned the weight 0.2.

7.7 Global and local spatial autocorrelation

The row-standardized spatial weights object named `weights` provides the needed information to perform exploratory spatial data analysis of median age in the Dallas-Fort Worth metropolitan area. In many cases, an analyst may be interested in understanding how the attributes of geographic features relate to those of their neighbors. Formally, this concept is called *spatial autocorrelation*. The concept of spatial autocorrelation relates to Waldo Tobler's famous "first law of geography," which reads (Tobler, 1970):

> Everything is related to everything else, but near things are more related than distant things.

This formulation informs much of the theory behind spatial data science and geographical inquiry more broadly. With respect to the exploratory spatial analysis of Census data, we might be interested in the degree to which a given Census variable clusters spatially, and subsequently where those clusters are found. One such way to assess clustering is to assess the degree to which ACS estimates are similar to or differ from those of their neighbors as defined by a weights matrix. Patterns can in turn be explained as follows:

- *Spatial clustering*: data values tend to be similar to neighboring data values;
- *Spatial uniformity*: data values tend to differ from neighboring data values;
- *Spatial randomness*: there is no apparent relationship between data values and those of their neighbors.

Given Tobler's first law of geography, we tend to expect that most geographic phenomena exhibit some degree of spatial clustering. This section introduces a variety of methods available in R to evaluate spatial clustering using ESDA and the **spdep** package (Bivand and Wong, 2018).

7.7.1 Spatial lags and Moran's I

Spatial weights matrices can be used to calculate the *spatial lag* of a given attribute for each observation in a dataset. The spatial lag refers to the neighboring values of an observation given a spatial weights matrix. As discussed above, row-standardized weights matrices will produce lagged means, and binary weights matrices will produce lagged sums. Spatial lag calculations are implemented in the function lag.listw(), which requires a spatial weights list object and a numeric vector from which to compute the lag.

```
dfw_tracts$lag_estimate <- lag.listw(weights, dfw_tracts$estimate)
```

The code above creates a new column in dfw_tracts, lag_estimate, that represents the average median age for the neighbors of each Census tract in the Dallas-Fort Worth metropolitan area. Using this information, we can draw a scatterplot of the ACS estimate vs. its lagged mean to do a preliminary assessment of spatial clustering in the data, shown in Figure 7.21.

```
ggplot(dfw_tracts, aes(x = estimate, y = lag_estimate)) + 
  geom_point(alpha = 0.3) + 
  geom_abline(color = "red") + 
  theme_minimal() + 
  labs(title = "Median age by Census tract, Dallas-Fort Worth TX",
       x = "Median age",
       y = "Spatial lag, median age",
       caption = "Data source: 2016-2020 ACS via the tidycensus R package.
```

The scatterplot suggests a positive correlation between the ACS estimate and its spatial lag, representative of spatial autocorrelation in the data. This relationship can be evaluated further by using a test of global spatial autocorrelation. The most common method used for spatial autocorrelation evaluation is Moran's I, which can be interpreted similar to a correlation coefficient but for the relationship between observations and their neighbors. The statistic is computed as:

$$I = \frac{N}{W} \frac{\sum_i \sum_j w_{ij}(x_i - \bar{x})(x_j - \bar{x})}{\sum_i (x_i - \bar{x})^2}$$

where w_{ij} represents the spatial weights matrix, N is the number of spatial units denoted by i and j, and W is the sum of the spatial weights.

Moran's I is implemented in **spdep** with the moran.test() function, which requires a numeric vector and a spatial weights list object.

```
moran.test(dfw_tracts$estimate, weights)
```

```
## 
##  Moran I test under randomisation
## 
## data:  dfw_tracts$estimate
## weights: weights
## 
## Moran I statistic standard deviate = 21.264, p-value < 2.2e-16
## alternative hypothesis: greater
```

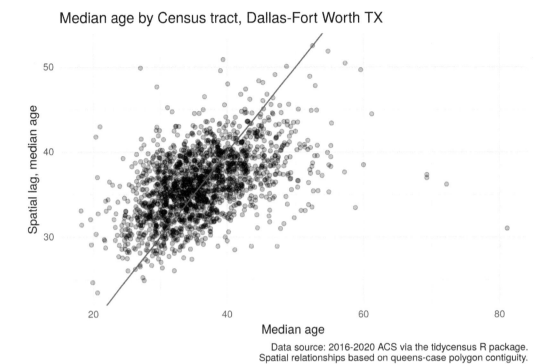

FIGURE 7.21 Scatterplot of median age relative to its spatial lag

```
## sample estimates:
## Moran I statistic       Expectation          Variance
##       0.2924898552    -0.0005889282      0.0001899598
```

The Moran's I statistic of 0.292 is positive, and the small p-value suggests that we reject the null hypothesis of spatial randomness in our dataset. (See Section 8.2.4 for additional discussion of p-values). As the statistic is positive, it suggests that our data are *spatially clustered*; a negative statistic would suggest spatial uniformity. In a practical sense, this means that Census tracts with older populations tend to be located near one another, and Census tracts with younger populations also tend to be found in the same areas.

7.7.2 Local spatial autocorrelation

We can explore this further with *local spatial autocorrelation analysis*. Local measures of spatial autocorrelation disaggregate global results to identify "hot spots" of similar values within a given spatial dataset. One such example is the Getis-Ord local G statistic (Getis and Ord, 1992), which is computed as follows:

$$G_i = \frac{\sum\limits_{j} w_{ij} x_j}{\sum\limits_{j=1}^{n} x_j} \text{ for all } i \neq j$$

In summary, the equation computes a ratio of the weighted average of the neighborhood values to the total sum of values for the dataset. While the default version of the local

7.7 Global and local spatial autocorrelation

FIGURE 7.22 Map of local Gi* scores

G (represented in the equation above) omits the location i from its calculation, a variant of the local G statistic, G_i*, includes this location. Results are returned as z-scores, and implemented in the `localG()` function in **spdep**.

The code below calculates the local G variant G_i* by re-generating the weights matrix with `include.self()`, then passing this weights matrix to the `localG()` function.

```r
# For Gi*, re-compute the weights with `include.self()`
localg_weights <- nb2listw(include.self(neighbors))

dfw_tracts$localG <- localG(dfw_tracts$estimate, localg_weights)

ggplot(dfw_tracts) +
  geom_sf(aes(fill = localG), color = NA) +
  scale_fill_distiller(palette = "RdYlBu") +
  theme_void() +
  labs(fill = "Local Gi* statistic")
```

Given that the returned results are z-scores, an analyst can choose hot spot thresholds in the statistic, calculate them with `case_when()`, then plot them accordingly.

```r
dfw_tracts <- dfw_tracts %>%
  mutate(hotspot = case_when(
    localG >= 2.56 ~ "High cluster",
    localG <= -2.56 ~ "Low cluster",
    TRUE ~ "Not significant"
  ))
```

FIGURE 7.23 Map of local Gi* scores with significant clusters highlighted

```
ggplot(dfw_tracts) +
  geom_sf(aes(fill = hotspot), color = "grey90", size = 0.1) +
  scale_fill_manual(values = c("red", "blue", "grey")) +
  theme_void()
```

The red areas on the resulting map in Figure 7.23 are representative of "high" clustering of median age, where neighborhoods with older populations are surrounded by other older-age neighborhoods. "Low" clusters are represented in blue, which reflect clustering of Census tracts with comparatively youthful populations.

7.7.3 Identifying clusters and spatial outliers with local indicators of spatial association (LISA)

An alternative method for the calculation of local spatial autocorrelation is the local indicators of spatial association statistic, commonly referred to as LISA or the local form of Moran's I (Anselin, 1995). As an extension of the Global Moran's I statistic, the local statistic I_i for a given local feature i with neighbors j is computed as follows:

$$I_i = z_i \sum_j w_{ij} z_j,$$

where z_i and z_j are expressed as deviations from the mean.

LISA is a popular method for exploratory spatial data analysis in the spatial social sciences implemented in a variety of software packages. ArcGIS implements LISA in its Cluster

7.7 Global and local spatial autocorrelation

and Outlier Analysis geoprocessing tool[11]; Anselin's open-source GeoDa[12] software has a graphical interface for calculating LISA statistics; and Python users can compute LISA using the PySAL library[13].

In R, LISA can be computed using the `localmoran()` family of functions in the spdep package. For users familiar with using LISA in other software packages, the `localmoran_perm()` function implements LISA where statistical significance is calculated based on a conditional permutation-based approach.

The example below calculates local Moran's I statistics in a way that resembles the output from GeoDa, which returns a cluster map and a Moran scatterplot. One of the major benefits of using LISA for exploratory analysis is its ability to identify both *spatial clusters*, where observations are surrounded by similar values, and *spatial outliers*, where observations are surrounded by dissimilar values. We'll use this method to explore clustering and the possible presence of spatial outliers with respect to Census tract median age in Dallas-Fort Worth.

```
set.seed(1983)

dfw_tracts$scaled_estimate <- as.numeric(scale(dfw_tracts$estimate))

dfw_lisa <- localmoran_perm(
  dfw_tracts$scaled_estimate,
  weights,
  nsim = 999L,
  alternative = "two.sided"
) %>%
  as_tibble() %>%
  set_names(c("local_i", "exp_i", "var_i", "z_i", "p_i",
              "p_i_sim", "pi_sim_folded", "skewness", "kurtosis"))

dfw_lisa_df <- dfw_tracts %>%
  select(GEOID, scaled_estimate) %>%
  mutate(lagged_estimate = lag.listw(weights, scaled_estimate)) %>%
  bind_cols(dfw_lisa)
```

The above code uses the following steps:

1. First, a random number seed is set given that we are using the conditional permutation approach to calculating statistical significance. This will ensure reproducibility of results when the process is re-run.
2. The ACS estimate for median age is converted to a z-score using `scale()`, which subtracts the mean from the estimate then divides by its standard deviation. This follows convention from GeoDa.
3. LISA is computed with `localmoran_perm()` for the scaled value for median age, using the contiguity-based spatial weights matrix. 999 conditional permutation simulations are used to calculate statistical significance, and the argument

---

[11]https://pro.arcgis.com/en/pro-app/latest/tool-reference/spatial-statistics/cluster-and-outlier-analysis-anselin-local-moran-s.htm
[12]https://geodacenter.github.io/
[13]https://pysal.org/esda/generated/esda.Moran_Local.html

TABLE 7.8 Local Moran's I results

GEOID	scaled_estimate	lagged_estimate	local_i	exp_i	var_i	z_i	p_i	geometry
48113018205	-1.2193253	-0.3045734	0.3715928	0.0200412	0.2808648	0.6633464	0.5071087	MULTIPOLYGON (((775616.1 21...
48113018508	-0.4251677	-0.7634200	0.3247727	0.0072807	0.0472315	1.4608877	0.1440463	MULTIPOLYGON (((766435.2 21...
48439105008	-0.8516598	-0.5222314	0.4450254	0.0015109	0.1479757	1.1529556	0.2489286	MULTIPOLYGON (((708333.6 21...
48113014409	-1.2634452	-0.8590131	1.0859551	-0.0138130	0.4009882	1.7367421	0.0824327	MULTIPOLYGON (((737242.6 21...
48113018135	-0.3810478	0.6042959	-0.2304013	-0.0018729	0.0216931	-1.5515979	0.1207585	MULTIPOLYGON (((782974 2136...

alternative = "two.sided" will identify both statistically significant clusters and statistically significant spatial outliers.

4. The LISA data frame is attached to the Census tract shapes after computing the lagged value for median age.

The result is shown in Table 7.8.

The information returned by localmoran_perm() can be used to compute both a GeoDa-style LISA quadrant plot as well as a cluster map. The LISA quadrant plot is similar to a Moran scatterplot, but also identifies "quadrants" of observations with respect to the spatial relationships identified by LISA. The code below uses case_when() to recode the data into appropriate categories for the LISA quadrant plot, using a significance level of $p = 0.05$.

```
dfw_lisa_clusters <- dfw_lisa_df %>%
  mutate(lisa_cluster = case_when(
    p_i >= 0.05 ~ "Not significant",
    scaled_estimate > 0 & local_i > 0 ~ "High-high",
    scaled_estimate > 0 & local_i < 0 ~ "High-low",
    scaled_estimate < 0 & local_i > 0 ~ "Low-low",
    scaled_estimate < 0 & local_i < 0 ~ "Low-high"
  ))
```

A LISA quadrant plot is visualized in Figure 7.24:

```
color_values <- c(`High-high` = "red",
                  `High-low` = "pink",
                  `Low-low` = "blue",
                  `Low-high` = "lightblue",
                  `Not significant` = "white")

ggplot(dfw_lisa_clusters, aes(x = scaled_estimate,
                              y = lagged_estimate,
                              fill = lisa_cluster)) +
  geom_point(color = "black", shape = 21, size = 2) +
  theme_minimal() +
  geom_hline(yintercept = 0, linetype = "dashed") +
  geom_vline(xintercept = 0, linetype = "dashed") +
  scale_fill_manual(values = color_values) +
  labs(x = "Median age (z-score)",
       y = "Spatial lag of median age (z-score)",
       fill = "Cluster type")
```

7.7 Global and local spatial autocorrelation

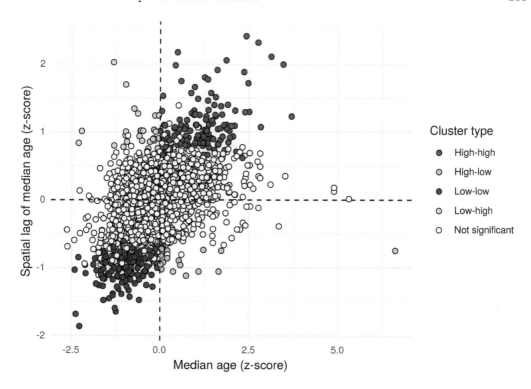

FIGURE 7.24 LISA quadrant scatterplot

Observations falling in the top-right quadrant represent "high-high" clusters, where Census tracts with higher median ages are also surrounded by Census tracts with older populations. Statistically significant clusters – those with a p-value less than or equal to 0.05 – are colored red on the chart. The bottom-left quadrant also represents spatial clusters, but instead includes lower-median-age tracts that are also surrounded by tracts with similarly low median ages. The top-left and bottom-right quadrants are home to the spatial outliers, where values are dissimilar from their neighbors.

GeoDa also implements a "cluster map" where observations are visualized in relationship to their cluster membership and statistical significance. The code below reproduces the GeoDa cluster map using **ggplot2** and geom_sf().

```
ggplot(dfw_lisa_clusters, aes(fill = lisa_cluster)) + 
  geom_sf(size = 0.1) + 
  theme_void() + 
  scale_fill_manual(values = color_values) + 
  labs(fill = "Cluster type")
```

The map illustrates distinctive patterns of spatial clustering by age in the Dallas-Fort Worth region. Older clusters are colored red; this includes areas like the wealthy Highland Park community north of downtown Dallas. Younger clusters are colored dark blue, and found in areas like east Fort Worth, east Dallas, and Arlington in the center of the metropolitan area. Spatial outliers appear scattered throughout the map as well; in the Dallas area, low-high clusters are Census tracts with large quantities of multifamily housing that are adjacent to predominantly single-family neighborhoods.

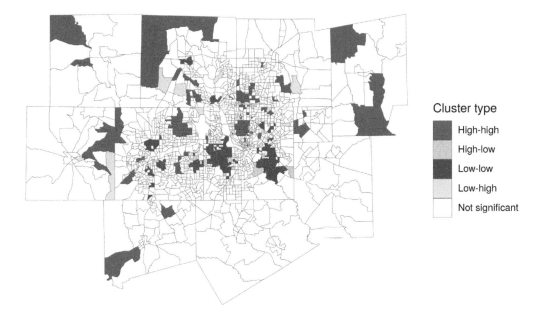

FIGURE 7.25 LISA cluster map

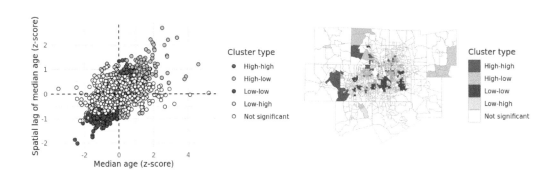

FIGURE 7.26 View of LISA Shiny app with linked brushing enabled

One very useful feature of GeoDa for exploratory spatial data analysis is the ability to perform linked brushing between the LISA quadrant plot and cluster map. This allows users to click and drag on either plot and highlight the corresponding observations on the other plot. Building on the chart linking example using ggiraph introduced in Section 6.6.2, a linked brushing approach similar to GeoDa can be implemented in Shiny and is represented in the image below and available at https://walkerke.shinyapps.io/linked-brushing/.

Using the lasso select tool, you can click and drag on either the scatterplot or the map and view the corresponding observations highlighted on the other chart panel. Code to reproduce this Shiny app is available in scripts/linked_brushing in the book's GitHub repository.

7.8 Exercises

1. Identify a different core-based statistical area of interest and use the methods introduced in this chapter to extract Census tracts or block groups for that CBSA.
2. Replicate the `erase_water()` cartographic workflow for a different county with significant water area; a good choice is King County, Washington. Be sure to transform your data to an appropriate projected coordinate system (selected with `suggest_crs()`) first. If the operation is too slow, try re-running with a higher area threshold and seeing what you get back.
3. Acquire a spatial dataset with **tidycensus** for a region of interest and a variable of interest to you. Follow the instructions in this chapter to generate a spatial weights matrix, then compute a hot-spot analysis with `localG()`.

8

Modeling US Census data

The previous chapter included a range of examples illustrating methods for analyzing and exploring spatial datasets. Census data can also be used to derive models for explaining patterns that occur across regions or within cities. These models draw from concepts introduced in prior chapters, but can also be used as part of explanatory frameworks or within broader analytic pipelines for statistical inference or machine learning. This chapter introduces a series of such frameworks. The first section looks at *segregation and diversity indices* which are widely used across the social sciences to explain demographic patterns. The second section explores topics in statistical modeling, including methods for *spatial regression* that take into account the spatial autocorrelation inherent in most Census variables. The third and final section explores concepts such as *classification, clustering*, and *regionalization* which are common in both unsupervised and supervised machine learning. Examples will illustrate how to use Census data to generate neighborhood typologies, which are widely used for business and marketing applications, and how to generate spatially coherent sales territories from Census data with regionalization.

8.1 Indices of segregation and diversity

A large body of research in the social sciences is concerned with neighborhood *segregation* and *diversity*. Segregation as addressed here generally refers to the measurement of the extent to which two or more groups live apart from each other; diversity as a companion metric measures neighborhood heterogeneity among groups. A wide range of indices have been developed by social scientists to measure segregation and diversity, and in many cases are inherently linked with spatial Census data which are often the best way to measure these concepts. Segregation and diversity indices are implemented in a variety of different R packages; the package recommended by this book is the **segregation** package (Elbers, 2021), which includes R functions for a variety of regional and local indices.

8.1.1 Data setup with spatial analysis

Much of the segregation and diversity literature focuses on race and ethnicity, which will be explored in the example below. The data setup code uses spatial methods covered in the previous three chapters to acquire Census tract-level data on population estimates for non-Hispanic white, non-Hispanic black, non-Hispanic Asian, and Hispanic populations in California, then filters those Census tracts those that intersect the largest urbanized areas by population in the state using an inner spatial join. In turn, it is an illustrative example of how spatial analysis tools can be important parts of data setup workflows for analysis. As urbanized areas for the 2020 Census are not yet defined at the time of this writing, we'll be using urbanized areas for 2019 and data from the 2015-2019 ACS.

DOI: 10.1201/9780203711415-8

```r
library(tidycensus)
library(tidyverse)
library(segregation)
library(tigris)
library(sf)

# Get California tract data by race/ethnicity
ca_acs_data <- get_acs(
  geography = "tract",
  variables = c(
    white = "B03002_003",
    black = "B03002_004",
    asian = "B03002_006",
    hispanic = "B03002_012"
  ),
  state = "CA",
  geometry = TRUE,
  year = 2019
)

# Use tidycensus to get urbanized areas by population with geometry,
# then filter for those that have populations of 750,000 or more
us_urban_areas <- get_acs(
  geography = "urban area",
  variables = "B01001_001",
  geometry = TRUE,
  year = 2019,
  survey = "acs1"
) %>%
  filter(estimate >= 750000) %>%
  transmute(urban_name = str_remove(NAME,
                                    fixed(", CA Urbanized Area (2010)")))

# Compute an inner spatial join between the California tracts and the
# urbanized areas, returning tracts in the largest California urban
# areas with the urban_name column appended
ca_urban_data <- ca_acs_data %>%
  st_join(us_urban_areas, left = FALSE) %>%
  select(-NAME) %>%
  st_drop_geometry()
```

To summarize, the spatial analysis workflow detailed above uses the following steps:

1. Data on race & ethnicity from the 2015-2019 5-year ACS for the four largest demographic groups in California is acquired with **tidycensus**'s get_acs() at the Census tract level with feature geometry included. Depending on the goals of the study, other racial/ethnic groups (e.g. native American, native Hawaiian/Pacific Islander) should be added or removed as needed.
2. As urban areas as defined by the Census Bureau often cross state boundaries, urban areas must be obtained for the entire US with get_acs(). Once obtained,

8.1 Indices of segregation and diversity

TABLE 8.1 Prepared data for segregation analysis

GEOID	variable	estimate	moe	urban_name
06013370000	white	1235	166	San Francisco–Oakland
06013370000	black	371	149	San Francisco–Oakland
06013370000	asian	540	88	San Francisco–Oakland
06013370000	hispanic	557	121	San Francisco–Oakland
06001442301	white	969	197	San Francisco–Oakland
06001442301	black	214	129	San Francisco–Oakland
06001442301	asian	3000	259	San Francisco–Oakland
06001442301	hispanic	1010	272	San Francisco–Oakland

urban areas are filtered to only those areas with populations of 750,000 or greater, and then `transmute()` is used to retain only a new column representing the area name (along with the simple feature geometry column).
3. A spatial join between the Census tract data and the urban area data is computed with `st_join()`. The argument `left = FALSE` computes an *inner spatial join*, which retains only those Census tracts that intersect the urban area boundaries, and appends the corresponding `urban_name` column to each Census tract.

The data structure appears as follows:

The data are in long (tidy) form, the default used by **tidycensus**; this data structure is ideal for computing indices in the **segregation** package.

8.1.2 The dissimilarity index

The dissimilarity index is widely used to assess neighborhood segregation between two groups within a region. It is computed as follows:

$$D = \frac{1}{2} \sum_{i=1}^{N} \left| \frac{a_i}{A} - \frac{b_i}{B} \right|$$

where a_i represents the population of group A in a given areal unit i; A is the total population of that group in the study region (e.g. a metropolitan area); and b_i and B are the equivalent metrics for the second group. The index ranges from a low of 0 to a high of 1, where 0 represents perfect integration between the two groups and 1 represents complete segregation. This index is implemented in the **segregation** package with the `dissimilarity()` function.

The example below computes the dissimilarity index between non-Hispanic white and Hispanic populations for the San Francisco/Oakland urbanized area. The data are filtered for only those rows that represent the target populations in the San Francisco/Oakland area, which is then piped to the `dissimilarity()` function. The function requires identification of a group column, for which we'll use `variable`; a unit column representing the neighborhood unit, for which we'll use `GEOID` to represent the Census tract; and a weight column that tells the function how many people are in each group.

```
ca_urban_data %>%
  filter(variable %in% c("white", "hispanic"),
         urban_name == "San Francisco--Oakland") %>%
```

TABLE 8.2 Dissimilarity indices for Hispanic and non-Hispanic white populations, large California urbanized areas

urban_name	stat	est
Los Angeles–Long Beach–Anaheim	D	0.5999229
San Francisco–Oakland	D	0.5135526
San Jose	D	0.4935633
San Diego	D	0.4898184
Riverside–San Bernardino	D	0.4079863
Sacramento	D	0.3687927

```
dissimilarity(
  group = "variable",
  unit = "GEOID",
  weight = "estimate"
)
```

```
##   stat       est
## 1:   D 0.5135526
```

The D index of segregation between non-Hispanic white and Hispanic populations in the San Francisco-Oakland area is 0.51. This statistic, however, is more meaningful in comparison with other cities. To compute dissimilarity for each urban area, we can creatively apply tidyverse techniques covered in earlier chapters and introduce a new function, group_modify(), for group-wise calculation. This example follows the recommended workflow in the **segregation** package documentation[1]. The code below filters the data for non-Hispanic white and Hispanic populations by Census tract, then groups the dataset by values in the urban_name column. The group_modify() function from **dplyr** then allows for the calculation of dissimilarity indices *by group*, which in this example is Census tracts within each respective urban area. It returns a combined dataset that is sorted in descending order with arrange() to make comparisons.

```
ca_urban_data %>%
  filter(variable %in% c("white", "hispanic")) %>%
  group_by(urban_name) %>%
  group_modify(~
    dissimilarity(.x,
      group = "variable",
      unit = "GEOID",
      weight = "estimate"
    )
  ) %>%
  arrange(desc(est))
```

The Los Angeles area is the most segregated of the large urbanized areas in California with respect to non-Hispanic white and Hispanic populations at the Census tract level, followed by

---

[1] https://elbersb.github.io/segregation/articles/faq.html#how-can-i-compute-indices-for-different-areas-at-once-

8.1 Indices of segregation and diversity

TABLE 8.3 Multi-group segregation results for California urban areas

urban_name	M	p	H	ent_ratio
Los Angeles–Long Beach–Anaheim	0.3391033	0.5016371	0.2851662	0.9693226
Riverside–San Bernardino	0.1497129	0.0867808	0.1408461	0.8664604
Sacramento	0.1658898	0.0736948	0.1426804	0.9477412
San Diego	0.2290891	0.1256072	0.2025728	0.9218445
San Francisco–Oakland	0.2685992	0.1394522	0.2116127	1.0346590
San Jose	0.2147445	0.0728278	0.1829190	0.9569681

San Francisco/Oakland. Riverside/San Bernardino and Sacramento are the least segregated of the large urban areas in the state.

8.1.3 Multi-group segregation indices

One disadvantage of the dissimilarity index is that it only measures segregation between two groups. For a state as diverse as California, we may be interested in measuring segregation and diversity between multiple groups at a time. The **segregation** package implements two such indices: the Mutual Information Index M, and Theil's Entropy Index H (Mora and Ruiz-Castillo, 2011). Following Elbers (2021), M is computed as follows for a dataset T:

$$M(\mathbf{T}) = \sum_{u=1}^{U} \sum_{g=1}^{G} p_{ug} \log \frac{p_{ug}}{p_u p_g}$$

where U is the total number of units u, G is the total number of groups g, and p_{ug} is the joint probability of being in unit u and group g, with p_u and p_g referring to unit and group probabilities. Theil's H for the same dataset T can then be written as:

$$H(\mathbf{T}) = \frac{M(\mathbf{T})}{E(\mathbf{T})}$$

where $E(T)$ is the entropy of T, normalizing H to range between values of 0 and 1.

Computing these indices is straightforward with the **segregation** package. The `mutual_total()` function computes both indices; when different regions are to be considered (like multiple urban areas, as in this example) the `mutual_within()` function will compute M and H by urban area with the `within` argument appropriately specified. We'll be using the full `ca_urban_data` dataset, which includes population estimates for non-Hispanic white, non-Hispanic Black, non-Hispanic Asian, and Hispanic populations.

```
mutual_within(
  data = ca_urban_data,
  group = "variable",
  unit = "GEOID",
  weight = "estimate",
  within = "urban_name",
  wide = TRUE
)
```

GEOID	ls	p
06037101110	0.2821846	0.0003363
06037101122	0.7790480	0.0002690
06037101210	0.1012193	0.0005088
06037101220	0.1182334	0.0002917
06037101300	0.6538220	0.0003094
06037101400	0.3951408	0.0002655
06037102103	0.3039904	0.0001384
06037102104	0.4462187	0.0002836
06037102105	0.1284913	0.0001496
06037102107	0.2389721	0.0003454

When multi-group segregation is considered using these indices, Los Angeles remains the most segregated urban area, whereas Riverside/San Bernardino is the least segregated.

The **segregation** package also offers a function for local segregation analysis, mutual_local(), which decomposes M into unit-level segregation scores, represented by ls. In the example below, we will use mutual_local() to examine patterns of segregation across the most segregated urban area, Los Angeles.

```
la_local_seg <- ca_urban_data %>%
  filter(urban_name == "Los Angeles--Long Beach--Anaheim") %>%
  mutual_local(
    group = "variable",
    unit = "GEOID",
    weight = "estimate",
    wide = TRUE
  )
```

The results can be mapped by joining the data to a dataset of Census tracts from **tigris**; the inner_join() function is used to retain tracts for the Los Angeles area only.

```
la_tracts_seg <- tracts("CA", cb = TRUE, year = 2019) %>%
  inner_join(la_local_seg, by = "GEOID")

la_tracts_seg %>%
  ggplot(aes(fill = ls)) +
  geom_sf(color = NA) +
  coord_sf(crs = 26946) +
  scale_fill_viridis_c(option = "inferno") +
  theme_void() +
  labs(fill = "Local\nsegregation index")
```

8.1.4 Visualizing the diversity gradient

The *diversity gradient* is a concept that uses scatterplot smoothing to visualize how neighborhood diversity varies by distance or travel-time from the core of an urban region (Walker, 2016a). Historically, literature on suburbanization in the social sciences assumes a more heterogeneous urban core relative to segregated and homogeneous suburban neighborhoods.

8.1 Indices of segregation and diversity

FIGURE 8.1 Map of local multi-group segregation scores in Los Angeles

The diversity gradient is a visual heuristic used to evaluate the validity of this demographic model.

The entropy index for a given geographic unit is calculated as follows:

$$E = \sum_{r=1}^{n} Q_r ln \frac{1}{Q_r}$$

Q_r in this calculation represents group r's proportion of the population in the geographic unit.

This statistic is implemented in the `entropy()` function in the **segregation** package. As the `entropy()` function calculates this statistic for a specific unit at a time, we will group the data by tract, and then use `group_modify()` to calculate the entropy for each tract separately. The argument `base = 4` is set by convention to the number of groups in the calculation; this sets the maximum value of the statistic to 1, which represents perfect evenness between the four groups in the area. Once computed, the indices are joined to a dataset of Census tracts from California; `inner_join()` is used to retain only those tracts in the Los Angeles urbanized area.

```
la_entropy <- ca_urban_data %>%
  filter(urban_name == "Los Angeles--Long Beach--Anaheim") %>%
  group_by(GEOID) %>%
  group_modify(~data.frame(entropy = entropy(
    data = .x,
    group = "variable",
    weight = "estimate",
    base = 4)))
```

```
la_entropy_geo <- tracts("CA", cb = TRUE, year = 2019) %>%
  inner_join(la_entropy, by = "GEOID")
```

Visualization of the diversity gradient then requires a relative measurement of how far each Census tract is from the urban core. The travel-time methods available in the **mapboxapi** package introduced in Chapter 7 are again used here to calculate driving distance to Los Angeles City Hall for all Census tracts in the Los Angeles urbanized area.

```
library(mapboxapi)

la_city_hall <- mb_geocode("City Hall, Los Angeles CA")

minutes_to_downtown <- mb_matrix(la_entropy_geo, la_city_hall)
```

Once computed, the travel times are stored in a vector `minutes_to_downtown`, then assigned to a new column `minutes` in the entropy data frame. The tract diversity index is visualized using **ggplot2** relative to its travel time to downtown Los Angeles, with a LOESS smoother superimposed over the scatterplot to represent the diversity gradient.

```
la_entropy_geo$minutes <- as.numeric(minutes_to_downtown)

ggplot(la_entropy_geo, aes(x = minutes_to_downtown, y = entropy)) +
  geom_point(alpha = 0.5) +
  geom_smooth(method = "loess") +
  theme_minimal() +
  scale_x_continuous(limits = c(0, 80)) +
  labs(title = "Diversity gradient, Los Angeles urbanized area",
       x = "Travel-time to downtown Los Angeles in minutes, Census tracts",
       y = "Entropy index")
```

The visualization of the diversity gradient in Figure 8.2 shows that neighborhood diversity increases with driving time from the urban core in Los Angeles, peaking at about 35 minutes in free-flowing traffic from the urban core then leveling off after that. The structure of the diversity gradient suggests that Census tracts near to downtown tend to be segregated, and suburban tracts more likely to be integrated.

8.2 Regression modeling with US Census data

Regression modeling is widely used in industry and the social sciences to understand social processes. In the social sciences, the goal of regression modeling is commonly to understand the relationships between a variable under study, termed an *outcome variable*, and one or

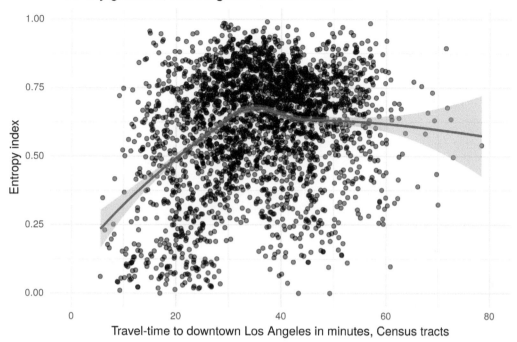

FIGURE 8.2 Diversity gradient visualization for the Los Angeles, CA urbanized area

more *predictors* that are believed to have some influence on the outcome variable. Following James et al. (2013), a model can be represented with the following general notation:

$$Y = f(X) + \epsilon$$

where Y represents the outcome variable; X represents one or more predictors hypothesized to have some influence on the outcome variable; f is a function that represents the relationships between X and Y; and ϵ represents the *error terms* or residuals, the differences between the modeled values of Y and the actual values. The function f will be estimated using a method appropriate for the structure of the data and selected by the analyst.

A complete treatment of regression modeling is beyond the scope of this book; recommended resources include James et al. (2013), Boehmke and Greenwell (2019), Çetinkaya-Rundel and Hardin (2021), and Matloff (2017). The purpose of this section is to illustrate an example workflow using regression modeling to analyze data from the American Community Survey. The section will start with a simple linear model and extend its discussion from there. In doing so, some problems with the application of the linear model to aggregated Census data will be discussed. First, demographic statistics are often highly correlated with one another, meaning that Census data-based models risk *collinearity* where predictors are not independent of one another. Second, spatial demographic data commonly exhibit *spatial autocorrelation*, which may lead to a violation of the assumption of independent and identically distributed error terms ($i.i.d$) in the linear model. Suggested approaches for addressing these problems discussed in this section include dimension reduction and spatial regression.

TABLE 8.4 Data acquired from tidycensus for regression modeling

GEOID	median_valueE	median_valueM	median_roomsE	median_roomsM	total_populationE	total_populationM	median_ageE	geometry
48085030101	183600	11112	6.0	0.3	2296	420	31.9	MULTIPOLYGON (((787828.3 21...
48085030102	198500	88036	6.0	0.7	2720	524	45.5	MULTIPOLYGON (((784027.9 21...
48085030201	324200	47203	7.4	0.6	3653	509	35.1	MULTIPOLYGON (((774843.6 21...
48085030202	366900	20014	7.3	0.4	3530	418	42.1	MULTIPOLYGON (((764886.2 21...
48085030204	217200	14152	6.2	0.8	6592	1193	32.7	MULTIPOLYGON (((778099.2 21...
48085030205	231200	41519	6.3	0.8	5257	1273	30.4	MULTIPOLYGON (((777818.4 21...

8.2.1 Data setup and exploratory data analysis

The topic of study in this illustrative applied workflow will be median home value by Census tract in the Dallas-Fort Worth metropolitan area. To get started, we'll define several counties in north Texas that we'll use to represent the DFW region, and use a named vector of variables to acquire data that will represent both our outcome variable and predictors. Data are returned from **tidycensus** with the argument output = "wide", giving one column per variable. The geometry is also transformed to an appropriate coordinate reference system for North Texas, EPSG code 32138 (NAD83 / Texas North Central with meters for measurement units).

```
library(tidycensus)
library(sf)

dfw_counties <- c("Collin County", "Dallas", "Denton",
                  "Ellis", "Hunt", "Kaufman", "Rockwall",
                  "Johnson", "Parker", "Tarrant", "Wise")

variables_to_get <- c(
  median_value = "B25077_001",
  median_rooms = "B25018_001",
  median_income = "DP03_0062",
  total_population = "B01003_001",
  median_age = "B01002_001",
  pct_college = "DP02_0068P",
  pct_foreign_born = "DP02_0094P",
  pct_white = "DP05_0077P",
  median_year_built = "B25037_001",
  percent_ooh = "DP04_0046P"
)

dfw_data <- get_acs(
  geography = "tract",
  variables = variables_to_get,
  state = "TX",
  county = dfw_counties,
  geometry = TRUE,
  output = "wide",
  year = 2020
) %>%
  select(-NAME) %>%
  st_transform(32138) # NAD83 / Texas North Central
```

The ACS estimates we've acquired include:

8.2 Regression modeling with US Census data

- `median_valueE`: The median home value of the Census tract (our outcome variable);
- `median_roomsE`: The median number of rooms for homes in the Census tract;
- `total_populationE`: The total population;
- `median_ageE`: The median age of the population in the Census tract;
- `median_year_builtE`: The median year built of housing structures in the tract;
- `median_incomeE`: The median income of households in the Census tract;
- `pct_collegeE`: The percentage of the population age 25 and up with a 4-year college degree;
- `pct_foreign_bornE`: The percentage of the population born outside the United States;
- `pct_whiteE`: The percentage of the population that identifies as non-Hispanic white;
- `percent_oohE`: The percentage of housing units in the tract that are owner-occupied.

8.2.2 Inspecting the outcome variable with visualization

To get started, we will examine both the geographic and data distributions of our outcome variable, median home value, with a quick map with `geom_sf()` and a histogram in Figure 8.3.

```
library(tidyverse)
library(patchwork)

mhv_map <- ggplot(dfw_data, aes(fill = median_valueE)) +
  geom_sf(color = NA) +
  scale_fill_viridis_c(labels = scales::label_dollar()) +
  theme_void() +
  labs(fill = "Median home value ")

mhv_histogram <- ggplot(dfw_data, aes(x = median_valueE)) +
  geom_histogram(alpha = 0.5, fill = "navy", color = "navy",
                 bins = 100) +
  theme_minimal() +
  scale_x_continuous(labels = scales::label_number_si(accuracy = 0.1)) +
  labs(x = "Median home value")

mhv_map + mhv_histogram
```

As is common with home values in metropolitan regions, the data distribution is right-skewed with a clustering of Census tracts on the lower end of the distribution of values and a long tail of very expensive areas, generally located north of downtown Dallas. This can lead to downstream violations of normality in model residuals. In turn, we might consider log-transforming our outcome variable, which will make its distribution closer to normal and will better capture the geographic variations in home values that we are trying to model.

```
library(tidyverse)
library(patchwork)
```

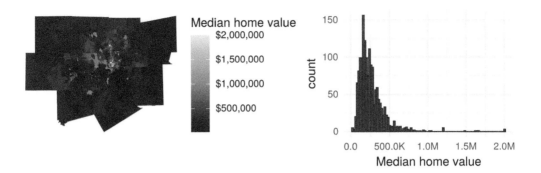

FIGURE 8.3 Median home value charts

```
mhv_map_log <- ggplot(dfw_data, aes(fill = log(median_valueE))) +
  geom_sf(color = NA) +
  scale_fill_viridis_c() +
  theme_void() +
  labs(fill = "Median home\nvalue (log)")

mhv_histogram_log <- ggplot(dfw_data, aes(x = log(median_valueE))) +
  geom_histogram(alpha = 0.5, fill = "navy", color = "navy",
                 bins = 100) +
  theme_minimal() +
  scale_x_continuous() +
  labs(x = "Median home value (log)")

mhv_map_log + mhv_histogram_log
```

The expensive areas of north Dallas still stand out in Figure 8.4, but the log-transformation makes the distribution of values more normal and better shows geographic variation of home values on the map. This suggests that we require some data preparation prior to fitting the model.

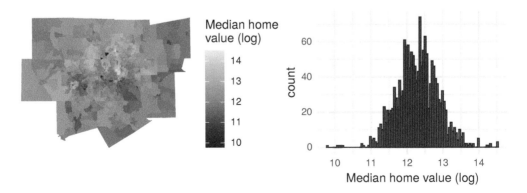

FIGURE 8.4 Logged median home value charts

8.2 Regression modeling with US Census data

TABLE 8.5 Engineered predictors for regression modeling

GEOID	median_value	median_rooms	total_population	median_age	median_year_built	median_income	pct_college	geometry
48085030101	183600	6.0	2296	31.9	1994	63036	8.6	MULTIPOLYGON (((787828.3 21...
48085030102	198500	6.0	2720	45.5	1995	65234	19.2	MULTIPOLYGON (((784027.9 21...
48085030201	324200	7.4	3653	35.1	2002	85938	41.3	MULTIPOLYGON (((774843.6 21...
48085030202	366900	7.3	3530	42.1	2001	134097	40.9	MULTIPOLYGON (((764886.2 21...
48085030204	217200	6.2	6592	32.7	2006	98622	51.8	MULTIPOLYGON (((778099.2 21...

8.2.3 "Feature engineering"

A common term used when preparing data for regression modeling is "feature engineering," which refers to the transformation of predictors in ways that better represent the relationships between those predictors and the outcome variable. Many of the variables acquired from the ACS in the steps above are already "pre-engineered" as they were returned as percentages from the ACS data profile, saving some steps. However, some variables would benefit from additional transformation.

The code below creates two new variables: `pop_density`, which represents the number of people in each Census tract per square kilometer, and `median_structure_age`, which represents the median age of housing structures in the tract.

```
library(sf)
library(units)

dfw_data_for_model <- dfw_data %>%
  mutate(pop_density = as.numeric(set_units(total_populationE / st_area(.),
                                            "1/km2")),
         median_structure_age = 2018 - median_year_builtE) %>%
  select(!ends_with("M")) %>%
  rename_with(.fn = ~str_remove(.x, "E$")) %>%
  na.omit()
```

The calculation of the `pop_density` column appears more complicated, so it is helpful to read it from the inside out. The `st_area()` function from the sf package calculates the area of the Census tract; by default this will be in square meters, using the base measurement unit of the data's coordinate reference system. The `total_population` column is then divided by the area of the tract. Next, the `set_units()` function is used to convert the measurement to population per square kilometer using `"1/km2"`. Finally, the calculation is converted from a units vector to a numeric vector with `as.numeric()`. Calculating median structure age is more straightforward, as the `median_year_builtE` column is subtracted from 2017, the mid-point of the 5-year ACS period from which our data are derived. Finally, to simplify the dataset, margin of error columns are dropped, the E at the end of the estimate columns is removed with `rename_with()`, and tracts with NA values are dropped as well with `na.omit()`.

8.2.4 A first regression model

After inspecting the distribution of the outcome variable and completing feature engineering with respect to the predictors, we are ready to fit a first linear model. Our linear model with a log-transformed outcome variable can be written as follows:

$$\begin{aligned}\log(\text{median\_value}) = \alpha &+ \beta_1(\text{median\_rooms}) + \beta_2(\text{median\_income}) + \\ &\beta_3(\text{pct\_college}) + \beta_4(\text{pct\_foreign\_born}) + \\ &\beta_5(\text{pct\_white}) + \beta_6(\text{median\_age}) + \\ &\beta_7(\text{median\_structure\_age}) + \beta_8(\text{percent\_ooh}) + \\ &\beta_9(\text{pop\_density}) + \beta_{10}(\text{total\_population}) + \epsilon\end{aligned}$$

where α is the model intercept, β_1 is the change in the log of median home value with a 1-unit increase in the median number of rooms (and so forth for all the model predictors) while holding all other predictors constant, and ϵ is the error term.

Model formulas in R are generally written as `outcome ~ predictor_1 + predictor_2 + ... + predictor_k`, where k is the number of model predictors. The formula can be supplied as a character string to the model function (as shown below) or supplied unquoted in the call to the function. We use `lm()` to fit the linear model, and then check the results with `summary()`.

```r
formula <- paste0("log(median_value) ~ median_rooms + median_income + ",
                  "pct_college + pct_foreign_born + pct_white + ",
                  "median_age + median_structure_age + ",
                  "percent_ooh + pop_density + total_population")

model1 <- lm(formula = formula, data = dfw_data_for_model)

summary(model1)
```

```
## 
## Call:
## lm(formula = formula, data = dfw_data_for_model)
## 
## Residuals:
##      Min       1Q   Median       3Q      Max 
## -2.03015 -0.14250  0.00033  0.14794  1.45712 
## 
## Coefficients:
##                        Estimate Std. Error t value Pr(>|t|)    
## (Intercept)           1.123e+01  6.199e-02 181.093  < 2e-16 ***
## median_rooms          8.800e-03  1.058e-02   0.832 0.405711    
## median_income         5.007e-06  4.202e-07  11.915  < 2e-16 ***
## pct_college           1.325e-02  5.994e-04  22.108  < 2e-16 ***
## pct_foreign_born      2.877e-03  8.005e-04   3.594 0.000336 ***
## pct_white             3.961e-03  4.735e-04   8.365  < 2e-16 ***
## median_age            4.782e-03  1.372e-03   3.485 0.000507 ***
## median_structure_age  1.202e-05  2.585e-05   0.465 0.642113    
## percent_ooh          -4.761e-03  5.599e-04  -8.504  < 2e-16 ***
## pop_density          -7.946e-06  6.160e-06  -1.290 0.197216    
## total_population      8.960e-06  4.460e-06   2.009 0.044733 *  
## ---
## Signif. codes:  0 '***' 0.001 '**' 0.01 '*' 0.05 '.' 0.1 ' ' 1
## 
```

8.2 Regression modeling with US Census data

```
## Residual standard error: 0.2695 on 1548 degrees of freedom
## Multiple R-squared:  0.7818, Adjusted R-squared:  0.7804
## F-statistic: 554.6 on 10 and 1548 DF,  p-value: < 2.2e-16
```

The printed summary gives us information about the model fit. The `Estimate` column represents the model parameters (the β values), followed by the standard error, the t-value test statistic, and the *p*-value which helps us assess relative statistical significance. James et al. (2013) (p. 67) provides a concise summary of how *p*-values should be interpreted:

> Roughly speaking, we interpret the *p*-value as follows: a small *p*-value indicates that it is unlikely to observe such a substantial association between the predictor and the response (outcome variable) due to chance, in the absence of any real association between the predictor and the response. Hence, if we see a small *p*-value, then we can infer that there is an association between the predictor and the response. We *reject the null hypothesis* – that is, we declare a relationship to exist between X and Y – if the *p*-value is small enough.

By convention, researchers will use *p*-value cutoffs of 0.05, 0.01, or 0.001, depending on the topic under study; these values are highlighted with asterisks in the model summary printout. Examining the model parameters and *p*-values suggests that median household income, bachelor's degree attainment, the percentage non-Hispanic white population, and median age are positively associated with median home values, whereas the percentage owner-occupied housing is negatively associated with median home values. The R-squared value is 0.78, suggesting that our model explains around 78 percent of the variance in `median_value`.

Somewhat surprisingly, `median_rooms` does not appear to have a significant relationship with median home value as per the model. On the one hand, this can be interpreted as "the effect of median rooms on median home value with all other predictors held constant" – but also could be suggestive of model mis-specification. As mentioned earlier, models using ACS data as predictors are highly vulnerable to *collinearity*. Collinearity occurs when two or more predictors are highly correlated with one another, which can lead to misinterpretation of the actual relationships between predictors and the outcome variable.

One way to inspect for collinearity is to visualize the *correlation matrix* of the predictors, in which correlations between all predictors are calculated with one another. The `correlate()` function in the **corrr** package (Kuhn et al., 2020) offers a straightforward method for calculating a correlation matrix over a rectangular data frame.

```
library(corrr)

dfw_estimates <- dfw_data_for_model %>%
  select(-GEOID, -median_value, -median_year_built) %>%
  st_drop_geometry()

correlations <- correlate(dfw_estimates, method = "pearson")
```

One calculated, the correlation matrix is visualized with `network_plot()` in Figure 8.5:

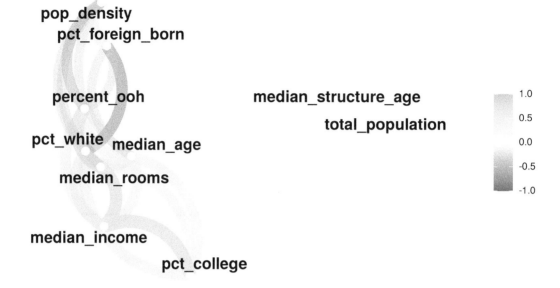

FIGURE 8.5 Network plot of correlations between model predictors

```
network_plot(correlations)
```

We notice that most of the predictors are correlated with one another to some degree, which is unsurprising given that they all represent social demographic data. Collinearity can be diagnosed further by calculating the *variance inflation factor* (VIF) for the model, which takes into account not just pairwise correlations but the extent to which predictors are collinear with all other predictors. A VIF value of 1 indicates no collinearity; VIF values above 5 suggest a level of collinearity that has a problematic influence on model interpretation (James et al., 2013). VIF is implemented by the vif() function in the **car** package (Fox and Weisberg, 2019).

```
library(car)

vif(model1)
```

```
##        median_rooms        median_income         pct_college
##            5.450436             6.210615            3.722434
##    pct_foreign_born            pct_white          median_age
##            2.013411             3.233142            1.833625
## median_structure_age         percent_ooh         pop_density
##            1.055760             3.953587            1.537508
##    total_population
##            1.174613
```

The most problematic variable is median_income, with a VIF value of over 6. A potential solution involves removing this variable and re-running the model; as it is highly correlated

8.2 Regression modeling with US Census data

with other predictors in the model, the effect of median household income would in theory be captured by the remaining predictors.

```
formula2 <- paste0("log(median_value) ~ median_rooms + pct_college + ",
                   "pct_foreign_born + pct_white + median_age + ",
                   "median_structure_age + percent_ooh + pop_density + ",
                   "total_population")

model2 <- lm(formula = formula2, data = dfw_data_for_model)

summary(model2)
```

```
## 
## Call:
## lm(formula = formula2, data = dfw_data_for_model)
## 
## Residuals:
##      Min       1Q   Median       3Q      Max 
## -1.91753 -0.15318 -0.00224  0.16192  1.58948 
## 
## Coefficients:
##                        Estimate Std. Error t value Pr(>|t|)    
## (Intercept)           1.101e+01  6.203e-02 177.566  < 2e-16 ***
## median_rooms          7.326e-02  9.497e-03   7.713 2.18e-14 ***
## pct_college           1.775e-02  4.862e-04  36.506  < 2e-16 ***
## pct_foreign_born      4.170e-03  8.284e-04   5.034 5.38e-07 ***
## pct_white             4.996e-03  4.862e-04  10.274  < 2e-16 ***
## median_age            3.527e-03  1.429e-03   2.468   0.0137 *  
## median_structure_age  2.831e-05  2.696e-05   1.050   0.2939    
## percent_ooh          -3.888e-03  5.798e-04  -6.705 2.81e-11 ***
## pop_density          -5.474e-06  6.430e-06  -0.851   0.3947    
## total_population      9.711e-06  4.658e-06   2.085   0.0373 *  
## ---
## Signif. codes:  0 '***' 0.001 '**' 0.01 '*' 0.05 '.' 0.1 ' ' 1
## 
## Residual standard error: 0.2815 on 1549 degrees of freedom
## Multiple R-squared:  0.7618, Adjusted R-squared:  0.7604 
## F-statistic: 550.4 on 9 and 1549 DF,  p-value: < 2.2e-16
```

The model R-squared drops slightly but not substantially to 0.76. Notably, the effect of `median_rooms` on median home value now comes through as strongly positive and statistically significant, suggesting that collinearity with median household income was suppressing this relationship in the first model. As a diagnostic, we can re-compute the VIF for the second model:

```
vif(model2)
```

```
##         median_rooms          pct_college     pct_foreign_born 
##             4.025411             2.245227             1.976425 
##            pct_white           median_age median_structure_age 
##             3.124400             1.822825             1.052805 
```

```
##     percent_ooh      pop_density  total_population
##        3.885779         1.535763          1.174378
```

The VIF values for all predictors in the model are now below 5.

8.2.5 Dimension reduction with principal components analysis

In the example above, dropping median household income from the model had a fairly negligible impact on the overall model fit and significantly improved the model's problems with collinearity. However, this will not always be the best solution for analysts, especially when dropping variables has a more significant impact on the model fit. An alternative approach to resolving problems with collinearity is *dimension reduction*, which transforms the predictors into a series of dimensions that represent the variance in the predictors but are uncorrelated with one another. Dimension reduction is also a useful technique when an analyst is dealing with a massive number of predictors (hundreds or even thousands) and needs to reduce the predictors in the model to a more manageable number while still retaining the ability to explain the variance in the outcome variable.

One of the most popular methods for dimension reduction is *principal components analysis*. Principal components analysis (PCA) reduces a higher-dimensional dataset into a lower-dimensional representation based linear combinations of the variables used. The *first principal component* is the linear combination of variables that explains the most overall variance in the data; the *second principal component* explains the second-most overall variance but is also constrained to be uncorrelated with the first component; and so forth.

PCA can be computed with the `prcomp()` function. We will use the `dfw_estimates` object that we used to compute the correlation data frame here as it includes only the predictors in the regression model, and use the notation `formula = ~.` to compute PCA over all the predictors. By convention, `scale.` and `center` should be set to `TRUE` as this normalizes the variables in the dataset before computing PCA given that they are measured differently.

```
pca <- prcomp(
  formula = ~.,
  data = dfw_estimates,
  scale. = TRUE,
  center = TRUE
)

summary(pca)
```

```
## Importance of components:
##                            PC1    PC2    PC3    PC4     PC5     PC6     PC7
## Standard deviation       2.020 1.1832 1.1307 1.0093 0.89917 0.70312 0.67686
## Proportion of Variance   0.408 0.1400 0.1278 0.1019 0.08085 0.04944 0.04581
## Cumulative Proportion    0.408 0.5481 0.6759 0.7778 0.85860 0.90803 0.95385
##                              PC8     PC9    PC10
## Standard deviation       0.48099 0.36127 0.31567
## Proportion of Variance   0.02314 0.01305 0.00997
## Cumulative Proportion    0.97698 0.99003 1.00000
```

8.2 Regression modeling with US Census data

TABLE 8.6 PCA variable loadings

predictor	PC1	PC2	PC3	PC4	PC5	PC6	PC7
median_rooms	-0.4077852	0.1323021	-0.3487253	0.0586440	-0.2076430	0.2123836	-0.1609126
total_population	-0.0032288	0.4604789	-0.5213433	-0.1272021	0.5458555	-0.4347310	0.0936404
median_age	-0.3474694	-0.1380008	0.2393805	-0.0781325	-0.2960241	-0.7849378	-0.1486493
median_income	-0.4149988	-0.2277427	-0.3213461	-0.0367450	0.0067868	0.2005786	0.1163622
pct_college	-0.3116212	-0.5358812	-0.1567328	-0.1562603	0.2061776	0.0588802	0.3496196
pct_foreign_born	0.2812618	-0.2409164	-0.4667693	0.0769341	-0.4514510	-0.2779627	0.3495310
pct_white	-0.3910590	-0.0924262	0.2855863	-0.1708498	0.3472563	-0.0631469	-0.0042362
percent_ooh	-0.3812989	0.3031944	-0.1930025	0.1680217	-0.3006266	0.0753161	-0.3374642
pop_density	0.2571344	-0.4253855	-0.2876627	-0.3247835	0.1261611	-0.0196740	-0.7385667
median_structure_age	-0.0095808	-0.2700238	-0.0450257	0.8829973	0.3131336	-0.1369883	-0.1609821

Printing the summary() of the PCA model shows 10 components that collectively explain 100% of the variance in the original predictors. The first principal component explains 40.8 percent of the overall variance; the second explains 14 percent; and so forth.

To understand what the different principal components now mean, it is helpful to plot the variable *loadings*. This represents the relationships between the original variables in the model and the derived components. This approach is derived from Julia Silge's blog post on the topic (Silge, 2021).

First, the variable loading matrix (stored in the rotation element of the pca object) is converted to a tibble so we can view it easier.

```
pca_tibble <- pca$rotation %>%
  as_tibble(rownames = "predictor")
```

Positive values for a given row mean that the original variable is *positively loaded* onto a given component, and negative values mean that the variable is *negatively loaded*. Larger values in each direction are of the most interest to us; values near 0 mean the variable is not meaningfully explained by a given component. To explore this further, we can visualize the first five components with **ggplot2** in Figure 8.6:

```
pca_tibble %>%
  select(predictor:PC5) %>%
  pivot_longer(PC1:PC5, names_to = "component", values_to = "value") %>%
  ggplot(aes(x = value, y = predictor)) + 
  geom_col(fill = "darkgreen", color = "darkgreen", alpha = 0.5) +
  facet_wrap(~component, nrow = 1) +
  labs(y = NULL, x = "Value") +
  theme_minimal()
```

With respect to PC1, which explains nearly 41 percent of the variance in the overall predictor set, the variables percent_ooh, pct_white, pct_college, median_rooms, median_income, and median_age load negatively, whereas pop_density and pct_foreign_born load positively. We can attach these principal components to our original data with predict() and cbind(), then make a map of PC1 for further exploration, shown in Figure 8.7.

```
components <- predict(pca, dfw_estimates)
```

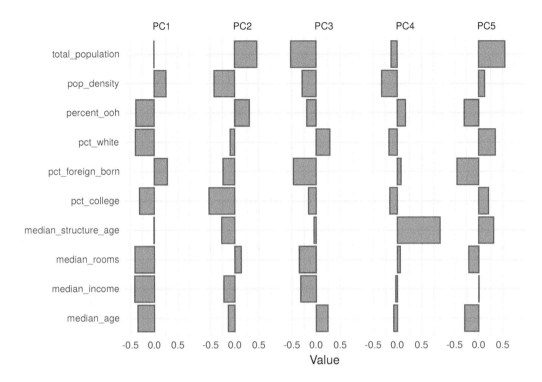

FIGURE 8.6 Loadings for first five principal components

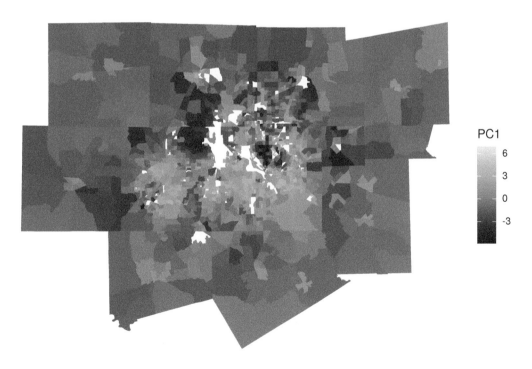

FIGURE 8.7 Map of principal component 1

8.2 Regression modeling with US Census data

```
dfw_pca <- dfw_data_for_model %>%
  select(GEOID, median_value) %>%
  cbind(components)

ggplot(dfw_pca, aes(fill = PC1)) +
  geom_sf(color = NA) +
  theme_void() +
  scale_fill_viridis_c()
```

The map, along with the bar chart, helps us understand how the multiple variables represent latent social processes at play in Dallas-Fort Worth. The brighter yellow areas, which have higher values for PC1, are located in communities like east Fort Worth, east Arlington, Grand Prairie, and south Dallas. Generally speaking, these are low-to-middle income areas with larger nonwhite populations. The locations with the lowest values for PC1 are Southlake (northeast of Fort Worth) and Highland Park (north of downtown Dallas); these communities are segregated, predominantly non-Hispanic white, and are among the wealthiest neighborhoods in the entire United States. In turn, PC1 captures the gradient that represents these social differences, with which multiple demographic characteristics will be associated.

These principal components can be used for *principal components regression*, in which the derived components themselves are used as model predictors. Generally, components should be chosen that account for at least 90 percent of the original variance in the predictors, though this will often be up to the discretion of the analyst. In the example below, we will fit a model using the first six principal components and the log of median home value as the outcome variable.

```
pca_formula <- paste0("log(median_value) ~ ",
                      paste0('PC', 1:6, collapse = ' + '))

pca_model <- lm(formula = pca_formula, data = dfw_pca)

summary(pca_model)

## 
## Call:
## lm(formula = pca_formula, data = dfw_pca)
## 
## Residuals:
##      Min       1Q   Median       3Q      Max 
## -1.78888 -0.16854 -0.00726  0.16941  1.60089 
## 
## Coefficients:
##              Estimate Std. Error  t value Pr(>|t|)
## (Intercept) 12.301439   0.007483 1643.902   <2e-16 ***
## PC1         -0.180706   0.003706  -48.765   <2e-16 ***
## PC2         -0.247181   0.006326  -39.072   <2e-16 ***
## PC3         -0.077097   0.006621  -11.645   <2e-16 ***
## PC4         -0.084417   0.007417  -11.382   <2e-16 ***
## PC5          0.111525   0.008325   13.397   <2e-16 ***
## PC6          0.003787   0.010646    0.356    0.722
```

```
## ---
## Signif. codes:  0 '***' 0.001 '**' 0.01 '*' 0.05 '.' 0.1 ' ' 1
## 
## Residual standard error: 0.2955 on 1552 degrees of freedom
## Multiple R-squared:  0.737,  Adjusted R-squared:  0.736
## F-statistic: 724.9 on 6 and 1552 DF,  p-value: < 2.2e-16
```

The model fit, as represented by the R-squared value, is similar to the models fit earlier in this chapter. One possible disadvantage of principal components regression, however, is the interpretation of the results as the different variables which are comprehensible on their own are now spread across the components. It can be helpful to think of the different components as *indices* in this sense.

As discussed above, PC1 represents a gradient from segregated, older, wealthy, white communities on the low end to more diverse, lower-income, and younger communities on the high end; this PC is negatively associated with median home values, with tracks with expectations. Reviewing the plot above, PC2 is associated with lower population densities and levels of educational attainment; in turn, it can be thought of as an urban (on the low end) to rural (on the high end) gradient. A negative association with median home value is then expected as home values are higher in the urban core than on the rural fringe of the metropolitan area.

8.3 Spatial regression

A core assumption of the linear model is that the errors are independent of one another and normally distributed. Log-transforming the right-skewed outcome variable, median home values, was indented to resolve the latter; we can check this by adding the residuals for `model2` to our dataset and drawing a histogram to check its distribution, shown in Figure 8.8.

```
dfw_data_for_model$residuals <- residuals(model2)

ggplot(dfw_data_for_model, aes(x = residuals)) + 
  geom_histogram(bins = 100, alpha = 0.5, color = "navy",
                 fill = "navy") + 
  theme_minimal()
```

The former assumption of independence of residuals is commonly violated in models that use spatial data, however. This is because models of spatial processes commonly are characterized by *spatial autocorrelation* in the error term, meaning that the model's performance itself depends on geographic location. We can assess this using techniques learned in the previous chapter such as Moran's I.

```
library(spdep)

wts <- dfw_data_for_model %>%
  poly2nb() %>%
  nb2listw()
```

8.3 Spatial regression

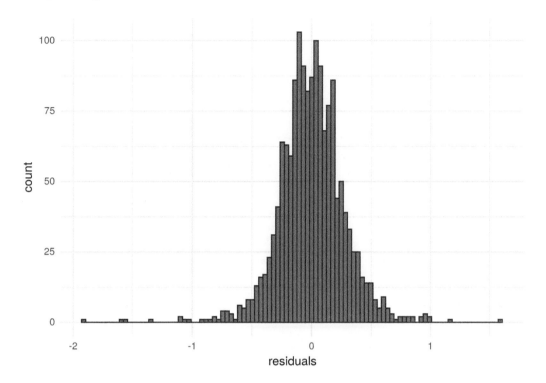

FIGURE 8.8 Distribution of model residuals with a ggplot2 histogram

```
moran.test(dfw_data_for_model$residuals, wts)
```

```
## 
##  Moran I test under randomisation
## 
## data:  dfw_data_for_model$residuals
## weights: wts
## 
## Moran I statistic standard deviate = 14.023, p-value < 2.2e-16
## alternative hypothesis: greater
## sample estimates:
## Moran I statistic       Expectation          Variance
##      0.2101748515       -0.0006418485      0.0002259981
```

The Moran's *I* test statistic is modest and positive (0.21) but is statistically significant. This can be visualized with a Moran scatterplot in Figure 8.9:

```
dfw_data_for_model$lagged_residuals <- lag.listw(wts, dfw_data_for_model$residuals)

ggplot(dfw_data_for_model, aes(x = residuals, y = lagged_residuals)) +
  theme_minimal() +
  geom_point(alpha = 0.5) +
  geom_smooth(method = "lm", color = "red")
```

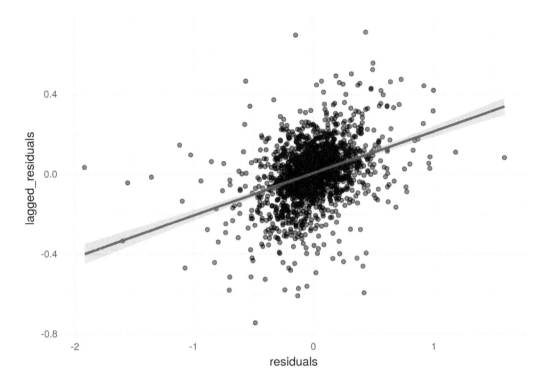

FIGURE 8.9 Moran scatterplot of residual spatial autocorrelation

The plot illustrates the positive spatial autocorrelation in the residuals, suggesting that the assumption of independence in the model error term is violated. To resolve this issue, we can turn to *spatial regression methods*.

8.3.1 Methods for spatial regression

The field of *spatial econometrics* is broadly concerned with the estimation and specification of models that are appropriate for handling spatial dependence in statistical processes. In general, two families of models are used to address these concerns with respect to regression: *spatial lag models* and *spatial error models*.

8.3.1.1 Spatial lag models

Spatial lag models account for spatial dependence by including a spatial lag of the outcome variable in the model. In doing so, it accounts for *spatial spillover effects* – the possibility that values in neighboring areas have an influence on values in a given location. A spatial lag model can be written as follows (Rey et al., 2020):

$$Y_i = \alpha + \rho Y_{lag-i} + \sum_k \beta_k X_{ki} + \epsilon_i,$$

where

$$Y_{lag-i} = \sum_j w_{ij} Y_j$$

8.3 Spatial regression

with w_{ij} representing the spatial weights. In this notation, ρ is the parameter measuring the effect of the spatial lag in the outcome variable, and k is the number of predictors in the model. However, the inclusion of a spatially lagged outcome variable on the right-hand side of the equation violates the exogeneity assumption of the linear model. In turn, special methods are required for estimating the spatial lag model, which are implemented in R in the **spatialreg** package (Bivand et al., 2013). Below, we use the function lagsarlm() to estimate the relationship between logged median home value and its predictors as a spatial lag model:

```
library(spatialreg)

lag_model <- lagsarlm(
  formula = formula2,
  data = dfw_data_for_model,
  listw = wts
)

summary(lag_model, Nagelkerke = TRUE)

##
## Call:lagsarlm(formula = formula2, data = dfw_data_for_model, listw = wts)
##
## Residuals:
##        Min         1Q      Median         3Q        Max
## -2.0647421 -0.1377312 -0.0032552  0.1386914  1.4820482
##
## Type: lag
## Coefficients: (asymptotic standard errors)
##                       Estimate  Std. Error z value  Pr(>|z|)
## (Intercept)          7.0184e+00  2.6898e-01 26.0927 < 2.2e-16
## median_rooms         6.2027e-02  8.8554e-03  7.0045 2.480e-12
## pct_college          1.2858e-02  5.4696e-04 23.5083 < 2.2e-16
## pct_foreign_born     2.0118e-03  7.7482e-04  2.5964  0.009420
## pct_white            2.7112e-03  4.7183e-04  5.7461 9.133e-09
## median_age           3.4421e-03  1.3163e-03  2.6150  0.008922
## median_structure_age 2.6093e-05  2.4827e-05  1.0510  0.293267
## percent_ooh         -3.0428e-03  5.4316e-04 -5.6021 2.118e-08
## pop_density         -1.3573e-05  5.9323e-06 -2.2879  0.022143
## total_population     8.3762e-06  4.2928e-06  1.9512  0.051031
##
## Rho: 0.35319, LR test value: 210.86, p-value: < 2.22e-16
## Asymptotic standard error: 0.023376
##     z-value: 15.109, p-value: < 2.22e-16
## Wald statistic: 228.29, p-value: < 2.22e-16
##
## Log likelihood: -125.2882 for lag model
## ML residual variance (sigma squared): 0.067179, (sigma: 0.25919)
## Nagelkerke pseudo-R-squared: 0.79193
## Number of observations: 1559
## Number of parameters estimated: 12
```

```
## AIC: 274.58, (AIC for lm: 483.43)
## LM test for residual autocorrelation
## test value: 6.9225, p-value: 0.0085118
```

The general statistical relationships observed in the non-spatial model are preserved in the spatial lag model, though the effect sizes (the model parameters) are smaller, illustrating the importance of controlling for the spatial lag. Additionally, the ρ parameter is positive and statistically significant, suggesting the presence of spatial spillover effects. This finding makes practical sense, as median home values may be influenced by the values of homes in neighboring Census tracts along with the characteristics of the neighborhood itself. The argument `Nagelkerke = TRUE` computes a pseudo-R-squared value, which is slightly higher than the corresponding value for the non-spatial model.

8.3.1.2 Spatial error models

In contrast with spatial lag models, spatial error models include a spatial lag in a model's error term. This is designed to capture latent spatial processes that are not currently being accounted for in the model estimation and in turn show up in the model's residuals. The spatial error model can be written as follows:

$$Y_i = \alpha + \sum_k \beta_k X_{ki} + u_i,$$

where

$$u_i = \lambda u_{lag-i} + \epsilon_i$$

and

$$u_{lag-i} = \sum_j w_{ij} u_j$$

Like the spatial lag model, estimating the spatial error model requires special methods, implemented with the `errorsarlm()` function in **spatialreg**.

```
error_model <- errorsarlm(
  formula = formula2,
  data = dfw_data_for_model,
  listw = wts
)

summary(error_model, Nagelkerke = TRUE)

##
## Call:errorsarlm(formula = formula2, data = dfw_data_for_model, listw = wts)
##
## Residuals:
##        Min         1Q     Median         3Q        Max
## -1.97990245 -0.13702534 -0.00030105  0.13933507  1.54937871
##
## Type: error
## Coefficients: (asymptotic standard errors)
```

8.3 Spatial regression

```
##                            Estimate  Std. Error   z value   Pr(>|z|)
## (Intercept)              1.1098e+01  6.6705e-02  166.3753  < 2.2e-16
## median_rooms             8.2815e-02  9.7089e-03    8.5298  < 2.2e-16
## pct_college              1.5857e-02  5.7427e-04   27.6120  < 2.2e-16
## pct_foreign_born         3.6601e-03  9.6570e-04    3.7901  0.0001506
## pct_white                4.6754e-03  6.1175e-04    7.6426  2.132e-14
## median_age               3.9346e-03  1.4130e-03    2.7845  0.0053605
## median_structure_age     2.6093e-05  2.5448e-05    1.0254  0.3051925
## percent_ooh             -4.7538e-03  5.6726e-04   -8.3803  < 2.2e-16
## pop_density             -1.4999e-05  6.8731e-06   -2.1823  0.0290853
## total_population         1.0497e-05  4.4668e-06    2.3499  0.0187796
##
## Lambda: 0.46765, LR test value: 164.17, p-value: < 2.22e-16
## Asymptotic standard error: 0.031997
##     z-value: 14.615, p-value: < 2.22e-16
## Wald statistic: 213.61, p-value: < 2.22e-16
##
## Log likelihood: -148.6309 for error model
## ML residual variance (sigma squared): 0.067878, (sigma: 0.26053)
## Nagelkerke pseudo-R-squared: 0.7856
## Number of observations: 1559
## Number of parameters estimated: 12
## AIC: 321.26, (AIC for lm: 483.43)
```

The λ (lambda) value is large and statistically significant, again illustrating the importance of accounting for spatial autocorrelation in the model.

8.3.2 Choosing between spatial lag and spatial error models

The spatial lag and spatial error models offer alternative approaches to accounting for processes of spatial autocorrelation when fitting models. This raises the question: which one of the two models should the analyst choose? On the one hand, this should be thought through in the context of the topic under study. For example, if spatial spillover effects are related to the hypotheses being evaluated by the analysts (e.g. the effect of neighboring home values on focal home values), a spatial lag model may be preferable; alternatively, if there are spatially autocorrelated factors that likely influence the outcome variable but are difficult to measure quantitatively (e.g. discrimination and racial bias in the housing market), a spatial error model might be preferred.

The two types of models can also be evaluated with respect to some quantitative metrics. For example, we can re-compute Moran's I over the model residuals to see if the spatial model has resolved our problems with spatial dependence. First, we'll check the spatial lag model:

```
moran.test(lag_model$residuals, wts)
```

```
##
##  Moran I test under randomisation
##
## data:  lag_model$residuals
## weights: wts
```

```
## 
## Moran I statistic standard deviate = 2.0436, p-value = 0.0205
## alternative hypothesis: greater
## sample estimates:
## Moran I statistic      Expectation         Variance
##      0.0300648348    -0.0006418485     0.0002257748
```

Next, the spatial error model:

```
moran.test(error_model$residuals, wts)
```

```
## 
##  Moran I test under randomisation
## 
## data:  error_model$residuals
## weights: wts
## 
## Moran I statistic standard deviate = -1.6126, p-value = 0.9466
## alternative hypothesis: greater
## sample estimates:
## Moran I statistic      Expectation         Variance
##     -0.0248732656    -0.0006418485     0.0002257849
```

Both models reduce Moran's I; however, the error model does a better job of eliminating spatial autocorrelation in the residuals entirely. We can also use *Lagrange multiplier tests* to evaluate the appropriateness of these models together (Anselin et al., 1996). These tests check for spatial error dependence, whether a spatially lagged dependent variable is missing, and the robustness of each in the presence of the other.

The `lm.LMtests()` function can be used with an input linear model to compute these tests. We'll use `model2`, the home value model with median household income omitted, to compute the tests.

```
lm.LMtests(
  model2,
  wts,
  test = c("LMerr", "LMlag", "RLMerr", "RLMlag")
)
```

```
## 
##  Lagrange multiplier diagnostics for spatial dependence
## 
## data:
## model: lm(formula = formula2, data = dfw_data_for_model)
## weights: wts
## 
## LMerr = 194.16, df = 1, p-value < 2.2e-16
## 
## 
##  Lagrange multiplier diagnostics for spatial dependence
## 
```

```
## data:
## model: lm(formula = formula2, data = dfw_data_for_model)
## weights: wts
##
## LMlag = 223.37, df = 1, p-value < 2.2e-16
##
##
##   Lagrange multiplier diagnostics for spatial dependence
##
## data:
## model: lm(formula = formula2, data = dfw_data_for_model)
## weights: wts
##
## RLMerr = 33.063, df = 1, p-value = 8.921e-09
##
##
##   Lagrange multiplier diagnostics for spatial dependence
##
## data:
## model: lm(formula = formula2, data = dfw_data_for_model)
## weights: wts
##
## RLMlag = 62.276, df = 1, p-value = 2.998e-15
```

All test statistics are large and statistically significant; in this case, the robust versions of the statistics should be compared. While both the lag and error models would be appropriate for this data, the test statistic for the robust version of the lag model is larger, suggesting that the spatial lag model should be preferred over the spatial error model in this example.

8.4 Geographically weighted regression

The models addressed in the previous sections – both the regular linear model and its spatial adaptations – estimate *global* relationships between the outcome variable, median home values, and its predictors. This lends itself to conclusions like "In the Dallas-Fort Worth metropolitan area, higher levels of educational attainment are associated with higher median home values." However, metropolitan regions like Dallas-Fort Worth are diverse and multifaceted. It is possible that a relationship between a predictor and the outcome variable that is observed for the entire region *on average* may vary significantly from neighborhood to neighborhood. This type of phenomenon is called *spatial non-stationarity*, and can be explored with *geographically weighted regression*, or GWR (Brunsdon et al., 1996).

GWR is a technique designed to evaluate local variations in the results of regression models given a kernel (distance-decay) weighting function. Following Lu et al. (2014), the basic form of GWR for a given location i can be written as:

$$Y_i = \alpha_i + \sum_{k=1}^{m} \beta_{ik} X_{ik} + \epsilon_i$$

where the model intercept, parameters, and error term are all location-specific. Notably, β_{ik} represents a *local regression coefficient* for predictor k (of the total number of predictors m) that is specific to location i.

GWR is implemented in the **GWmodel** R package (Gollini et al., 2015) as well as the spgwr package (Bivand and Yu, 2020). These packages offer an interface to a wider family of geographically weighted methods, such as binomial, generalized linear, and robust geographically weighted regression; geographically weighted PCA; and geographically weighted summary statistics. The example below will adapt the regression model used in earlier examples to a locally-variation model with GWR.

8.4.1 Choosing a bandwidth for GWR

GWR relies on the concept of a "kernel bandwidth" to compute the local regression model for each location. A kernel bandwidth is based on the kernel type (fixed or adaptive) and a distance-decay function. A fixed kernel uses a cutoff distance to determine which observations will be included in the local model for a given location i, whereas an adaptive kernel uses the nearest neighbors to a given location. In most circumstances with Census tract data where the size of tracts in a region will vary widely, an adaptive kernel will be preferred to a fixed kernel to ensure consistency of neighborhoods across the region. The distance-decay function then governs how observations will be weighted in the local model relative to their distance from location i. Closer tracts to i will have a greater influence on the results for location i, with influence falling with distance.

Bandwidth sizes (either a distance cutoff or number of nearest neighbors) can be selected directly by the user; in the **GWmodel** R package, the `bw.gwr()` function also helps analysts choose an appropriate kernel bandwidth using cross-validation. The code below computes a bandwidth bw using this method. Note that we must first convert our data to a legacy `SpatialPolygonsDataFrame` object from the sp package as **GWmodel** does not yet support sf objects.

```
library(GWmodel)
library(sf)

dfw_data_sp <- dfw_data_for_model %>%
  as_Spatial()

bw <- bw.gwr(
  formula = formula2,
  data = dfw_data_sp,
  kernel = "bisquare",
  adaptive = TRUE
)
```

8.4 Geographically weighted regression

`bw.gwr()` chose 187 as the number of nearest neighbors based on cross-validation. This means that for each Census tract, the nearest 187 of the total 1559 Census tracts in the Dallas-Fort Worth region will be used to estimate the local model, with weights calculated using the bisquare distance-decay function as follows:

$$w_{ij} = 1 - (\frac{d_{ij}^2}{h^2})^2$$

where d_{ij} is the distance between local observation i and neighbor j, and h is the kernel bandwidth. As we are using an adaptive kernel, h will vary for each observation and take on the distance between location i and its "neighbor" furthest from that location.

8.4.2 Fitting and evaluating the GWR model

The basic form of GWR can be fit with the `gwr.basic()` function, which uses a similar argument structure to other models fit in this chapter. The formula can be passed to the `formula` parameter as a character string; we'll use the original formula with median household income omitted and include it below for a refresher. The derived bandwidth from `bw.gwr()` will be used with a bisquare, adaptive kernel.

```
formula2 <- paste0("log(median_value) ~ median_rooms + pct_college + ",
                  "pct_foreign_born + pct_white + median_age + ",
                  "median_structure_age + percent_ooh + pop_density + ",
                  "total_population")

gw_model <- gwr.basic(
  formula = formula2,
  data = dfw_data_sp,
  bw = bw,
  kernel = "bisquare",
  adaptive = TRUE
)
```

Printing the object `gw_model` will show both the results of the global model and the ranges of the locally varying parameter estimates. The model object itself has the following elements:

```
names(gw_model)
```

```
## [1] "GW.arguments"  "GW.diagnostic" "lm"           "SDF"
## [5] "timings"       "this.call"     "Ftests"
```

Each element provides information about the model fit, but perhaps most interesting to the analyst is the `SDF` element, in this case a `SpatialPolygonsDataFrame` containing mappable model results. We will extract that element from the model object and convert it to simple features, then take a look at the columns in the object.

```
gw_model_results <- gw_model$SDF %>%
  st_as_sf()

names(gw_model_results)
```

```
##  [1] "Intercept"              "median_rooms"
##  [3] "pct_college"            "pct_foreign_born"
##  [5] "pct_white"              "median_age"
##  [7] "median_structure_age"   "percent_ooh"
##  [9] "pop_density"            "total_population"
## [11] "y"                      "yhat"
## [13] "residual"               "CV_Score"
## [15] "Stud_residual"          "Intercept_SE"
## [17] "median_rooms_SE"        "pct_college_SE"
## [19] "pct_foreign_born_SE"    "pct_white_SE"
## [21] "median_age_SE"          "median_structure_age_SE"
## [23] "percent_ooh_SE"         "pop_density_SE"
## [25] "total_population_SE"    "Intercept_TV"
## [27] "median_rooms_TV"        "pct_college_TV"
## [29] "pct_foreign_born_TV"    "pct_white_TV"
## [31] "median_age_TV"          "median_structure_age_TV"
## [33] "percent_ooh_TV"         "pop_density_TV"
## [35] "total_population_TV"    "Local_R2"
## [37] "geometry"
```

The sf object includes columns with local parameter estimates, standard errors, and t-values for each predictor, along with local diagnostic elements such as the Local R-squared, giving information about how well a model performs in any particular location. In Figure 8.11, we use ggplot2 and geom_sf() to map the local R-squared, though it may be useful as well to use mapview::mapview(gw_model_results, zcol = "Local_R2") and explore these results interactively.

```
ggplot(gw_model_results, aes(fill = Local_R2)) + 
  geom_sf(color = NA) + 
  scale_fill_viridis_c() + 
  theme_void()
```

The map suggests that the model performs very well in Fort Worth, Collin County, and the eastern edge of the metropolitan area, with local R-squared values exceeding 0.9. It performs worse in northwestern Denton County and in other more rural areas to the west of Fort Worth.

We can examine locally varying parameter estimates in much the same way. The first map below visualizes the local relationships between the percentage owner-occupied housing and median home values. Recall from the global model that this coefficient was negative and statistically significant.

```
ggplot(gw_model_results, aes(fill = percent_ooh)) + 
  geom_sf(color = NA) + 
  scale_fill_viridis_c() + 
  theme_void() + 
  labs(fill = "Local ? for \npercent_ooh")
```

The dark purple areas on the map in Figure 8.10 are those areas where the global relationship in the model reflects the local relationship, as local parameter estimates are negative. The areas that stick out include the high-density area of uptown Dallas, where renter-occupied

8.4 Geographically weighted regression

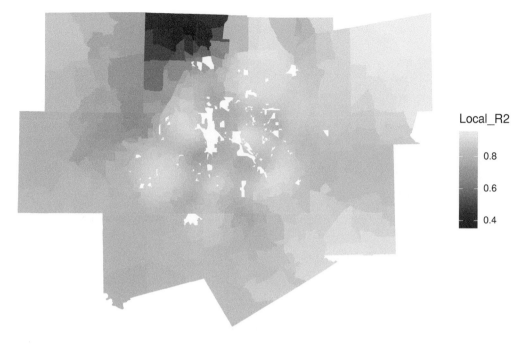

FIGURE 8.10 Local R-squared values from the GWR model

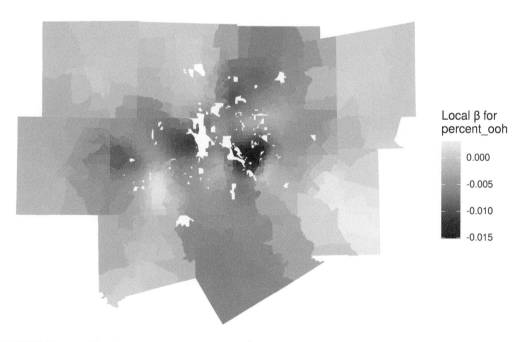

FIGURE 8.11 Local parameter estimates for percent owner-occupied housing

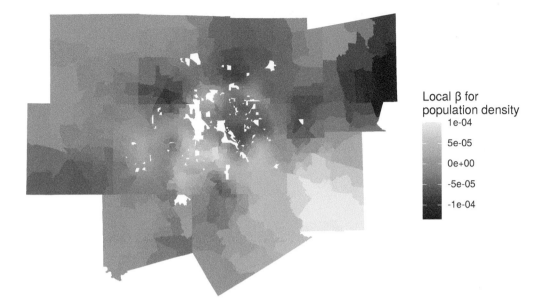

FIGURE 8.12 Local parameter estimates for population density

housing is common and median home values are very high. However, in rural areas on the fringe of the metropolitan area this relationship reverses, returning in some cases positive parameter estimates (the yellow parts of the map). This means that for those local areas, a greater percentage of owner-occupied housing is associated with higher home values.

We can explore this further by investigating the local parameter estimates for population density, which was not significant in the global model:

```
ggplot(gw_model_results, aes(fill = pop_density)) + 
  geom_sf(color = NA) + 
  scale_fill_viridis_c() + 
  theme_void() + 
  labs(fill = "Local ? for \npopulation density")
```

For large portions of the metropolitan area, the relationship between population density and median home values is negligible, which is what drives the global relationship. However, we do observe variations in different parts of the metropolitan area. Bright yellow locations are those where high population densities are associated with higher home values. Conversely, the darker blue and purple areas represent several affluent enclaves or suburbs of Dallas and Fort Worth where lower densities are associated with higher home values.

8.4.3 Limitations of GWR

While GWR is an excellent method for exploring spatial non-stationarity in regression model results, it does have some limitations. When the **spgwr** package itself *is loaded into your R environment*, it prints the following warning:

> NOTE: This package does not constitute approval of GWR as a method of spatial analysis.

Why would an R package itself warn the user about its use? GWR is particularly susceptible to problems that plague other regression models using spatial data. Earlier sections in this chapter covered the topic of *collinearity*, where parameter estimates are biased due to high correlations between predictors. Given that predictor values tend to cluster spatially, GWR models often suffer from *local multicollinearity* where predictors are highly correlated in local areas.

Additionally, the impact of *edge effects* can be acute in GWR models. Edge effects – which are present in most spatial models and analysis techniques – refer to misleading results for observations on the edge of a dataset. In the example used in this chapter, the Dallas-Fort Worth metropolitan area represents the region under study. This artificially restricts the neighborhoods around Census tracts on the edge of the metropolitan area to only those tracts *that are also within the metropolitan area*, and omits the rural tracts that border them. For local models, this is a particular problem as results for observations on the edge of an area are based on incomplete information that may not reflect the true circumstances of that location.

With these concerns in mind, GWR is generally recommended as an *exploratory technique* that serves as a useful companion to the estimation of a global model. For example, a global parameter estimate may suggest that in Dallas-Fort Worth, median home values tend to be higher in areas with lower percentages of owner-occupied housing, controlling for other predictors in the model. GWR, used as a follow-up, helps the analyst understand that this global relationship is driven by higher home values near to urban cores such as in the uptown Dallas area, and may not necessarily characterize the dynamics of rural and exurban areas elsewhere in the metropolitan region.

8.5 Classification and clustering of ACS data

The statistical models discussed earlier in this chapter were fit for the purpose of understanding relationships between an outcome variable and a series of predictors. In the social sciences, such models are generally used for *inference*, where a researcher tests those hypotheses with respect to those relationships to understand social processes. In industry, regression models are commonly used instead for *prediction*, where a model is trained on the relationship between an observed outcome and predictors then used to make predictions on out-of-sample data. In machine learning terminology, this is referred to as *supervised learning*, where the prediction target is known.

In other cases, the researcher is interested in discovering the structure of a dataset and generating meaningful labels for it rather than making predictions based on a known outcome. This type of approach is termed *unsupervised machine learning*. This section will explore two common applications of unsupervised machine learning with respect to demographic data: *geodemographic classification*, which identifies "clusters" of similar areas based on common demographic characteristics, and *regionalization*, which partitions an area into salient *regions* that are both spatially contiguous and share common demographic attributes.

8.5.1 Geodemographic classification

Geodemographic classification refers to the grouping of geographic observations based on similar demographic (or other) characteristics (Singleton and Spielman, 2013). It is commonly used to generate neighborhood "typologies" that can help explain general similarities and differences among neighborhoods in a broader region. While the geodemographic approach has been criticized for essentializing neighborhoods (Goss, 1995), it is also widely used to understand dynamics of urban systems (Vicino et al., 2011) and has been proposed as a possible solution to problems with large margins of error for individual variables in the ACS (Spielman and Singleton, 2015). In industry, geodemographics are widely used for marketing and customer segmentation purposes. Popular frameworks include Esri's Tapestry Segmentation[2] and Experian's Mosaic product[3].

While the exact methodology to produce a geodemographic classification system varies from implementation to implementation, the general process used involves *dimension reduction* applied to a high-dimensional input dataset of model features, followed by a *clustering algorithm* to partition observations into groups based on the derived dimensions. As we have already employed principal components analysis for dimension reduction on the Dallas-Fort Worth dataset, we can re-use those components for this purpose.

The k-means clustering algorithm is one of the most common unsupervised algorithms used to partition data in this way. K-means works by attempting to generate K clusters that are internally similar but dissimilar from other clusters. Following James et al. (2013) and Boehmke and Greenwell (2019), the goal of K-means clustering can be written as:

$$\underset{C_1 \ldots C_k}{\text{minimize}} \left\{ \sum_{k=1}^{K} W(C_k) \right\}$$

where the within-cluster variation $W(C_k)$ is computed as

$$W(C_k) = \sum_{x_i \in C_k} (x_i - \mu_k)^2$$

with x_i representing an observation in the cluster C_k and μ_k representing the mean value of all observations in cluster C_k.

To compute k-means, the analyst must first choose the number of desired clusters, represented with k. The analyst then specifies k initial "centers" from the data (this is generally done at random) to seed the algorithm. The algorithm then iteratively assigns observations to clusters until the total within-cluster variation is minimized, returning a cluster solution.

In R, k-means can be computed with the `kmeans()` function. This example solution will generate 6 cluster groups. Given that the algorithm relies on random seeding of the cluster centers, `set.seed()` should be used to ensure stability of the solution.

---

[2] https://www.esri.com/en-us/arcgis/products/data/data-portfolio/tapestry-segmentation
[3] https://www.experian.com/marketing-services/consumer-segmentation

8.5 Classification and clustering of ACS data

```
set.seed(1983)

dfw_kmeans <- dfw_pca %>%
  st_drop_geometry() %>%
  select(PC1:PC8) %>%
  kmeans(centers = 6)

table(dfw_kmeans$cluster)
```

```
##
##   1   2   3   4   5   6
## 456 193 172  83 228 427
```

The algorithm has partitioned the data into six clusters; the smallest (Cluster 4) has 83 Census tracts, whereas the largest (Cluster 3) has 456 Census tracts. At this stage, it is useful to explore the data in both *geographic space* and *variable space* to understand how the clusters differ from one another. We can assign the cluster ID to the original dataset as a new column and map it with geom_sf().

```
dfw_clusters <- dfw_pca %>%
  mutate(cluster = as.character(dfw_kmeans$cluster))

ggplot(dfw_clusters, aes(fill = cluster)) +
  geom_sf(size = 0.1) +
  scale_fill_brewer(palette = "Set1") +
  theme_void() +
  labs(fill = "Cluster ")
```

Some notable geographic patterns in the clusters are evident from the map in Figure 8.13, even if the viewer does not have local knowledge of the Dallas-Fort Worth region. Cluster 1 represents more rural communities on the edges of the metropolitan area, whereas Cluster 6 tends to be located in the core counties of Tarrant and Dallas as well as higher-density tracts in outer counties. Cluster 2 covers both big-city downtowns of Fort Worth and Dallas along with a scattering of suburban tracts.

A useful companion visualization to the map is a color-coded scatterplot using two of the principal components in the PCA dataset, shown in Figure 8.14. We will use the two components discussed as "indices" in Section 8.2.5: PC1, which is a gradient from affluent/older/white to lower-income/younger/nonwhite, and PC2, which represents areas with high population densities and educational attainment on the low end to lower-density, less educated areas on the high end. Given the data density, ggplotly() from the **plotly** package will convert the scatterplot to a graphic with an interactive legend, allowing the analyst to turn cluster groups on and off.

```
library(plotly)

cluster_plot <- ggplot(dfw_clusters,
                       aes(x = PC1, y = PC2, color = cluster)) +
  geom_point() +
  scale_color_brewer(palette = "Set1") +
```

FIGURE 8.13 Map of geodemographic clusters in Dallas-Fort Worth

```
  theme_minimal()

ggplotly(cluster_plot) %>%
  layout(legend = list(orientation = "h", y = -0.15,
                       x = 0.2, title = "Cluster"))
```

While all clusters overlap one another to some degree, each occupies distinct feature space. Double-click any cluster in the legend to isolate it. Cluster 1, which covers much of the rural fringe of Dallas-Fort Worth, scores very high on the "rurality" index PC2 (low density / educational attainment) and modestly negative on the "diversity" index PC1. Cluster 2, which includes the two downtowns, scores lower on the "rurality" index PC2 but scores higher on the "diversity" index PC1. A geodemographic analyst may adopt these visualization approaches to explore the proposed typology in greater depth, and aim to produce informative "labels" for each cluster.

8.5.2 Spatial clustering & regionalization

The geodemographic classification outlined in the previous section offers a useful methodology for identifying similar types of Census tracts in varying parts of a metropolitan region. However, this approach was *aspatial* in that it did not take the geographic properties of the Census tracts into account. In other applications, an analyst may want to generate meaningful clusters that are constrained to be neighboring or contiguous areas. An application of this workflow might be sales territory generation, where sales representatives will be assigned to communities in which they have local market knowledge but also want to minimize overall travel time.

8.5 Classification and clustering of ACS data

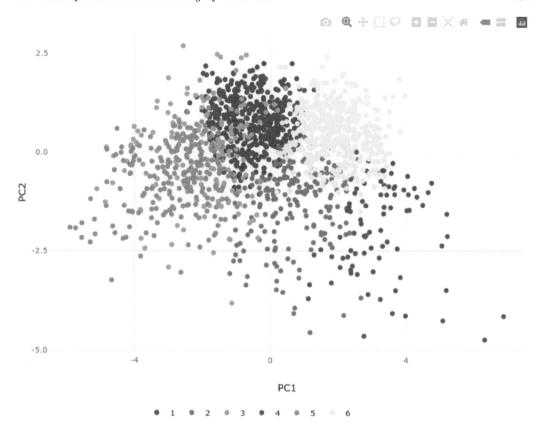

FIGURE 8.14 Interactive scatterplot of PC1 and PC2 colored by cluster

A suite of *regionalization* algorithms are available that adapt the clustering approach by introducing spatial constraints. A spatial constraint might be an additional requirement that the algorithm minimize overall geographic distance between observations, or even that the derived clusters must be geographically contiguous. The latter scenario is explored in this subsection.

The workflow below illustrates the SKATER algorithm (Assunção et al., 2006), an acronym that stands for "Spatial 'K'luster Analysis by Tree Edge Removal." This algorithm is implemented in R with the skater() function in the **spdep** package and is also available in PySAL, GeoDa, and ArcGIS as the "Spatially Constrained Multivariate Clustering[4]" tool.

SKATER relies on the concept of *minimum spanning trees*, where a connectivity graph is drawn between all observations in the dataset with graph edges weighted by the attribute similarity between observations. The graph is then "pruned" by removing edges that connect observations that are not similar to one another.

The setup for SKATER involves similar steps to clustering algorithms used in Chapter 7, as a queens-case contiguity weights matrix is generated. The key differences include the use of *costs* – which represent the differences between neighbors based on an input set of variables,

---
[4]https://pro.arcgis.com/en/latest/tool-reference/spatial-statistics/how-spatially-constrained-multivariate-clustering-works.htm

which in this example will be principal components 1 through 8 – and the use of a binary weights matrix with style = "B".

```
library(spdep)

input_vars <- dfw_pca %>%
  select(PC1:PC8) %>%
  st_drop_geometry() %>%
  as.data.frame()

skater_nbrs <- poly2nb(dfw_pca, queen = TRUE)
costs <- nbcosts(skater_nbrs, input_vars)
skater_weights <- nb2listw(skater_nbrs, costs, style = "B")
```

Once the weights have been generated, a minimum spanning tree is created with mstree(), and used in the call to skater().

```
mst <- mstree(skater_weights)

regions <- skater(
  mst[,1:2],
  input_vars,
  ncuts = 7,
  crit = 10
)
```

The ncuts parameter dictates how many times the algorithm should prune the minimum spanning tree; a value of 7 will create 8 groups. The crit parameter is used to determine the minimum number of observations per group; above, we have set this value to 10, requiring that each region will have at least 10 Census tracts.

The solution can be extracted and assigned to the spatial dataset, then visualized with geom_sf():

```
dfw_clusters$region <- as.character(regions$group)

ggplot(dfw_clusters, aes(fill = region)) +
  geom_sf(size = 0.1) +
  scale_fill_brewer(palette = "Set1") +
  theme_void()
```

Figure 8.15 shows how the algorithm has partitioned the data into eight contiguous regions. These regions are largely geographic in nature (region 5 covers the northwestern portion of the metropolitan area, whereas region 2 covers the east and south), but they also incorporate demographic variations in the data. For example, region 6 covers downtown and uptown Dallas along with the Bishop Arts neighborhood; these represent the highest-density and most traditionally "urban" parts of the metropolitan area. Additionally, region 4 represents the "northeast Tarrant County" suburban community along with similar suburbs in Denton County, which is a socially meaningful sub-region in north Texas.

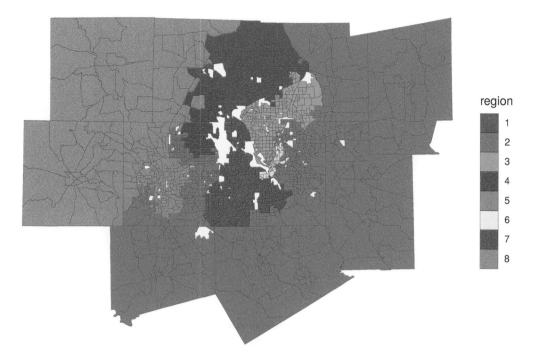

FIGURE 8.15 Map of contiguous regions derived with the SKATER algorithm

8.6 Exercises

Identify a different region of the United States of interest to you. Complete the following tasks:

1. Acquire race/ethnicity data from **tidycensus** for your chosen region and compute the dissimilarity index. How does segregation in your chosen region compare with urban areas in California?
2. Reproduce the regression modeling workflow outlined in this chapter for your chosen region. Is residual spatial autocorrelation more, or less, of an issue for your region than in Dallas-Fort Worth?
3. Create a geodemographic classification for your region using the sample code in this chapter. Does the typology you've generated resemble that of Dallas-Fort Worth, or does it differ?

9

Introduction to Census microdata

The previous chapters in this book focus on aggregate-level analysis of US Census Bureau data. However, such analyses are limited to the pre-tabulated estimates provided by the Census Bureau. While these estimates are voluminous, they may not include the level of detail required by researchers, and they are limited to analyses appropriate for aggregate-level data. In turn, many researchers turn to Census *microdata*, which are anonymized individual-level Census records, to help answer demographic questions. In 2020, **tidycensus** added support for American Community Survey microdata along with a series of tools to assist with analysis of these datasets. The next two chapters provide an overview of this functionality in **tidycensus** and help users get started analyzing and modeling ACS microdata appropriately.

9.1 What is "microdata?"

Microdata refer to individual-level data made available to researchers. In many cases, microdata reflect responses to surveys that are de-identified and anonymized, then prepared in datasets that include rich detail about survey responses. US Census microdata are available for both the decennial Census and the American Community Survey; these datasets, named the Public Use Microdata Series (PUMS)[1], allow for detailed cross-tabulations not available in aggregated data.

The ACS PUMS is available, like the aggregate data, in both 1-year and 5-year versions. The 1-year PUMS covers about 1 percent of the US population, whereas the 5-year PUMS covers about 5 percent; this means that microdata represent a smaller subset of the US population than the regular ACS. Public use microdata downloads available in bulk from the Census FTP server[2] or from data.census.gov's MDAT tool[3].

The Census Bureau also operates a network of Federal Statistical Research Data Centers[4] (FSRDCs) around the country that grant access to microdata with larger sample sizes and greater demographic detail. To work at one of these centers, researchers must get special government clearance and have an approved proposal with the US Census Bureau. This and the following chapter focus on the public use microdata product, which is much more accessible to researchers and analysts.

---

[1] https://www.census.gov/programs-surveys/acs/microdata.html
[2] https://www2.census.gov/programs-surveys/acs/data/pums/2019/5-Year/
[3] https://data.census.gov/mdat/#/search?ds=ACSPUMS5Y2019
[4] https://www.census.gov/about/adrm/fsrdc.html

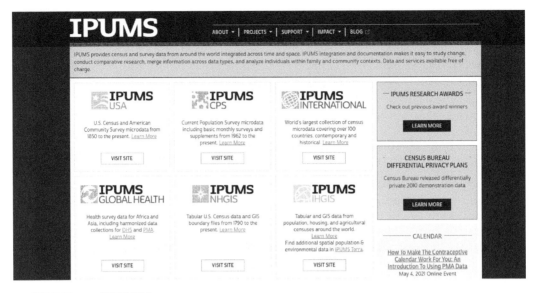

FIGURE 9.1 IPUMS home page

9.1.1 Microdata resources: IPUMS

One of the most popular and comprehensive repositories for research microdata is the University of Minnesota's IPUMS project (Ruggles et al., 2020). IPUMS includes Decennial US Census and ACS microdata (IPUMS USA), microdata from the Current Population Survey (IPUMS CPS), and over 100 countries around the world (IPUMS International).

IPUMS releases microdata that are *harmonized,* which means that changing variable definitions over time are aligned by the IPUMS team to allow for coherent longitudinal analysis. Using IPUMS requires signing up for an account and making a request through their web interface, then downloading a data extract; an API is under development[5]. IPUMS data products will be covered in more detail in Chapter 11, which will also introduce the **ipumsr** R package (Ellis and Burk, 2020) for working with IPUMS data in R.

9.1.2 Microdata and the Census API

The migration of US Census data from American FactFinder to the data.census.gov tool integrated the Census Bureau's data download interface with its API. The Census Bureau's MDAT tool allows for flat file downloads of microdata along with API queries for microdata, marking the first time that microdata are available via the API.

This means that microdata can be accessed with `httr::GET()` requests in R, but also made ACS microdata accessible to **tidycensus**. In 2020, tidycensus released a range of features to support ACS microdata for R users; this functionality is covered in the remainder of this chapter.

---

[5] https://developer.ipums.org/docs/apiprogram/

9.2 Using microdata in tidycensus

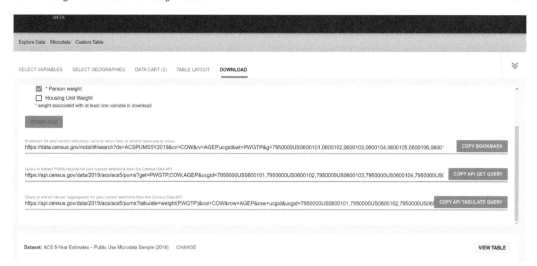

FIGURE 9.2 MDAT tool from data.census.gov

9.2 Using microdata in tidycensus

American Community Survey microdata are available in **tidycensus** by using the `get_pums()` function, which communicates with the Census API much like other **tidycensus** functions and returns PUMS data. Given the unique properties of Census microdata and the different structure of individual-level records as opposed to aggregate data, the data returned by `get_pums()` differs from other **tidycensus** functions. This section covers the basics of requesting microdata extracts with **tidycensus**.

9.2.1 Basic usage of `get_pums()`

`get_pums()` requires specifying one or more variables and the state for which you'd like to request data. For national-level analyses, `state = 'all'` *can* get data for the entire USA by iterating over all US states, but the data can take some time to download depending on the user's internet connection. The `get_pums()` function defaults to the 5-year ACS with `survey = "acs5"`; 1-year ACS data is available with `survey = "acs1"`. At the time of this writing, data are available from 2006 through 2019 for the 1-year ACS and 2005-2009 through 2016-2020 for the 5-year ACS.

Let's take a look at a first example using `get_pums()` to request microdata for Wyoming from the 1-year 2019 ACS with information on sex, age (`AGEP`), and household type (`HHT`).

```
library(tidycensus)

wy_pums <- get_pums(
  variables = c("SEX", "AGEP", "HHT"),
  state = "WY",
  survey = "acs1",
```

TABLE 9.1 1-year ACS PUMS data for Wyoming

SERIALNO	SPORDER	WGTP	PWGTP	AGEP	ST	HHT	SEX
2019GQ0000335	1	0	206	17	56	b	2
2019GQ0000958	1	0	50	37	56	b	1
2019GQ0009156	1	0	33	75	56	b	2
2019GQ0012426	1	0	23	21	56	b	2
2019GQ0015243	1	0	209	94	56	b	1
2019GQ0018773	1	0	273	26	56	b	1
2019GQ0019978	1	0	6	16	56	b	1
2019GQ0022931	1	0	67	18	56	b	2
2019GQ0032421	1	0	113	94	56	b	2
2019GQ0033045	1	0	10	44	56	b	1

```
  year = 2019
)
```

The function returns just under 6,000 rows of data with our requested variables in the columns. However, a few other variables are also returned that we did not request; these default variables are covered below.

9.2.2 Understanding default data from `get_pums()`

`get_pums()` returns some technical variables by default without the user needing to request them specifically. These technical variables are essential for uniquely identifying observations in the dataset and eventually performing any analysis and modeling. These default technical variables include:

- `SERIALNO`: a serial number that uniquely identifies households in the sample;
- `SPORDER`: the order of the person in the household, which when combined with `SERIALNO` uniquely identifies a person;
- `WGTP`: the household weight;
- `PWGTP`: the person weight;
- `ST`: the state FIPS code.

Given that PUMS data are a *sample* of the US population, the weights columns must be used for analysis. In general terms, we can interpret the weights as "the number of observations in the general population represented by this particular row in the dataset." In turn, a row with a `PWGTP` value of 50 represents about 50 people in Wyoming with the same demographic characteristics of the "person" in that row.

Inferences about population characteristics can be made by summing over the weights columns. For example, let's say we want to get an estimate of the number of people in Wyoming who are 50 years old in 2019, and compare this with the total population in Wyoming. We can filter the dataset for rows that match the condition `AGEP == 50`, then sum over the `PWGTP` column.

9.2 Using microdata in tidycensus

```
library(tidyverse)

wy_age_50 <- filter(wy_pums, AGEP == 50)

print(sum(wy_pums$PWGTP))
```

```
## [1] 578759
```

```
print(sum(wy_age_50$PWGTP))
```

```
## [1] 4756
```

Our data suggest that of the 578,759 people in Wyoming in 2019, about 4,756 were 50 years old. Of course, this estimate will be subject to a margin of error; the topic of error calculations from PUMS data will be covered in the next chapter.

It is important to note that `get_pums()` returns two separate weights columns: one for households and one for persons. Let's take a look at a single household in the Wyoming dataset to examine this further.

```
wy_hh_example <- filter(wy_pums, SERIALNO == "2019HU0456721")

wy_hh_example
```

```
## # A tibble: 4 x 8
##   SERIALNO      SPORDER  WGTP PWGTP  AGEP ST    HHT   SEX
##   <chr>           <dbl> <dbl> <dbl> <dbl> <chr> <chr> <chr>
## 1 2019HU0456721       1   146   146    40 56    1     2
## 2 2019HU0456721       2   146   132    45 56    1     1
## 3 2019HU0456721       3   146    94     8 56    1     2
## 4 2019HU0456721       4   146   154     5 56    1     1
```

This household includes a woman aged 40, a man aged 45, and two children: a girl aged 8 and a boy aged 5. The `HHT` value is 1, which tells us that this is a married-couple household. Notably, the household weight value, `WGTP`, is identical for all household members, whereas the person weight value, `PWGTP`, is not.

Microdata retrieved from the Census API are a hybrid of both household-level data and person-level data, which means that analysts need to take care to use appropriate weights and filters for household-level or person-level analyses. For example, to determine the number of households in Wyoming, the dataset should be filtered to records where the `SPORDER` column is equal to 1 then summed over the `WGTP` column. Persons living in group quarters will be excluded automatically as they have a household weight of 0.

```
wy_households <- filter(wy_pums, SPORDER == 1)

sum(wy_households$WGTP)
```

```
## [1] 233126
```

TABLE 9.2 1-year ACS PUMS data with vacant housing units

SERIALNO	SPORDER	WGTP	PWGTP	AGEP	ST	VACS	HHT
2019HU0094511	NA	157	NA	NA	56	1	b
2019HU0271067	NA	268	NA	NA	56	1	b
2019HU0466134	NA	158	NA	NA	56	1	b
2019HU0650356	NA	243	NA	NA	56	1	b
2019HU0707994	NA	30	NA	NA	56	1	b
2019HU0720346	NA	100	NA	NA	56	1	b
2019HU0885957	NA	170	NA	NA	56	1	b
2019HU0905994	NA	45	NA	NA	56	1	b
2019HU0928583	NA	25	NA	NA	56	1	b
2019HU0929765	NA	57	NA	NA	56	1	b

Housing unit rather than simply household-level analyses introduce an additional level of complexity, as housing units can be both occupied and vacant. Vacant housing units are returned in a different format from the Census API which makes them a special case in **tidycensus**. To return vacant housing units along with person and household records, use the argument return_vacant = TRUE.

```
wy_with_vacant <- get_pums(
  variables = c("SEX", "AGEP", "HHT"),
  state = "WY",
  survey = "acs1",
  year = 2019,
  return_vacant = TRUE
) %>%
  arrange(VACS)
```

Vacant housing units are included in the dataset, but as they do not have person-level characteristics (due to their lack of occupancy) all person-level variables like age and sex have values of NA.

9.3 Working with PUMS variables

While the ACS PUMS dataset does not include tens of thousands of variables choices like its aggregate counterpart, it nonetheless includes variables and variable codes that can be difficult to understand without a data dictionary. In the Wyoming example above, only interpretation of the AGEP column for age is straightforward. HHT, for household type, and SEX, for sex, are coded as integers represented as character strings. To help users understand the meanings of these codes, **tidycensus** includes a built-in dataset, pums_variables, that can be viewed, filtered, and browsed.

9.3 Working with PUMS variables

9.3.1 Variables available in the ACS PUMS

As with the data dictionaries for the decennial Census and aggregate ACS obtained with `load_variables()`, it is advisable to browse the PUMS data dictionary, `pums_variables`, with the `View()` function in RStudio.

```
View(pums_variables)
```

`pums_variables` is a long-form dataset that organizes specific *value codes* by variable so you know what you can get with `get_pums()`. You'll use information in the `var_code` column to fetch variables, but pay close attention to the `var_label`, `val_min`, `val_max`, `val_label`, and `data_type` columns. These columns should be interpreted as follows:

- `var_code` gives you the variable codes that should be supplied to the `variables` parameter (as a character vector) in `get_pums()`. These variables will be represented in the columns of your output dataset.
- `var_label` is a more informative description of the variable's topic.
- `data_type` is one of `"chr"`, for categorical variables that will be returned as R character strings, or `"num"`, for variables that will be returned as numeric.
- `val_min` and `val_max` provide information about the meaning of the data values. For categorical variables, these two columns will be the same; for numeric variables, they will give you the possible range of data values.
- `val_label` contains the value labels, which are particularly important for understanding the content of categorical variables.

9.3.2 Recoding PUMS variables

A typical **tidycensus** workflow covered earlier in this book involves browsing the appropriate data dictionary, choosing variable IDs, and using those IDs in your scripts and workflows. Analysts will likely follow this same process with `get_pums()`, but can also use the argument `recode = TRUE` to return additional contextual information with the requested data. `recode = TRUE` instructs `get_pums()` to append recoded columns to your returned dataset based on information available in `pums_variables`. Let's take a look at the Wyoming example with `recode = TRUE`.

```
wy_pums_recoded <- get_pums(
  variables = c("SEX", "AGEP", "HHT"),
  state = "WY",
  survey = "acs1",
  year = 2019,
  recode = TRUE
)
```

Note that the dataset returns three new columns: `ST_label`, `HHT_label`, and `SEX_label` which include longer and more informative descriptions of the value labels. These columns are returned as ordered factors that preserve the original ordering of the columns independent of their alphabetical order. Numeric columns like `AGEP` are not recoded as the data values reflect numbers, not a categorical label.

TABLE 9.3 Recoded PUMS data for Wyoming

SERIALNO	SPORDER	WGTP	PWGTP	AGEP	ST	HHT	SEX
2019GQ0000523	1	0	14	16	56	b	1
2019GQ0002701	1	0	58	36	56	b	1
2019GQ0002909	1	0	72	82	56	b	2
2019GQ0010377	1	0	68	18	56	b	2
2019GQ0016100	1	0	68	35	56	b	1
2019GQ0018538	1	0	11	54	56	b	2
2019GQ0019066	1	0	49	26	56	b	1
2019GQ0019370	1	0	61	18	56	b	1
2019GQ0020847	1	0	140	35	56	b	1
2019GQ0020895	1	0	75	18	56	b	1

9.3.3 Using variables filters

PUMS datasets, especially those from the 5-year ACS, can get quite large. Even users with speedy internet connections will need to be patient when downloading what could be millions of records from the Census API and potentially risk internet hiccups. When only subsets of data are required for an analysis, the `variables_filter` argument can return a subset of data from the API, reducing long download times.

The `variables_filter` argument should be supplied as a named list where variable names (which can be quoted or unquoted) are paired with a data value or vector of data values to be requested from the API. The "filter" works by passing a special query to the Census API which will only return a subset of data, meaning that the entire dataset does not need to be first downloaded then filtered on the R side. This leads to substantial time savings for targeted queries.

In the example below, the Wyoming request is modified with `variables_filter` to return only women (`SEX = 2`) between the ages of 30 and 49, but this time from the 5-year ACS PUMS.

```
wy_pums_filtered <- get_pums(
  variables = c("SEX", "AGEP", "HHT"),
  state = "WY",
  survey = "acs5",
  variables_filter = list(
    SEX = 2,
    AGEP = 30:49
  ),
  year = 2019
)
```

The returned dataset reflects the filter request, with data values in the AGEP column only ranging between 30 and 49 and the SEX column only including 2, for female.

TABLE 9.4 Filtered extract of Wyoming PUMS data

SERIALNO	SPORDER	WGTP	PWGTP	AGEP	ST	HHT	SEX
2015000002861	1	62	63	46	56	6	2
2015000007968	5	3	3	33	56	2	2
2015000009972	2	22	55	32	56	2	2
2015000010667	1	11	12	38	56	1	2
2015000022223	1	31	31	47	56	1	2
2015000025344	1	35	35	38	56	3	2
2015000027928	1	55	54	45	56	1	2
2015000037505	1	10	10	48	56	1	2
2015000054619	1	8	8	41	56	1	2
2015000065987	1	14	14	47	56	6	2

9.4 Public Use Microdata Areas (PUMAs)

One of the steps the Census Bureau takes to preserve anonymity in the PUMS datasets is limiting the geographical detail in the data. Granular Census geographical information like the Census tract or block group of residence for individuals in the PUMS samples are not available. That said, some geographical information is available in the PUMS samples in the form of the Public Use Microdata Area, or PUMA.

9.4.1 What is a PUMA?

Public Use Microdata Areas (PUMAs) are the smallest available geographies at which records are identifiable in the PUMS datasets. PUMAs are redrawn with each decennial US Census, and typically are home to between 100,000 and 200,000 people when drawn, although some may be much larger by the end of a Census cycle. In large cities, a PUMA will represent a collection of nearby neighborhoods; in rural areas, it might represent several counties across a large area of a state.

At the time of this writing, PUMA geographies correspond to the 2010 Census definitions. While PUMAs are redrawn with each decennial US Census, their release lags the decennial Census by a couple of years. To determine appropriate PUMA geographies, the Census Bureau consults with State Data Centers (SDCs) and incorporates their suggested revisions to these geographies. For 2020 PUMAs, this process took place through the 2021 calendar year. Updated 2020 PUMA geographies were released in mid-2022, and will be incorporated into ACS data products starting in 2023. For more information, visit https://www.census.gov/programs-surveys/geography/guidance/geo-areas/pumas/2020pumas.html.

PUMA geographies can be obtained and reviewed with the pumas() function in the **tigris** package. Let's take a look at PUMA geographies for the state of Wyoming in Figure 9.3:

```
library(tigris)
options(tigris_use_cache = TRUE)

wy_pumas <- pumas(state = "WY", cb = TRUE, year = 2019)
```

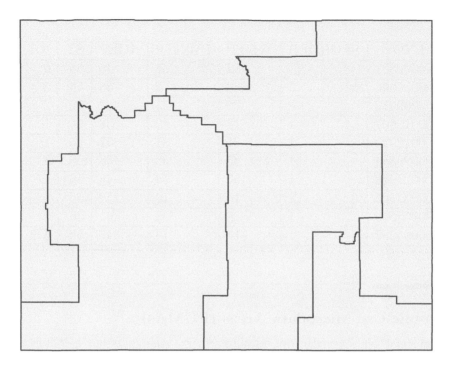

FIGURE 9.3 Basic map of PUMAs in Wyoming

```
ggplot(wy_pumas) +
  geom_sf() +
  theme_void()
```

There are five PUMAs in Wyoming, largely covering large rural areas of the state, although the smallest PUMA by area covers the more urban southeast corner of the state. The returned object includes a NAME10 column with an informative description of the PUMAs:

```
wy_pumas$NAME10
```

```
## [1] "Sheridan, Park, Teton, Lincoln & Big Horn Counties"
## [2] "Sweetwater, Fremont, Uinta, Sublette & Hot Springs Counties--
Wind River Reservation"
## [3] "Campbell, Goshen, Platte, Johnson, Washakie, Weston, Crook & Niobrara
    Counties"
## [4] "Natrona, Carbon & Converse Counties"
## [5] "Laramie & Albany Counties"
```

In a denser urban area, PUMAs will reflect subsections of major cities and are drawn in attempts to reflect meaningful local areas. In New York City, for example, PUMAs are drawn to align with recognized community districts in the city as shown in Figure 9.4.

9.4 Public Use Microdata Areas (PUMAs)

FIGURE 9.4 Map of PUMAs in New York City

```
nyc_pumas <- pumas(state = "NY", cb = TRUE, year = 2019) %>%
  filter(str_detect(NAME10, "NYC"))

ggplot(nyc_pumas) +
  geom_sf() +
  theme_void()
```

The names of NYC's PUMAs themselves reflect the community districts in the geographic data returned by **tigris**.

```
nyc_pumas$NAME10[1:5]
```

```
## [1] "NYC-Manhattan Community District 11--East Harlem"
## [2] "NYC-Manhattan Community District 8--Upper East Side"
## [3] "NYC-Bronx Community District 11--Pelham Parkway, Morris Park & Laconia"
## [4] "NYC-Manhattan Community District 9--Hamilton Heights, Manhattanville &
   West Harlem"
## [5] "NYC-Brooklyn Community District 12--Borough Park, Kensington & Ocean Parkway"
```

9.4.2 Working with PUMAs in PUMS data

PUMA information is available with the variable code `PUMA` in `get_pums()`. Use `PUMA` like any other variable to return information about the PUMA of residence for the individual records.

TABLE 9.5 Wyoming microdata with PUMA and age information

SERIALNO	SPORDER	WGTP	PWGTP	AGEP	PUMA	ST
2015000004982	1	28	29	39	00200	56
2015000004982	2	28	25	36	00200	56
2015000004982	3	28	29	10	00200	56
2015000004982	4	28	25	8	00200	56
2015000005881	1	27	28	74	00100	56
2015000007222	1	0	8	61	00100	56
2015000008820	1	0	12	28	00400	56
2015000008875	1	18	18	61	00300	56
2015000008875	2	18	19	58	00300	56
2015000008875	3	18	22	14	00300	56

```
wy_age_by_puma <- get_pums(
  variables = c("PUMA", "AGEP"),
  state = "WY",
  survey = "acs5",
  year = 2019
)
```

PUMA IDs are replicated across states, so the PUMA column should be combined with the ST column to uniquely identify PUMAs when performing multi-state analyses.

The puma argument in get_pums() can also be used to obtain data for a specific PUMA or multiple PUMAs. Like the variables_filter parameter, puma uses a query on the API side to reduce long download times for users only interested in a geographical subset of data.

```
wy_puma_subset <- get_pums(
  variables = "AGEP",
  state = "WY",
  survey = "acs5",
  puma = "00500",
  year = 2019
)
```

For multi-state geographical queries, the puma argument must be adapted slightly due to the aforementioned possibility that PUMA IDs will be replicated across states. To perform a multi-state query by PUMA, specify state = "multiple" and pass a named vector of state/PUMA pairs to the puma parameter.

```
twostate_puma_subset <- get_pums(
  variables = "AGEP",
  state = "multiple",
  survey = "acs5",
  puma = c("WY" = "00500", "UT" = "05001"),
```

TABLE 9.6 Microdata for PUMA 00500 in Wyoming

SERIALNO	SPORDER	WGTP	PWGTP	AGEP	PUMA	ST
2015000007382	1	17	17	33	00500	56
2015000014560	1	3	3	56	00500	56
2015000017683	1	17	16	54	00500	56
2015000017683	2	17	7	54	00500	56
2015000024415	1	13	13	74	00500	56
2015000024415	2	13	15	68	00500	56
2015000024421	1	6	7	47	00500	56
2015000035380	1	4	4	53	00500	56
2015000035380	2	4	3	62	00500	56
2015000035380	3	4	5	29	00500	56

TABLE 9.7 Data for PUMA 00500 in Wyoming and PUMA 05001 in Utah

SERIALNO	SPORDER	WGTP	PWGTP	AGEP	PUMA	ST
2015000000582	1	17	17	55	05001	49
2015000000582	2	17	14	51	05001	49
2015000000582	3	17	23	22	05001	49
2015000007011	1	71	71	26	05001	49
2015000007011	2	71	25	49	05001	49
2015000007011	3	71	32	26	05001	49
2015000007382	1	17	17	33	00500	56
2015000014560	1	3	3	56	00500	56
2015000017683	1	17	16	54	00500	56
2015000017683	2	17	7	54	00500	56

```
  year = 2019
)
```

The returned data include neighboring areas in Wyoming and Utah.

9.5 Exercises

- Try requesting PUMS data using `get_pums()` yourselves, but for a state other than Wyoming.
- Use the `pums_variables` dataset to browse the available variables in the PUMS. Create a custom query with `get_pums()` to request data for variables other than those we've used in the above examples.

10
Analyzing Census microdata

A major benefit of using the individual-level microdata returned by `get_pums()` is the ability to create detailed, granular estimates of ACS data. While the aggregate ACS data available with `get_acs()` includes tens of thousands of indicators to choose from, researchers and analysts still may be interested in cross-tabulations not available in the aggregate files. Additionally, microdata helps researchers design statistical models to assess demographic relationships at the individual level in a way not possible with aggregate data.

Analysts must pay careful attention to the structure of the PUMS datasets in order to produce accurate estimates and handle errors appropriately. PUMS datasets are *weighted* samples, in which each person or household is not considered unique or individual but rather representative of multiple other persons or households. In turn, analyses and tabulations using PUMS data must use appropriate tools for handling weighting variables to accurately produce estimates. Fortunately, tidyverse tools like **dplyr**, covered elsewhere in this book, are excellent for producing these tabulations and handling survey weights

As covered in Chapter 3, data from the American Community Survey are based on a sample and in turn characterized by error. This means that ACS data acquired with `get_pums()` are similarly characterized by error, which can be substantial when cross-tabulations are highly specific. Fortunately, the US Census Bureau provides *replicate weights* to help analysts generate standard errors around tabulated estimates with PUMS data as they take into account the complex structure of the survey sample. While working with replicate weights has traditionally been cumbersome for analysts, **tidycensus** with help from the **survey** (Lumley, 2010) and **srvyr** (Freedman Ellis and Schneider, 2021) R packages has integrated tools for handling replicate weights and correctly estimating standard errors when tabulating and modeling data. These workflows will be covered later in this chapter.

10.1 PUMS data and the tidyverse

As discussed in Chapter 9, `get_pums()` automatically returns data with both household (`WGTP`) and person (`PWGTP`) weights. These weights can loosely be interpreted as the number of households or persons represented by each individual row in the PUMS data. Appropriate use of these weights columns is essential for tabulating accurate estimates of population characteristics with PUMS data. Fortunately, weighted tabulations work quite well within familiar tidyverse workflows, such as those covered in Chapter 3.

10.1.1 Basic tabulation of weights with tidyverse tools

Let's get some basic sample PUMS data from the 2016-2020 ACS for Mississippi with information on sex and age.

TABLE 10.1 PUMS data for Mississippi

SERIALNO	SPORDER	WGTP	PWGTP	AGEP	ST	SEX	ST_label
2016000000411	1	54	54	30	28	1	Mississippi/MS
2016000000411	2	54	95	22	28	2	Mississippi/MS
2016000000739	1	27	26	51	28	1	Mississippi/MS
2016000000739	2	27	17	17	28	2	Mississippi/MS
2016000000803	1	3	3	30	28	2	Mississippi/MS
2016000000803	2	3	4	8	28	1	Mississippi/MS
2016000000858	1	9	9	90	28	1	Mississippi/MS
2016000000858	2	9	24	63	28	1	Mississippi/MS
2016000000901	1	16	16	70	28	1	Mississippi/MS
2016000000901	2	16	24	65	28	2	Mississippi/MS

```r
library(tidycensus)
library(tidyverse)

ms_pums <- get_pums(
  variables = c("SEX", "AGEP"),
  state = "MS",
  survey = "acs5",
  year = 2020,
  recode = TRUE
)
```

As we learned in Chapter 10, the number of people in Mississippi can be tabulated by summing over the person-weight column:

```r
sum(ms_pums$PWGTP)
```

```
## [1] 2981835
```

We can perform similar calculations with tidyverse tools. The count() function in the dplyr package performs a simple tabulation over a dataset. The optional wt argument in count() allows you to specify a weight column, which in this case will be our person-weight.

```r
ms_pums %>% count(wt = PWGTP)
```

```
## # A tibble: 1 x 1
##         n
##     <dbl>
## 1 2981835
```

count() has the additional benefit of allowing for the specification of one our more columns that will be grouped and tabulated. For example, we could tabulate data by unique values of age and sex in Mississippi. The wt argument in count() specifies the PWGTP column as the appropriate weight for data tabulation.

10.1 PUMS data and the tidyverse

```
ms_pums %>%
  count(SEX_label, AGEP, wt = PWGTP)
```

```
## # A tibble: 186 x 3
##    SEX_label  AGEP     n
##    <ord>     <dbl> <dbl>
##  1 Male          0 18111
##  2 Male          1 19206
##  3 Male          2 18507
##  4 Male          3 18558
##  5 Male          4 20054
##  6 Male          5 17884
##  7 Male          6 18875
##  8 Male          7 18775
##  9 Male          8 19316
## 10 Male          9 20866
## # ... with 176 more rows
```

We can also perform more custom analyses, such as tabulating the number of people over age 65 by sex in Mississippi. This involves specifying a filter condition to retain rows for records with an age of 65 and up, then tabulating by sex.

```
ms_pums %>%
  filter(AGEP >= 65) %>%
  count(SEX, wt = PWGTP)
```

```
## # A tibble: 2 x 2
##   SEX        n
##   <chr>  <dbl>
## 1 1     206504
## 2 2     267707
```

We can then use `get_acs()` to check our answer:

```
get_acs(geography = "state",
        state = "MS",
        variables = c("DP05_0030", "DP05_0031"),
        year = 2020)
```

```
## # A tibble: 2 x 5
##   GEOID NAME        variable  estimate   moe
##   <chr> <chr>       <chr>        <dbl> <dbl>
## 1 28    Mississippi DP05_0030   206518   547
## 2 28    Mississippi DP05_0031   267752   466
```

We notice that our tabulations are very close to the ACS estimates available in `get_acs()`, and well within the margin of error. When we are doing tabulations with microdata, it is important to remember that we are tabulating data based on a smaller subsample of information than is available to the aggregate ACS estimates. In turn, as the US Census Bureau reminds us (American Community Survey Office, 2021):

Because PUMS data consist of a subset of the full ACS sample, tabulations from the ACS PUMS will not match those from published tables of ACS data.

Analysts will often want to use PUMS data and the tabulated aggregate ACS data in tandem as appropriate, as each data type offers complimentary strengths. As the aggregate ACS data are based on a larger sample, its data aggregations will be preferable to those produced with PUMS data. However, PUMS data offer the ability to compute detailed cross-tabulations not available in aggregate ACS tables and to fit models of demographic relationships at the individual level. Examples of each follow in this chapter.

10.1.2 Group-wise data tabulation

When combined with tidyverse tools as introduced in Chapter 3, PUMS data can produce highly detailed estimates not available in the regular aggregate ACS. The example below acquires data on rent burden, family type, and race/ethnicity to examine intersections between these variables for households in Mississippi. The PUMA variable is also included for use later in this chapter.

Our guiding research question is as follows: how does rent burden vary by race/ethnicity and household type for Mississippi households? This requires obtaining data on rent burden (gross rent as percentage of household income) with variable GRPIP; race and ethnicity with variables RAC1P and HISP; and household type with variable HHT. The variables_filter argument is used to filter the sample to only renter-occupied households paying cash rent, speeding download times.

```
hh_variables <- c("PUMA", "GRPIP", "RAC1P",
                  "HISP", "HHT")

ms_hh_data <- get_pums(
  variables = hh_variables,
  state = "MS",
  year = 2020,
  variables_filter = list(
    SPORDER = 1,
    TEN = 3
  ),
  recode = TRUE
)
```

A subset of the data is shown in Table 10.2.

To analyze rent burdens with respect to the marital status and race/ethnicity of the householder, it will be useful to do some additional recoding using **dplyr**'s case_when() function. A new race_ethnicity column will identify householders by general categories, and a married column will identify whether or not the household is a married-couple household.

10.1 PUMS data and the tidyverse

TABLE 10.2 Household microdata for Mississippi

SERIALNO	SPORDER	WGTP	PWGTP	GRPIP	PUMA	ST	TEN
2019HU0266269	1	14	14	101	01100	28	3
2019HU0266529	1	18	17	27	02000	28	3
2019HU0266712	1	105	105	57	02000	28	3
2019HU0268193	1	13	13	18	00500	28	3
2019HU0268829	1	1	1	22	00400	28	3

```
ms_hh_recoded <- ms_hh_data %>%
  mutate(
    race_ethnicity = case_when(
      HISP != "01" ~ "Hispanic",
      HISP == "01" & RAC1P == "1" ~ "White",
      HISP == "01" & RAC1P == "2" ~ "Black",
      TRUE ~ "Other"
    ),
    married = case_when(
      HHT == "1" ~ "Married",
      TRUE ~ "Not married"
    )
  )
```

This information can then be summarized with respect to the household weight variable WGTP and the rent burden variable GRPIP within a group_by() %>% summarize() workflow. The dataset is filtered to only non-Hispanic white, non-Hispanic Black, and Hispanic householders to focus on those groups, then grouped by race/ethnicity and marital status. Within the summarize() call, the percentage of each subgroup paying 40 percent or more of their household incomes in rent is calculated by summing over the household weight column WGTP, but filtering for households with rent burdens of 40 percent or more in the numerator.

```
ms_hh_summary <- ms_hh_recoded %>%
  filter(race_ethnicity != "Other") %>%
  group_by(race_ethnicity, married) %>%
  summarize(
    prop_above_40 = sum(WGTP[GRPIP >= 40]) / sum(WGTP)
  )
```

TABLE 10.3 Tabulated PUMS data for Mississippi

race_ethnicity	married	prop_above_40
Black	Married	0.1625791
Black	Not married	0.4080033
Hispanic	Married	0.1716087
Hispanic	Not married	0.3569935
White	Married	0.1266644
White	Not married	0.3356546

FIGURE 10.1 Basic plot of PUMAs in Mississippi

The demographic group in this example with the largest rent burden is Black, Not married; nearly 41 percent of households in this group pay over 40 percent of their incomes in gross rent. The least rent-burdened group is White, Married, with a value under 13 percent. For each of the three racial/ethnic groups, there is a distinctive financial advantage for married-couple households over non-married households; this is particularly pronounced for Black householders.

10.2 Mapping PUMS data

In the previous example, we see that rent burdens for Black, unmarried households are particularly acute in Mississippi. A follow-up question may involve an examination of how this trend varies geographically. As discussed in the previous chapter, the most granular geography available in the PUMS data is the PUMA, which generally includes 100,000-200,000 people. PUMA geographies are available in the **tigris** package with the function pumas().

```
library(tigris)
library(tmap)
options(tigris_use_cache = TRUE)

ms_pumas <- pumas("MS", year = 2020)

plot(ms_pumas$geometry)
```

A geographical visualization of rent burdens in Mississippi requires a slight adaptation of the above code. Instead of returning a comparative table, the dataset should also be grouped by the PUMA column then filtered for the combination of variables that represent the group the analyst wants to visualize. In this case, the focus is on unmarried Black households by PUMA.

```
ms_data_for_map <- ms_hh_recoded %>%
  group_by(race_ethnicity, married, PUMA) %>%
  summarize(
    percent_above_40 = 100 * (sum(WGTP[GRPIP >= 40]) / sum(WGTP))
  ) %>%
  filter(race_ethnicity == "Black",
         married == "Not married")
```

The output dataset has one row per PUMA and is suitable for joining to the spatial dataset for visualization.

```
library(tmap)

joined_pumas <- ms_pumas %>%
  left_join(ms_data_for_map, by = c("PUMACE10" = "PUMA"))

tm_shape(joined_pumas) +
  tm_polygons(col = "percent_above_40",
              palette = "Reds",
              title = "% rent-burdened\nunmarried Black households") +
  tm_layout(legend.outside = TRUE,
            legend.outside.position = "right")
```

The map in Figure 10.2 illustrates geographic variations in our indicator of interest. In particular, unmarried Black households are particularly rent-burdened along the Gulf Coast, with over half of households paying at least 40 percent of their household incomes in gross rent. The least rent-burdened areas for this demographic group are in the suburban PUMAs around Jackson.

10.3 Survey design and the ACS PUMS

As earlier chapters have addressed, the American Community Survey is based on a *sample* of the US population and in turn subject to sampling error. This becomes particularly acute when dealing with small sub-populations like those explored at the PUMA level in the previous section. Given that PUMS data are individual-level records and not aggregates, standard errors and in turn margins of error must be computed by the analyst. Doing so correctly requires accounting for the complex sample design of the ACS. Fortunately, **tidycensus** with help from the **survey** and **srvyr** packages includes tools to assist with these tasks.

10.3.1 Getting replicate weights

The Census Bureau recommends using the *Successive Difference Replication* method to compute standard errors around derived estimates from PUMS data. To calculate standard errors, the Census Bureau publishes 80 "replicate weights" for each observation, representing either person (PWGTP1 through PWGTP80) or household (WGTP1 through WGTP80)

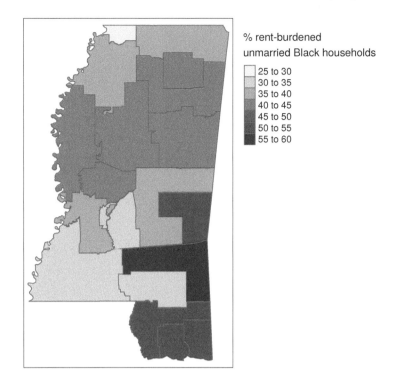

FIGURE 10.2 Map of rent-burdened unmarried Black household share by PUMA in Mississippi

weights (American Community Survey Office, 2021). The formula for computing the standard error SE for a derived PUMS estimate x is as follows:

$$SE(x) = \sqrt{\frac{4}{80}\sum_{r=1}^{80}(x_r - x)^2}$$

where x is the PUMS estimate and x_r is the rth weighted estimate.

With respect to SDR standard errors, the PUMS documentation acknowledges (p. 12),

---

Successive Difference Replication (SDR) standard errors and margins of error are expected to be more accurate than generalized variance formulas (GVF) standard errors and margins of error, although they may be more inconvenient for some users to calculate.

The "inconvenience" is generally due to the need to download 80 additional weighting variables and prepare the equation written above. The rep_weights parameter in get_pums() makes it easier for users to retrieve the replicate weights variables without having to request

10.3 Survey design and the ACS PUMS

all 80 directly. In a call to `get_pums()`, an analyst can use `rep_weights = "person"` for person-weights, `"housing"` for household weights, or `"both"` to get both sets.

The code below re-downloads the Mississippi rent burden dataset used above, but with household replicate weights included.

```
ms_hh_replicate <- get_pums(
  variables = c("TEN", hh_variables),
  state = "MS",
  recode = TRUE,
  year = 2020,
  variables_filter = list(
    SPORDER = 1
  ),
  rep_weights = "housing"
)
```

```
names(ms_hh_replicate)
```

```
##  [1] "SERIALNO"    "SPORDER"     "GRPIP"       "PUMA"        "ST"
##  [6] "TEN"         "HHT"         "HISP"        "RAC1P"       "ST_label"
## [11] "TEN_label"   "HHT_label"   "HISP_label"  "RAC1P_label" "WGTP"
## [16] "PWGTP"       "WGTP1"       "WGTP2"       "WGTP3"       "WGTP4"
## [21] "WGTP5"       "WGTP6"       "WGTP7"       "WGTP8"       "WGTP9"
## [26] "WGTP10"      "WGTP11"      "WGTP12"      "WGTP13"      "WGTP14"
## [31] "WGTP15"      "WGTP16"      "WGTP17"      "WGTP18"      "WGTP19"
## [36] "WGTP20"      "WGTP21"      "WGTP22"      "WGTP23"      "WGTP24"
## [41] "WGTP25"      "WGTP26"      "WGTP27"      "WGTP28"      "WGTP29"
## [46] "WGTP30"      "WGTP31"      "WGTP32"      "WGTP33"      "WGTP34"
## [51] "WGTP35"      "WGTP36"      "WGTP37"      "WGTP38"      "WGTP39"
## [56] "WGTP40"      "WGTP41"      "WGTP42"      "WGTP43"      "WGTP44"
## [61] "WGTP45"      "WGTP46"      "WGTP47"      "WGTP48"      "WGTP49"
## [66] "WGTP50"      "WGTP51"      "WGTP52"      "WGTP53"      "WGTP54"
## [71] "WGTP55"      "WGTP56"      "WGTP57"      "WGTP58"      "WGTP59"
## [76] "WGTP60"      "WGTP61"      "WGTP62"      "WGTP63"      "WGTP64"
## [81] "WGTP65"      "WGTP66"      "WGTP67"      "WGTP68"      "WGTP69"
## [86] "WGTP70"      "WGTP71"      "WGTP72"      "WGTP73"      "WGTP74"
## [91] "WGTP75"      "WGTP76"      "WGTP77"      "WGTP78"      "WGTP79"
## [96] "WGTP80"
```

All 80 household replicate weights are included in the dataset. A key distinction in the above code, however, is that the housing tenure variable TEN is not included in the `variables_filter` argument, instead returning the full sample of households in Mississippi. This is because standard error estimation for complex survey samples requires special methods for *subpopulations*, which will be covered below.

10.3.2 Creating a survey object

With replicate weights in hand, analysts can turn to a suite of tools in R for handling complex survey samples. The **survey** package (Lumley, 2010) is the standard for handling these types of datasets in R. The more recent **srvyr** package (Freedman Ellis and Schneider, 2021) wraps **survey** to allow the use of tidyverse functions on survey objects. Both packages return a survey class object that intelligently calculates standard errors when data are tabulated with appropriate functions. **tidycensus** includes a function, to_survey(), to convert ACS microdata to **survey** or **srvyr** objects in a way that incorporates the recommended formula for SDR standard error calculation with replicate weights.

```r
library(survey)
library(srvyr)

ms_hh_svy <- ms_hh_replicate %>%
  to_survey(type = "housing",
            design = "rep_weights") %>%
  filter(TEN == 3)

class(ms_hh_svy)
```

```
## [1] "tbl_svy"      "svyrep.design"
```

The to_survey() function returns the original dataset as an object of class tbl_svy and svyrep.design with minimal hassle.

Note the use of filter() after converting the replicate weights dataset to a survey object to subset the data to only renter-occupied households paying cash rent. When computing standard errors for derived estimates using complex survey samples, it is necessary to take the entire structure of the sample into account. In turn, it is important to *first* convert the dataset into a survey object *and then* identify the "subpopulation" for which the model will be fit.

For analysis of subpopulations, srvyr::filter() works like survey::subset() for appropriate standard error estimation. This data structure will then be taken into account when calculating standard errors.

10.3.3 Calculating estimates and errors with srvyr

srvyr's survey_*() family of functions automatically calculates standard errors around tabulated estimates using tidyverse-equivalent functions. For example, analogous to the use of count() to tabulate weighted data, survey_count() will do the same for a survey object but will also return appropriately-calculated standard errors.

```r
ms_hh_svy %>%
  survey_count(PUMA, HHT_label)
```

The survey_count() function returns tabulations for each household type by PUMA in Mississippi along with the estimate's standard error. The **srvyr** package can also accommodate more complex workflows. Below is an adaptation of the rent burden analysis computed above, but using the **srvyr** function survey_mean().

```r
ms_svy_summary <- ms_hh_svy %>%
  mutate(
    race_ethnicity = case_when(
```

10.3 Survey design and the ACS PUMS

TABLE 10.4 Tabulated PUMS data for household types in Mississippi by PUMA with standard errors

PUMA	HHT_label	n	n_se
00100	Married couple household	5579	494.9350
00100	Other family household: Male householder, no spouse present	1474	236.1260
00100	Other family household: Female householder, no spouse present	3684	312.6579
00100	Nonfamily household: Male householder: Living alone	1814	251.1036
00100	Nonfamily household: Male householder: Not living alone	710	162.1937

```
      HISP != "01" ~ "Hispanic",
      HISP == "01" & RAC1P == "1" ~ "White",
      HISP == "01" & RAC1P == "2" ~ "Black",
      TRUE ~ "Other"
    ),
    married = case_when(
      HHT == "1" ~ "Married",
      TRUE ~ "Not married"
    ),
    above_40 = GRPIP >= 40
  ) %>%
  filter(race_ethnicity != "Other") %>%
  group_by(race_ethnicity, married) %>%
  summarize(
    prop_above_40 = survey_mean(above_40)
  )
```

The derived estimates shown in Table 10.5 are the same as before, but the **srvyr** workflow also returns standard errors.

10.3.4 Converting standard errors to margins of error

To convert standard errors to *margins of error* around the derived PUMS estimates, analysts should multiply the standard errors by the following coefficients:

- 90 percent confidence level: 1.645
- 95 percent confidence level: 1.96

TABLE 10.5 Derived estimates for PUMS data with standard errors

race_ethnicity	married	prop_above_40	prop_above_40_se
Black	Married	0.1625791	0.0135852
Black	Not married	0.4080033	0.0081999
Hispanic	Married	0.1716087	0.0339661
Hispanic	Not married	0.3569935	0.0418501
White	Married	0.1266644	0.0118274
White	Not married	0.3356546	0.0078927

- 99 percent confidence level: 2.56

Computing margins of error around derived ACS estimates from PUMS data allows for familiar visualization of uncertainty in the ACS as shown earlier in this book. The example below calculates margins of error at a 90 percent confidence level for the rent burden estimates for Mississippi, then draws a margin of error plot as illustrated in Section 4.3.

```r
ms_svy_summary_moe <- ms_svy_summary %>%
  mutate(prop_above_40_moe = prop_above_40_se * 1.645,
         label = paste(race_ethnicity, married, sep = ", "))

ggplot(ms_svy_summary_moe, aes(x = prop_above_40,
                               y = reorder(label,
                                           prop_above_40))) +
  geom_errorbar(aes(xmin = prop_above_40 - prop_above_40_moe,
                    xmax = prop_above_40 + prop_above_40_moe)) +
  geom_point(size = 3, color = "navy") +
  labs(title = "Rent burdened-households in Mississippi",
       x = "2016-2020 ACS estimate (from PUMS data)",
       y = "",
       caption = paste0("Rent-burdened defined when gross rent ",
                        "is 40 percent or more\nof household income. ",
                        "Error bars represent a 90 ",
                        "percent confidence level.")) +
  scale_x_continuous(labels = scales::percent) +
  theme_grey(base_size = 12)
```

Figure 10.3 effectively represents the uncertainty associated with estimates for the relatively small Hispanic population in Mississippi.

10.4 Modeling with PUMS data

The rich complexity of demographic data available in the PUMS samples allow for the estimation of statistical models to study a wide range of social processes. Like the tabulation of summary statistics with PUMS data, however, statistical models that use complex survey samples require special methods. Fortunately, these methods are incorporated into the **srvyr** and **survey** packages.

Before estimating the model, data should be acquired with `get_pums()` along with appropriate replicate weights. The example below will model whether or not an individual in the labor force aged between 25 and 49 changed residences in the past year as a function of educational attainment, wages, age, class of worker, and family status in Rhode Island.

```r
ri_pums_to_model <- get_pums(
  variables = c("PUMA", "SEX", "MIG",
                "AGEP", "SCHL", "WAGP",
                "COW", "ESR", "MAR", "NOC"),
  state = "RI",
```

10.4 Modeling with PUMS data

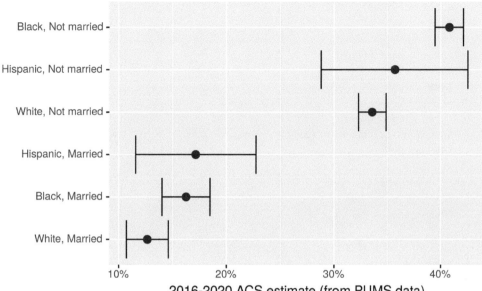

FIGURE 10.3 Margin of error plot for derived PUMS estimates

```
  survey = "acs5",
  year = 2020,
  rep_weights = "person"
)
```

Even though our model will focus on the population in the labor force aged 25 to 49, `variables_filter` should not be used here as the full dataset is needed for appropriate model estimation. This will be addressed in the next section.

10.4.1 Data preparation

Similar to Section 8.2.3, we will perform some feature engineering before fitting the model. This largely involves recoding both the outcome variable and the predictors to more general categories to assist with ease of interpretation. As with other recoding workflows in this book, `case_when()` collapses the categories.

```
ri_pums_recoded <- ri_pums_to_model %>%
  mutate(
    emp_type = case_when(
      COW %in% c("1", "2") ~ "private",
      COW %in% c("3", "4", "5") ~ "public",
      TRUE ~ "self"
    ),
```

```
    child = case_when(
      NOC > 0 ~ "yes",
      TRUE ~ "no"
    ),
    married = case_when(
      MAR == 1 ~ "yes",
      TRUE ~ "no"
    ),
    college = case_when(
      SCHL %in% as.character(21:24) ~ "yes",
      TRUE ~ "no"
    ),
    sex = case_when(
      SEX == 2 ~ "female",
      TRUE ~ "male"
    ),
    migrated = case_when(
      MIG == 1 ~ 0,
      TRUE ~ 1
    )
  )
```

Given that we will be estimating a logistic regression model with a binary outcome (whether or not an individual is a migrant, `migrated` is coded as either 0 or 1. The other recoded variables will be used as *categorical predictors*, in which parameter estimates refer to probabilities of having migrated relative to a reference category (e.g. college graduates relative to individuals who have not graduated college).

In the next step, the subpopulation for which the model will be estimated is identified using `filter()`. We will focus on individuals aged 25 to 49 who are employed and earned wages in the past year.

```
ri_model_svy <- ri_pums_recoded %>%
  to_survey() %>%
  filter(
    ESR == 1,    # civilian employed
    WAGP > 0,    # earned wages last year
    AGEP >= 25,
    AGEP <= 49
  ) %>%
  rename(age = AGEP, wages = WAGP)
```

10.4.2 Fitting and evaluating the model

The family of modeling functions in the **survey** package should be used for modeling data in survey design objects, as they will take into account the replicate weights, survey design, and subpopulation structure. In the example below, we use the `svyglm()` function for this purpose. The formula is written using standard R formula notation, the survey design object

10.4 Modeling with PUMS data

is passed to the design parameter, and family = quasibinomial() is used to fit a logistic regression model.

```
library(survey)

migration_model <- svyglm(
  formula = migrated ~ log(wages) + sex + age + emp_type +
    child + married + college + PUMA,
  design = ri_model_svy,
  family = quasibinomial()
)
```

Once fit, we can examine the results:

```
summary(migration_model)
```

```
## 
## Call:
## svyglm(formula = migrated ~ log(wages) + sex + age + emp_type +
##     child + married + college + PUMA, design = ri_model_svy,
##     family = quasibinomial())
## 
## Survey design:
## Called via srvyr
## 
## Coefficients:
##                  Estimate Std. Error t value Pr(>|t|)
## (Intercept)      1.489784   0.496118   3.003  0.00379 **
## log(wages)      -0.098089   0.047027  -2.086  0.04093 *
## sexmale          0.249331   0.056123   4.443 3.53e-05 ***
## age             -0.068431   0.008195  -8.350 6.98e-12 ***
## emp_typepublic  -0.057571   0.099477  -0.579  0.56477
## emp_typeself    -0.243479   0.196835  -1.237  0.22055
## childyes        -0.192214   0.105508  -1.822  0.07309 .
## marriedyes      -0.141021   0.115814  -1.218  0.22776
## collegeyes       0.256121   0.094649   2.706  0.00869 **
## PUMA00102        0.098035   0.150894   0.650  0.51818
## PUMA00103        0.102302   0.162798   0.628  0.53195
## PUMA00104        0.187429   0.184095   1.018  0.31240
## PUMA00201        0.190723   0.135870   1.404  0.16516
## PUMA00300        0.288592   0.179487   1.608  0.11271
## PUMA00400       -0.329335   0.202088  -1.630  0.10801
## ---
## Signif. codes:  0 '***' 0.001 '**' 0.01 '*' 0.05 '.' 0.1 ' ' 1
## 
## (Dispersion parameter for quasibinomial family taken to be 10453.18)
## 
## Number of Fisher Scoring iterations: 5
```

The model identifies some notable differences in recent migrants relative to non-migrants, controlling for other demographic factors. Males are more likely to have moved than females, as are younger people relative to older people in the subpopulation. Individuals with children are slightly more stationary, whereas college-educated individuals in the sample are more likely to have moved. The PUMAs are included as a categorical predictor largely to control for geographic differences in the state; the model does not identify any substantive differences among Rhode Island PUMAs in this analysis.

10.5 Exercises

- Using the dataset you acquired from the exercises in Chapter 9 (or the example Wyoming dataset in that chapter), tabulate a group-wise summary using the PWGTP column and dplyr functions as you've learned in this section.

- Advanced follow-up: using `get_acs()`, attempt to acquire the same aggregated data from the ACS. Compare your tabulated estimate with the ACS estimate.

- Second advanced follow-up: request the same data as before, but this time with replicate weights. Calculate the margin of error as you've learned in this section – and if you have time, compare with the posted ACS margin of error!

11

Other Census and government data resources

Most of the examples covered in the book to this point use data from recent US Census Bureau datasets such as the Decennial Census since 2000 and the American Community Survey. These datasets are available through the US Census Bureau's APIs and in turn accessible with **tidycensus** and related tools. However, analysts and historians may be interested in accessing data from much earlier – perhaps all the way back to 1790, the first US Census! Fortunately, these historical datasets are available to analysts through the National Historical Geographic Information System (NHGIS) project[1] and the Minnesota Population Center's[2] IPUMS project[3]. While both of these data repositories have typically attracted researchers using commercial software such as ArcGIS (for NHGIS) and Stata/SAS (for IPUMS), the Minnesota Population Center has developed an associated **ipumsr** R package to help analysts integrate these datasets into R-based workflows.

Additionally, the US Census Bureau publishes many other surveys and datasets besides the decennial US Census and American Community Survey. While **tidycensus** focuses on these core datasets, other R packages provide support for the wide range of datasets available from the Census Bureau and other government agencies.

The first part of this chapter provides an overview of how to access and use historical US Census datasets in R with NHGIS, IPUMS, and the **ipumsr** package. Due to the size of the datasets involved, these datasets are not provided with the sample data available in the book's data repository. To reproduce, readers should follow the steps provided to sign up for an IPUMS account and download the data themselves. The second part of this chapter covers R workflows for Census data resources outside the decennial US Census and American Community Survey. It highlights packages such as **censusapi**, which allows for programmatic access to all US Census Bureau APIs, and **lehdr**, which grants access to the LEHD LODES dataset for analyzing commuting flows and jobs distributions. Other government data resources are also addressed at the end of the chapter.

11.1 Mapping historical geographies of New York City with NHGIS

The National Historical Geographic Information System (NHGIS) project (Manson et al., 2021) is a tremendous resource for both contemporary and historical US Census data. While some datasets (e.g. the 2000 and 2010 decennial US Censuses, the ACS) can be accessed with both **tidycensus** and NHGIS, NHGIS is an excellent option for users who prefer browsing data menus to request data and/or who require historical information earlier than 2000. The example in this section will illustrate an applied workflow using NHGIS and its companion

---

[1] https://www.nhgis.org/
[2] https://pop.umn.edu/
[3] https://ipums.org/

FIGURE 11.1 NHGIS data browser interface

R package, **ipumsr** (Ellis and Burk, 2020) to map geographies of immigration in New York City from the 1910 Census.

11.1.1 Getting started with NHGIS

To get started with NHGIS, visit the NHGIS website[4] and click the "REGISTER" link at the top of the screen to register for an account. NHGIS asks for some basic information about your work and how you plan to use the data, and you'll agree to the NHGIS usage license agreement. Once registered, return to the NHGIS home page and click the "Get Data" button to visit the NHGIS data browser interface.

A series of options on the left-hand side of your screen. These options include:

- **Geographic levels**: the level of aggregation for your data. NHGIS includes a series of filters to help you choose the correct level of aggregation; click the plus sign to select it. Keep in mind that not all geographic levels will be available for all variables and all years. To reproduce the example in this section, click "CENSUS TRACT" then "SUBMIT."

- **Years**: The year(s) for which you would like to request data. Decennial, non-decennial, and 5-year ranges are available for Census tracts. Note that many years are greyed out – this means that no data are available for those years at the Census tract level. The earliest year for which Census tract-level data are available is 1910, which is the year we will choose; check the box next to "1910" then click SUBMIT.

- **Topics**: This menu helps you filter down to specific areas of interest in which you are searching for data to select. Data are organized into categories (e.g. race, ethnicity, and origins) and sub-topics (e.g. age, sex). Topics not available at your chosen geography/year combination are greyed out. Choose the "Nativity and Place of Birth" topic then click SUBMIT.

- **Datasets**: The specific datasets from which you would like to request data, which is particularly useful when there are multiple datasets available in a given year. In this applied example, there is only one dataset that aligns with our choices: Population Data for Census tracts in New York City in 1910. As such, there is no need to select anything here.

The **Select Data** menu shows which Census tables are available given your filter selections. Usefully, the menu includes embedded links that give additional information about the

---

[4] https://www.nhgis.org/

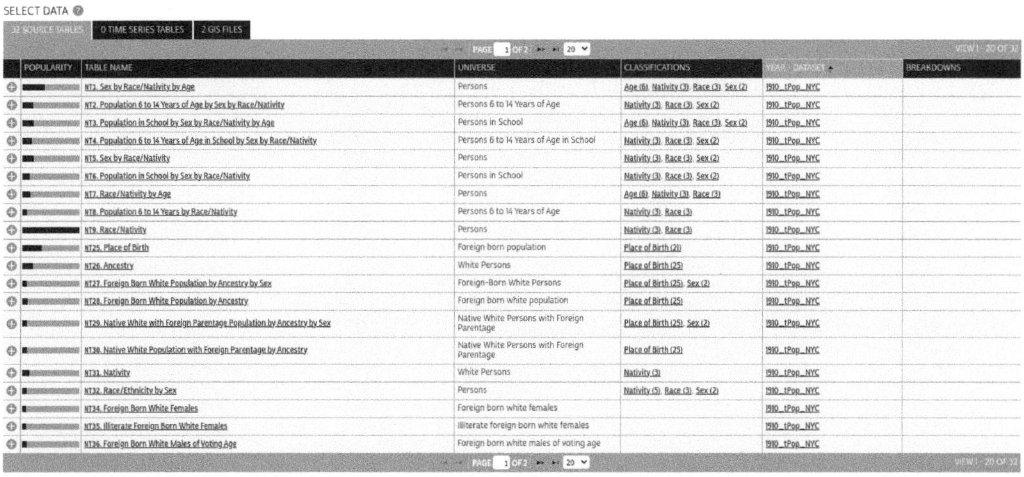

11.1 Mapping historical geographies of New York City with NHGIS

FIGURE 11.2 NHGIS select data screen

available data choices, along with a "popularity" bar graph showing the most-downloaded tables for that particular dataset.

For this example, choose the tables "NT26: Ancestry" and "NT45: Race/Ethnicity" by clicking the green plus signs next to the two to select them. Then, click the **GIS FILES** tab. This tab allows you to select companion shapefiles that can be merged to the demographic extracts for mapping in a desktop GIS or in software like R. Choose either of the Census Tract options then click "CONTINUE" to review your selection. On the "REVIEW AND SUBMIT" screen, keep the "Comma delimited" file structure selected and give your extract a description if you would like.

When finished, click SUBMIT. You'll be taken to the "EXTRACTS HISTORY" screen where you can download your data when it is ready; you'll receive a notification by email when your data can be downloaded. Once you receive this notification, return to NHGIS and download both the table data and the GIS data to the same directory on your computer.

11.1.2 Working with NHGIS data in R

Once acquired, NHGIS spatial and attribute data can be integrated seamlessly into R-based data analysis workflows thanks to the **ipumsr** package. **ipumsr** includes a series of NHGIS-specific functions: `read_nhgis()`, which reads in the tabular aggregate data; `read_nhgis_sf()`, which reads in the spatial data as a simple features object; and `read_nhgis_sp()`, which reads in the spatial data in legacy sp format.

`read_nhgis_sf()` has built-in features to make working with spatial and demographic data simpler for R users. If the `data_file` argument is pointed to the CSV file with the demographic data, and the `shape_file` argument is pointed to the shapefile, `read_nhgis_sf()` will read in both files simultaneously and join them correctly based on a common GISJOIN column found in both files. An additional perk is that `read_nhgis_sf()` can handle the zipped folders, removing an additional step from the analyst's data preparation workflow.

The example below uses `read_nhgis_sf()` to read in spatial and demographic data on immigrants in New York City in 1910. As the 1910 shapefile folder includes both NYC Census tracts and a separate dataset with US counties, the top-level folder should be unzipped, shape_file pointed to the second-level zipped folder, and the shape_layer argument used to exclusively read in the tracts. The filter() call will drop Census tracts that do not have corresponding data (so, outside NYC).

```
library(ipumsr)
library(tidyverse)

# Note: the NHGIS file name depends on a download number
# that is unique to each user, so your file names will be
# different from the code below
nyc_1910 <- read_nhgis_sf(
  data_file = "data/NHGIS/nhgis0099_csv.zip",
  shape_file = paste0("data/NHGIS/nhgis0099_shape/",
                      "nhgis0099_shapefile_tl2000_us_tract_1910.zip"),
  shape_layer = starts_with("US_tract_1910")
) %>%
  filter(str_detect(GISJOIN, "G36"))
```

```
## Use of data from NHGIS is subject to conditions including that users should
## cite the data appropriately. Use command `ipums_conditions()` for more details.
##
##
## Reading data file...
## Reading geography...
## options:               ENCODING=latin1
## Reading layer `US_tract_1910' from data source
##   `/tmp/Rtmp8Jqkeo/file5d13486bf918/US_tract_1910.shp' using driver
##   `ESRI Shapefile'
## Simple feature collection with 1989 features and 6 fields
## Geometry type: MULTIPOLYGON
## Dimension:     XY
## Bounding box:  xmin: 489737.4 ymin: 130629.6 xmax: 2029575 ymax: 816129.7
## Projected CRS: USA_Contiguous_Albers_Equal_Area_Conic
```

read_nhgis_sf() has read in the tracts shapefile as a simple features object then joined the corresponding CSV file to it *and* imported data labels from the data codebook. Note that if you are reproducing this example with data downloaded yourself, you will have a unique zipped folder & file name based on your unique download ID. The "99" in the example above reflects the 99th extract from NHGIS for a given user, not a unique dataset name.

The best way to review the variables and their labels is the View() command in RStudio, which is most efficient on sf objects when the geometry is first dropped with sf::st_drop_geometry().

```
View(sf::st_drop_geometry(nyc_1910))
```

As shown in Figure 11.3, the variable labels are particularly useful when using View() to understand what the different variables mean without having to reference the codebook.

11.1 Mapping historical geographies of New York City with NHGIS

A6G007 Finland	A6G008 France	A6G009 Germany	A6G010 Greece	A6G011 Holland	A6G012 Hungary	A6G013 Ireland	A6G014 Italy	A6G015 Norway	A6G016 Rumania	A6G017 Russia	A6... Sco...
153	70	4533	7	63	581	5546	864	91	141	2032	
117	14	1478	1	6	82	1315	220	72	15	315	
6	14	784	1	5	24	677	538	149	2	378	
16	14	1194	5	14	13	1531	320	342	1	225	
4	12	599	0	1	2	796	1801	133	1	447	
0	2	49	0	4	2	260	6590	0	2	49	
8	10	545	1	11	31	2406	309	110	8	249	
20	11	385	0	6	38	1969	314	40	1	197	
8	17	516	6	7	46	2185	589	230	10	171	
2	7	493	0	1	13	1847	383	378	3	84	
0	10	126	0	1	2	1034	453	111	0	78	
5	12	301	0	3	17	1163	1245	445	8	101	
13	9	247	0	1	2	1012	322	1059	2	101	
16	8	233	2	0	8	1046	1813	1486	0	71	
0	3	217	0	3	12	335	39	8	0	75	
5	5	82	0	0	2	839	3161	117	9	91	
4	11	203	6	1	9	538	9068	627	21	140	
2	7	229	3	1	3	2663	1793	139	6	89	
0	22	371	1	9	9	1375	68	125	1	86	
4	7	507	0	7	15	1061	487	83	1	102	
9	15	294	2	8	0	1823	677	1269	9	140	

FIGURE 11.3 NHGIS data in the RStudio Viewer

FIGURE 11.4 Plot of NYC Census tracts in 1910 using an Albers Equal Area CRS

11.1.3 Mapping NHGIS data in R

The message displayed when reading in the NHGIS shapefile above indicates that the Census tract data are in a projected coordinate reference system, `USA_Contiguous_Albers_Equal_Area_Conic`. The spatial data can be displayed with `plot()`:

```
plot(nyc_1910$geometry)
```

The data shown in Figure 11.4 reflect Census tracts for New York City, but appear rotated counter-clockwise. This is because the coordinate reference system used,

TABLE 11.1 Suggested CRS options for New York City

crs_code	crs_name	crs_type	crs_gcs	crs_units
6539	NAD83(2011) / New York Long Island (ftUS)	projected	6318	us-ft
6538	NAD83(2011) / New York Long Island	projected	6318	m
4456	NAD27 / New York Long Island	projected	4267	us-ft
3628	NAD83(NSRS2007) / New York Long Island (ftUS)	projected	4759	us-ft
3627	NAD83(NSRS2007) / New York Long Island	projected	4759	m
32118	NAD83 / New York Long Island	projected	4269	m
2908	NAD83(HARN) / New York Long Island (ftUS)	projected	4152	us-ft
2831	NAD83(HARN) / New York Long Island	projected	4152	m
2263	NAD83 / New York Long Island (ftUS)	projected	4269	us-ft
3748	NAD83(HARN) / UTM zone 18N	projected	4152	m

FIGURE 11.5 NYC Census tracts with an area-appropriate CRS

`ESRI:100023`, is appropriate for the entire United States (in fact, it is the base CRS used for `tigris::shift_geometry()`), but will not be appropriate for any specific small area. As covered in Chapter 5, the **crsuggest** package helps identify more appropriate projected coordinate reference system options.

```
library(crsuggest)
library(sf)

suggest_crs(nyc_1910)
```

Based on these suggestions, we'll select the CRS "NAD83(2011) / New York Long Island" which has an EPSG code of 6538.

```
nyc_1910_proj <- st_transform(nyc_1910, 6538)

plot(nyc_1910_proj$geometry)
```

Given the information in the two tables downloaded from NHGIS, there are multiple ways to visualize the demographics of New York City in 1910. The first example is a choropleth

11.1 Mapping historical geographies of New York City with NHGIS

map of the percentage of the total population born outside the United States. As there is no "total population" column in the dataset, the code below uses **dplyr**'s `rowwise()` and `c_across()` functions to perform row-wise calculations and sum across the columns `A60001` through `A60007`. The `transmute()` function then works like a combination of `mutate()` and `select()`: it calculates two new columns, then selects only those columns in the output dataset `nyc_pctfb`.

```
nyc_pctfb <- nyc_1910_proj %>%
  rowwise() %>%
  mutate(total = sum(c_across(A60001:A60007))) %>%
  ungroup() %>%
  transmute(
    tract_id = GISJOIN,
    pct_fb = A60005 / total
  )
```

The result can be visualized with any of the mapping packages covered in Chapter 6, such as ggplot2 and `geom_sf()`.

```
ggplot(nyc_pctfb, aes(fill = pct_fb)) +
  geom_sf(color = NA) +
  scale_fill_viridis_c(option = "magma", labels = scales::percent) +
  theme_void(base_family = "Verdana") +
  labs(title = "Percent foreign-born by Census tract, 1910",
       subtitle = "New York City",
       caption =  "Data source: NHGIS",
       fill = "Percentage")
```

Manhattan's Lower East Side stands out as the part of the city with the largest proportion of foreign-born residents in 1910, with percentages exceeding 60%.

An alternative view could focus on one of the specific groups represented in the columns in the dataset. For example, the number of Italy-born residents by Census tract is represented in the column `A6G014`; this type of information could be represented by either a graduated symbol map or a dot-density map. Using techniques learned in Section 6.3.4.3, we can use the `as_dot_density()` function in the **tidycensus** package to generate one dot for approximately every 100 Italian immigrants. Next, the Census tracts are dissolved with the `st_union()` function to generate a base layer on top of which the dots will be plotted.

```
library(tidycensus)

italy_dots <- nyc_1910_proj %>%
  as_dot_density(
    value = "A6G014",
    values_per_dot = 100
  )

nyc_base <- nyc_1910_proj %>%
  st_union()
```

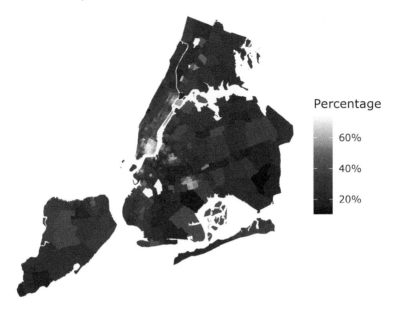

FIGURE 11.6 Percent foreign-born by Census tract in NYC in 1910, mapped with ggplot2

In Section 6.3.4.3, we used tmap::tm_dots() to create a dot-density map. ggplot2 and geom_sf() also work well for dot-density mapping; cartographers can either use geom_sf() with a very small size argument, or set shape = "." where each data point will be represented by a single pixel on the screen.

```
ggplot() +
  geom_sf(data = nyc_base, size = 0.1) +
  geom_sf(data = italy_dots, shape = ".", color = "darkgreen") +
  theme_void(base_family = "Verdana") +
  labs(title = "Italy-born population in New York City, 1910",
       subtitle = "1 dot = 100 people",
       caption = "Data source: NHGIS")
```

The map in Figure 11.7 highlights areas with large concentrations of Italian immigrants in New York City in 1910.

11.2 Analyzing complete-count historical microdata with IPUMS and R

Chapters 9 and 10 covered the process of acquiring and analyzing microdata from the American Community Survey with **tidycensus**. As noted, these workflows are only available

Italy-born population in New York City, 1910
1 dot = 100 people

Data source: NHGIS

FIGURE 11.7 Dot-density map of the Italy-born population in NYC in 1910, mapped with ggplot2

for recent demographics, reflecting the recent availability of the ACS. Historical researchers will need data that goes further back, and will likely turn to IPUMS-USA[5] for these datasets. IPUMS-USA makes available microdata samples all the way back to 1790, enabling historical demographic research not possible elsewhere.

A core focus of Chapter 10 was the use of sampling weights to appropriately analyze and model microdata. Historical Census datasets, however, are subject to the "72-year rule"[6], which states:

> The U.S. government will not release personally identifiable information about an individual to any other individual or agency until 72 years after it was collected for the decennial census. This "72-Year Rule" (92 Stat. 915; Public Law 95-416[7]; October 5, 1978) restricts access to decennial census records to all but the individual named on the record or their legal heir.

---

[5] https://usa.ipums.org/usa/
[6] https://www.census.gov/history/www/genealogy/decennial_census_records/the_72_year_rule_1.html
[7] https://www.census.gov/history/pdf/NARA_Legislation.pdf

FIGURE 11.8 IPUMS data browser

This means that decennial Census records that reflect periods 72 years ago or older can be made available to researchers by the IPUMS team. In fact, complete-count Census microdata can be downloaded from IPUMS at the person-level for the Census years 1850-1940, and at the household level for years earlier than 1850.

The availability of complete-count Census records offers a tremendous analytic opportunity for researchers, but also comes with some challenges. The largest ACS microdata sample – the 2016-2020 5-year ACS – has around 16 million records, which can be read into memory on a standard desktop computer with 16GB RAM. Complete-count Census data can have records exceeding 100 million, which will not be possible to read into memory in R on a standard computer. This chapter covers R-based workflows for handling massive Census microdata without needing to upgrade one's computer or set up a cloud computing instance. The solution presented involves setting up a local database with PostgreSQL and the DBeaver platform, then interacting with microdata in that database using R's tidyverse and database interface tooling.

11.2.1 Getting microdata from IPUMS

To get started, visit the IPUMS-USA website at https://usa.ipums.org/usa/. If you already signed up for an IPUMS account in Section 11.1.1, log in with your user name and password; otherwise follow those instructions to register for an account, which you can use for all of the IPUMS resources including NHGIS. Once you are logged in, click the "Get Data" button to visit the data selection menu.

You'll use the various options displayed in Figure 11.8 to define your extract. These options include:

- **Select samples**: choose one or more data samples to include in your microdata extract. You can choose a wide range of American Community Survey and Decennial US Census samples, or you can download full count Census data from the "USA FULL COUNT" tab. To reproduce this example, choose the 1910 100% dataset by first un-checking the "Default sample from each year" box, clicking the "USA FULL COUNT" tab, then choosing the 1910 dataset and clicking **SUBMIT SAMPLE SELECTIONS**.
- **Select harmonized variables**: One of the major benefits of using IPUMS for microdata analysis is that the IPUMS team has produced *harmonized variables* that aim to reconcile

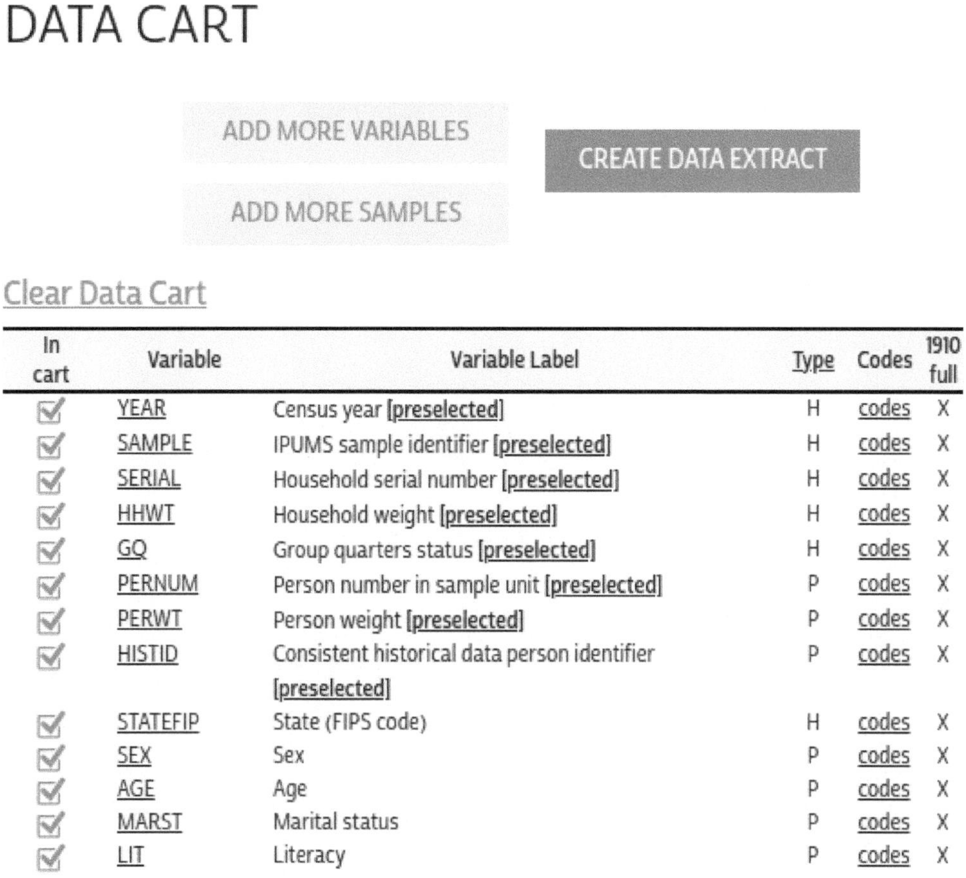

DATA CART

[ADD MORE VARIABLES] [CREATE DATA EXTRACT]

[ADD MORE SAMPLES]

Clear Data Cart

In cart	Variable	Variable Label	Type	Codes	1910 full
✓	YEAR	Census year [preselected]	H	codes	X
✓	SAMPLE	IPUMS sample identifier [preselected]	H	codes	X
✓	SERIAL	Household serial number [preselected]	H	codes	X
✓	HHWT	Household weight [preselected]	H	codes	X
✓	GQ	Group quarters status [preselected]	H	codes	X
✓	PERNUM	Person number in sample unit [preselected]	P	codes	X
✓	PERWT	Person weight [preselected]	P	codes	X
✓	HISTID	Consistent historical data person identifier [preselected]	P	codes	X
✓	STATEFIP	State (FIPS code)	H	codes	X
✓	SEX	Sex	P	codes	X
✓	AGE	Age	P	codes	X
✓	MARST	Marital status	P	codes	X
✓	LIT	Literacy	P	codes	X

FIGURE 11.9 IPUMS data cart

variable IDs and variable definitions over time allowing for easier longitudinal analysis. By default, users will browse and select from these harmonized variables. Choose from household-level variables and person-level variables by browsing the drop-down menus and selecting appropriate variable IDs; these will be added to your output extract. For users who want to work with variables as they originally were in the source dataset, click the **SOURCE VARIABLES** radio button to switch to source variables. To replicate this example, choose the STATEFIP (household > geographic), SEX, AGE, and MARST (person > demographic), and LIT (person > education) variables.

- **Display options:** This menu gives a number of options to modify the display of variables when browsing; try out the different options if you'd like.

When finished, click the "VIEW CART" link to go to your data cart. You'll see the variables that will be returned with your extract.

Notice that there will be more variables in your output extract than you selected; this is because a number of technical variables are *pre-selected*, which is similar to the approach taken by `get_pums()` in **tidycensus**. When you are ready to create the extract, click the **CREATE DATA EXTRACT** button to get a summary of your extract before submitting.

Note that the example shown reflects changing the output data format to CSV, which will be used to load the IPUMS data into a database. If using R directly for a smaller extract

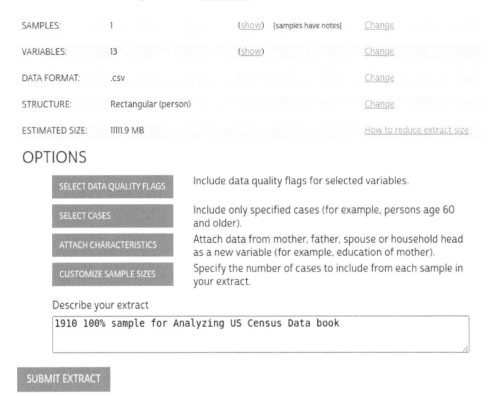

FIGURE 11.10 IPUMS extract request screen

and the **ipumsr** R package, the default fixed-width text file with the extension .dat can be selected and the data can be read into R with `ipumsr::read_ipums_micro()`. To replicate the workflow below, however, the output data format should be changed to CSV.

Click **SUBMIT EXTRACT** to submit your extract request to the IPUMS system. You'll need to agree to some special usage terms for the 100% data, then wait patiently for your extract to process. You'll get an email notification when your extract is ready; when you do, return to IPUMS and download your data extract to your computer. The output format will be a gzipped CSV with a prefix unique to your download ID, e.g. `usa_00032.csv.gz`. Use an appropriate utility to unzip the CSV file.

11.2.2 Loading microdata into a database

Census data analysts may feel comfortable working with .csv files, and reading them into R with `readr::read_csv()` or one of the many other options for loading data from comma-separated text files. The data extract created in the previous section presents additional challenges for the Census analyst, however. It contains approximately 92.4 million rows – one for each person in the 1910 US Census! This will fill up the memory of a standard laptop computer when read into R very quickly using standard tools.

An alternative solution that can be performed on a standard laptop computer is setting up a database. The solution proposed here uses PostgreSQL[8], a popular open-source database, and DBeaver[9], a free cross-platform tool for working with databases.

If PostgreSQL is not already installed on your computer, visit https://www.postgresql.org/download/ and follow the instructions for your operating system to install it. A full tutorial on PostgreSQL for each operating system is beyond the scope of this book, so this example will use standard defaults. When you are finished installing PostgreSQL, you will be prompted to set up a default database, which will be called `postgres` and will be associated with a user, also named `postgres`. You'll be asked to set a password for the database; the examples in this chapter also use `postgres` for the password, but you can choose whatever password you would like.

Once the default database has been set up, visit https://dbeaver.io/download/ and download/install the appropriate DBeaver Community Edition for your operating system. Launch DBeaver, and look for the "New Database Connection" icon. Click there to launch a connection to your PostgreSQL database. Choose "PostgreSQL" from the menu of database options and fill out your connection settings appropriately.

For most defaults, the host should be `localhost` running on port `5432` with the database name `postgres` and the username `postgres` as well. Enter the password you selected when setting up PostgreSQL and click **Finish**. You'll notice your database connection appear in the "Database Navigator" pane of DBeaver.

Within the database, you can create **schemas** to organize sets of related tables. To create a new schema, right-click the postgres database in the Database Navigator pane and choose "Create > Schema." This example uses a schema named "ipums." Within the new schema you'll be able to import your CSV file as a table. Expand the "ipums" schema and right-click on "Tables" then choose **Import Data.** Select the unzipped IPUMS CSV file and progress through the menus. This example changes the default "Target" name to `census1910`.

On the final menu, click the **Start** button to import your data. Given the size of the dataset, this will likely take some time; it took around 2 hours to prepare the example on my machine. Once the data has imported successfully to the new database table in DBeaver, you'll be able to inspect your dataset using the DBeaver interface.

11.2.3 Accessing your microdata database with R

In a typical R workflow with flat files, an analyst will read in a dataset (such as a CSV file) entirely in-memory into R then use their preferred toolset to interact with the data. The alternative discussed in this section involves connecting to the 1910 Census database from R then using R's database interface toolkit through the tidyverse to analyze the data. A major benefit to this workflow is that it allows an analyst to perform tidyverse operations on datasets *in the database*. This means that tidyverse functions are translated to Structured Query Language (SQL) queries and passed to the database, with outputs displayed in R, allowing the analyst to interact with data without having to read it into memory.

---

[8] https://www.postgresql.org/
[9] https://dbeaver.io/

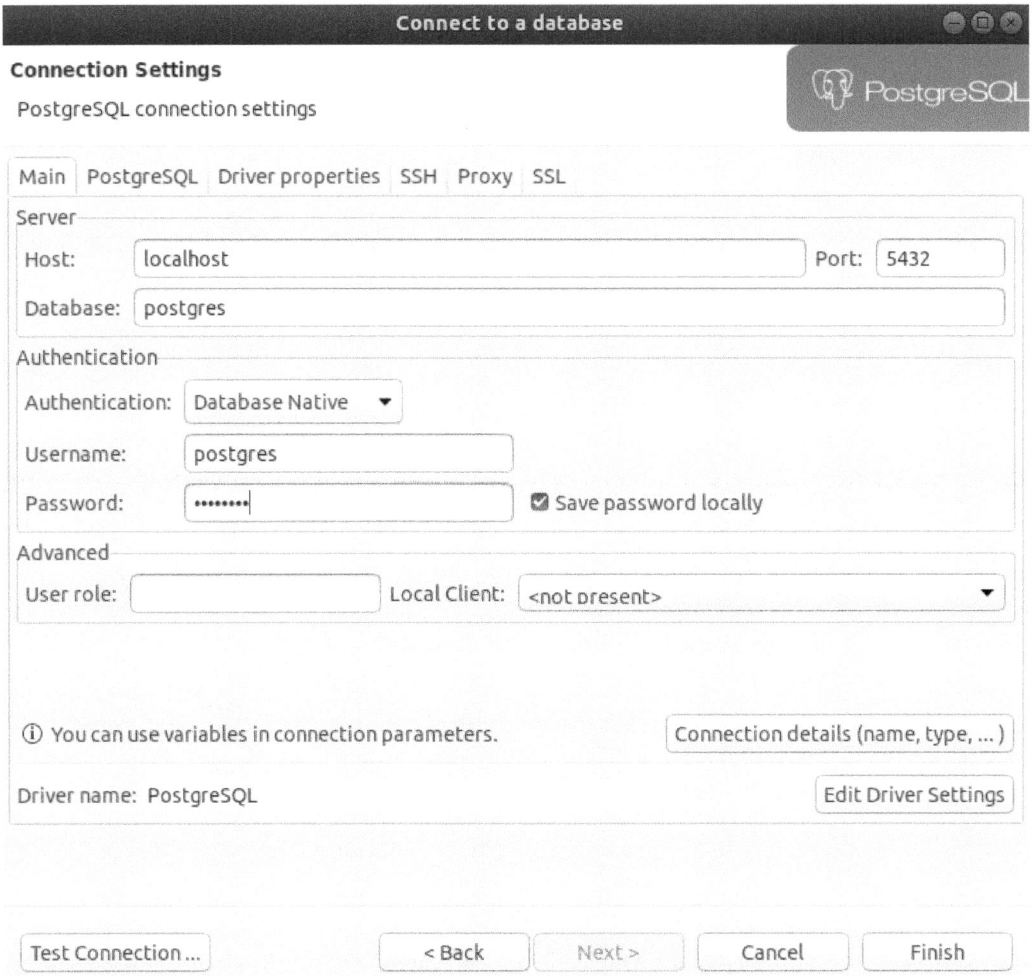

FIGURE 11.11 DBeaver connection screen

The first step requires making a database connection from R. The DBI infrastructure for R[10] allows users to make connections to a wide range of databases using a consistent interface (R Special Interest Group on Databases (R-SIG-DB) et al., 2021). The PostgreSQL interface is handled with the **RPostgres** package (Wickham et al., 2021c).

The dbConnect() function is used to make a database connection, which will be assigned to the object conn. Arguments include the database driver Postgres() and the username, password, database name, host, and port, which are all familiar from the database setup process.

```
library(tidyverse)
library(RPostgres)
library(dbplyr)
```

---

[10] https://www.r-dbi.org/

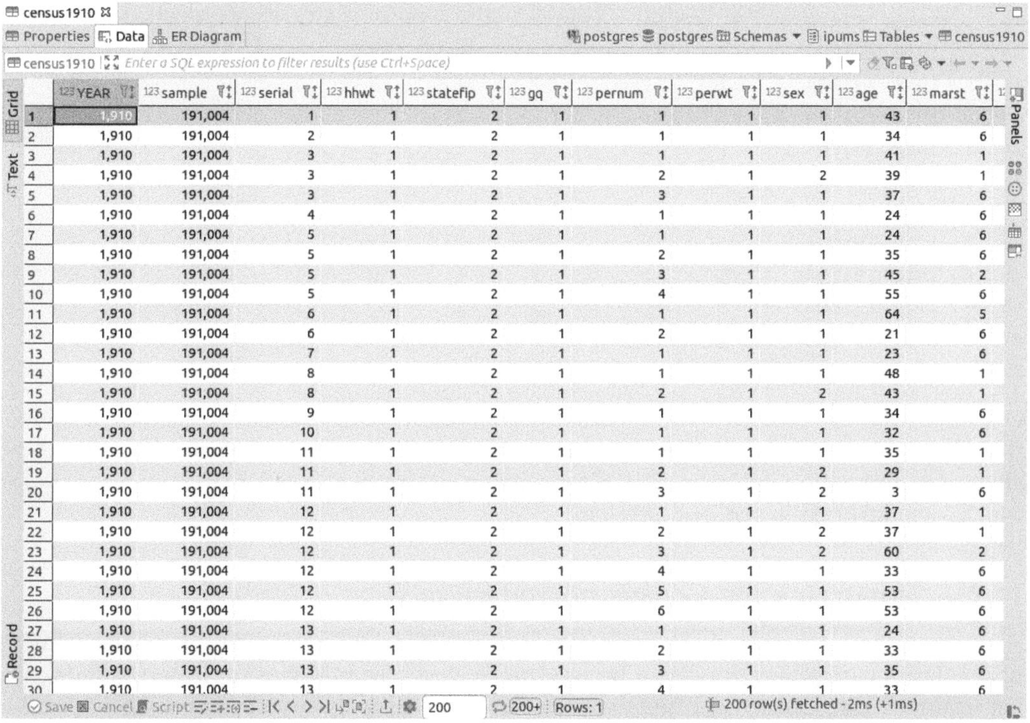

11.2 Analyzing complete-count historical microdata with IPUMS and R

FIGURE 11.12 DBeaver data view

```
conn <- dbConnect(
  drv = Postgres(),
  user = "postgres",
  password = "postgres",
  host = "localhost",
  port = "5432",
  dbname = "postgres"
)
```

Once connected to the database, the database extension to **dplyr**, **dbplyr**, facilitates interaction with database tables (Wickham et al., 2021b). tbl() links to the connection object conn to retrieve data from the database; the in_schema() function points tbl() to the census1910 table in the ipums schema.

Printing the new object census1910 shows the general structure of the 1910 Census microdata:

```
census1910 <- tbl(conn, in_schema("ipums", "census1910"))

census1910

## # Source:   table<"ipums"."census1910"> [?? x 13]
## # Database: postgres [postgres@localhost:5432/postgres]
##      YEAR sample serial  hhwt statefip    gq pernum perwt   sex   age marst   lit
```

```
##      <int>  <int> <int> <dbl>   <int> <int>  <int> <dbl> <int> <int> <int> <int>
## 1     1910 191004     1     1       2     1      1     1     1    43     6     4
## 2     1910 191004     2     1       2     1      1     1     1    34     6     4
## 3     1910 191004     3     1       2     1      1     1     1    41     1     4
## 4     1910 191004     3     1       2     1      2     1     2    39     1     4
## 5     1910 191004     3     1       2     1      3     1     1    37     6     4
## 6     1910 191004     4     1       2     1      1     1     1    24     6     4
## 7     1910 191004     5     1       2     1      1     1     1    24     6     4
## 8     1910 191004     5     1       2     1      2     1     1    35     6     4
## 9     1910 191004     5     1       2     1      3     1     1    45     2     4
## 10    1910 191004     5     1       2     1      4     1     1    55     6     4
## # ... with more rows, and 1 more variable: histid <chr>
```

Our data have 13 columns and an unknown number of rows; the database table is so large that **dbplyr** won't calculate this automatically. However, the connection to the database allows for interaction with the 1910 Census microdata using familiar tidyverse workflows, which are addressed in the next section.

11.2.4 Analyzing big Census microdata in R

While the default printing of the microdata shown above does not reveal the number of rows in the dataset, tidyverse tools can be used to request this information from the database. For example, the `summarize()` function will generate summary tabulations as shown earlier in this book; without a companion `group_by()`, it will do so over the whole dataset.

```
census1910 %>% summarize(n())
```

```
## # Source:   lazy query [?? x 1]
## # Database: postgres [postgres@localhost:5432/postgres]
##     `n()`
##    <int64>
## 1 92404011
```

There are 92.4 million rows in the dataset, reflecting the US population size in 1910. Given that we are working with complete-count data, the workflow for using microdata differs from the sample data covered in Chapters 9 and 10. While IPUMS by default returns a person-weight column, the value for all rows in the dataset for this column is 1, reflecting a 1:1 relationship between the records and actual people in the United States at that time.

dplyr's database interface can accommodate more complex examples as well. Let's say we want to tabulate literacy statistics in 1910 for the population age 18 and up in Texas by sex. A straightforward tidyverse pipeline can accommodate this request to the database. For reference, male is coded as 1 and female as 2, and the literacy codes are as follows:

- 1: No, illiterate (cannot read or write)
- 2: Cannot read, can write
- 3: Cannot write, can read
- 4: Yes, literate (reads and writes)

```
census1910 %>%
  filter(age > 17, statefip == "48") %>%
```

```
  group_by(sex, lit) %>%
  summarize(num = n())
```

```
## # Source:    lazy query [?? x 3]
## # Database:  postgres [postgres@localhost:5432/postgres]
## # Groups:    sex
##     sex   lit    num
##   <int> <int> <int64>
## 1    1     1  115952
## 2    1     2      28
## 3    1     3   14183
## 4    1     4  997844
## 5    2     1  111150
## 6    2     2      16
## 7    2     3   14531
## 8    2     4  883636
```

dbplyr also includes some helper functions to better understand how it is working and to work with the derived results. For example, the `show_query()` function can be attached to the end of a tidyverse pipeline to show the SQL query that the R code is translated to in order to perform operations in-database:

```
census1910 %>%
  filter(age > 17, statefip == "48") %>%
  group_by(sex, lit) %>%
  summarize(num = n()) %>%
  show_query()
```

```
## <SQL>
## SELECT "sex", "lit", COUNT(*) AS "num"
## FROM "ipums"."census1910"
## WHERE (("age" > 17.0) AND ("statefip" = '48'))
## GROUP BY "sex", "lit"
```

If an analyst wants the result of a database operation to be brought into R as an R object rather than as a database view, the `collect()` function can be used at the end of a pipeline to load data directly. A companion function from **ipumsr**, `ipums_collect()`, will add variable and value labels to the collected data based on an IPUMS codebook.

The aforementioned toolsets can now be used for robust analyses of historical microdata based on complete-count Census data. The example below illustrates how to extend the literacy by sex example above to visualize literacy gaps by sex by state in 1910.

```
literacy_props <- census1910 %>%
  filter(age > 18) %>%
  group_by(statefip, sex, lit) %>%
  summarize(num = n()) %>%
  group_by(statefip, sex) %>%
  mutate(total = sum(num, na.rm = TRUE)) %>%
  ungroup() %>%
```

```
  mutate(prop = num / total) %>%
  filter(lit == 4) %>%
  collect()

state_names <- tigris::fips_codes %>%
  select(state_code, state_name) %>%
  distinct()

literacy_props_with_name <- literacy_props %>%
  mutate(statefip = str_pad(statefip, 2, "left", "0")) %>%
  left_join(state_names, by = c("statefip" = "state_code")) %>%
  mutate(sex = ifelse(sex == 1, "Male", "Female"))

ggplot(literacy_props_with_name, aes(x = prop, y = reorder(state_name, prop),
                                     color = sex)) +
  geom_line(aes(group = state_name), color = "grey10") +
  geom_point(size = 2.5) +
  theme_minimal() +
  scale_color_manual(values = c(Male = "navy", Female = "darkred")) +
  scale_x_continuous(labels = scales::percent) +
  labs(x = "Percent fully literate, 1910",
       color = "",
       y = "")
```

11.3 Other US government datasets

To this point, this book has focused on a smaller number of US Census Bureau datasets, with a primary focus on the decennial US Census and the American Community Survey. However, many more US government datasets are available to researchers, some from the US Census Bureau and others from different US government agencies. This section covers a series of R packages to help analysts access these resources, and illustrates some applied workflows using those datasets.

11.3.1 Accessing Census data resources with censusapi

censusapi (Recht, 2021) is an R package designed to give R users access to *all* of the US Census Bureau's API endpoints. Unlike **tidycensus**, which only focuses on a select number of datasets, **censusapi**'s getCensus() function can be widely applied to the hundreds of possible datasets the Census Bureau makes available. **censusapi** requires some knowledge of how to structure Census Bureau API requests to work; however, this makes the package more flexible than **tidycensus** and may be preferable for users who want to submit highly customized queries to the decennial Census or ACS APIs.

censusapi uses the same Census API key as **tidycensus**, though references it with the R environment variable CENSUS_KEY. If this environment variable is set in a user's .Renviron file, functions in **censusapi** will pick up the key without having to supply it directly.

11.3 Other US government datasets

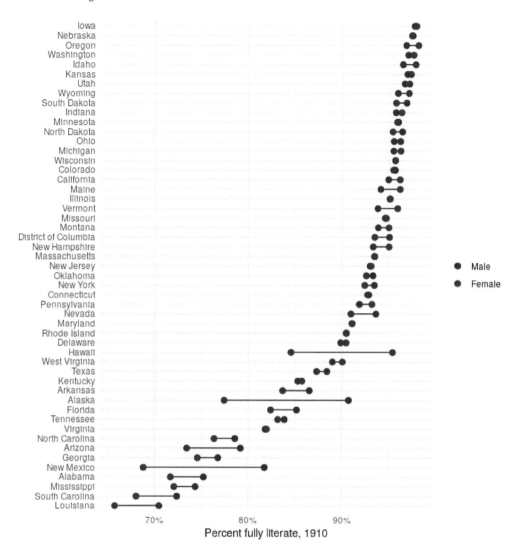

FIGURE 11.13 Literacy gaps by sex and state in 1910

The **usethis** package (Wickham and Bryan, 2021) is the most user-friendly way to work with the .Renviron file in R with its function edit_r_environ(). Calling this function will bring up the .Renviron file in a text editor, allowing users to set environment variables that will be made available to R when R starts up.

```
library(usethis)
```

```
edit_r_environ()
```

Add the line CENSUS_KEY='YOUR KEY HERE' to your .Renviron file, replacing the YOUR KEY HERE text with your API key, then restart R to get access to your key.

TABLE 11.2 Data from the 2017 Economic Census acquired with the censusapi package

state	county	EMP	PAYANN	GEO_ID	NAICS2017
48	373	1041	15256	0500000US48373	72
48	391	246	3242	0500000US48391	72
48	467	1254	19187	0500000US48467	72
48	055	910	15247	0500000US48055	72
48	383	140	2178	0500000US48383	72
48	501	147	1867	0500000US48501	72
48	387	0	0	0500000US48387	72
48	399	170	1595	0500000US48399	72
48	407	0	0	0500000US48407	72
48	435	195	4063	0500000US48435	72

censusapi's core function is getCensus(), which translates R code to Census API queries. The name argument references the API name; the censusapi documentation[11] or the function listCensusApis() helps you understand how to format this.

The example below makes a request to the Economic Census API[12], getting data on employment and payroll for NAICS code 72 (accommodation and food services businesses) in counties in Texas in 2017.

```
library(censusapi)

tx_econ17 <- getCensus(
  name = "ecnbasic",
  vintage = 2017,
  vars = c("EMP", "PAYANN", "GEO_ID"),
  region = "county:*",
  regionin = "state:48",
  NAICS2017 = 72
)
```

censusapi can also be used in combination with other packages covered in this book such as **tigris** for mapping. The example below uses the Small Area Health Insurance Estimates API[13], which delivers modeled estimates of health insurance coverage at the county level with various demographic breakdowns. Using **censusapi** and **tigris**, we can retrieve data on the percent of the population below age 19 without health insurance for all counties in the United States. This information will be joined to a counties dataset from **tigris** with shifted geometry, then mapped with **ggplot2**.

```
library(tigris)
library(tidyverse)

us_youth_sahie <- getCensus(
  name = "timeseries/healthins/sahie",
```

---

[11] https://www.hrecht.com/censusapi/articles/example-masterlist.html
[12] https://www.census.gov/programs-surveys/economic-census/data/api.html
[13] https://www.census.gov/programs-surveys/sahie/data/api.html

11.3 Other US government datasets

```
  vars = c("GEOID", "PCTUI_PT"),
  region = "county:*",
  regionin = "state:*",
  time = 2019,
  AGECAT = 4
)

us_counties <- counties(cb = TRUE, resolution = "20m", year = 2019) %>%
  shift_geometry(position = "outside") %>%
  inner_join(us_youth_sahie, by = "GEOID")

ggplot(us_counties, aes(fill = PCTUI_PT)) +
  geom_sf(color = NA) +
  theme_void() +
  scale_fill_viridis_c() +
  labs(fill = "% uninsured ",
       caption = "Data source: SAHIE via the censusapi R package",
       title = "  Percent uninsured under age 19 by county, 2019")
```

As these are *modeled* estimates, state-level influences on the county-level estimates are apparent on the map in Figure 11.14.

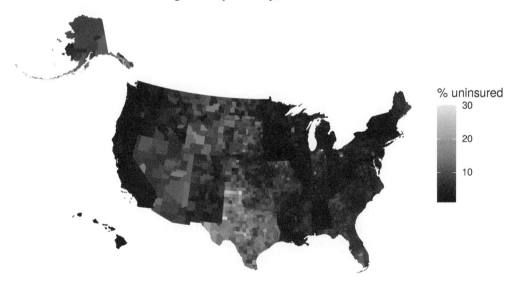

FIGURE 11.14 Map of youth without health insurance using censusapi and tigris

11.3.2 Analyzing labor markets with lehdr

Another very useful package for working with Census Bureau data is the **lehdr** R package (Green et al., 2021), which access the Longitudinal and Employer-Household Dynamics (LEHD) Origin-Destination Employment Statistics (LODES) data. LODES is not available from the Census API, meriting an alternative package and approach. LODES includes synthetic estimates of residential, workplace, and residential-workplace links at the Census block level, allowing for highly detailed geographic analysis of jobs and commuter patterns over time.

The core function implemented in **lehdr** is `grab_lodes()`, which downloads a LODES file of a specified `lodes_type` (either `"rac"` for residential, `"wac"` for workplace, or `"od"` for origin-destination) for a given state and year. While the raw LODES data are available at the Census block level, the `agg_geo` parameter offers a convenient way to roll up estimates to higher levels of aggregation. For origin-destination data, the `state_part = "main"` argument below captures within-state commuters; use `state_part = "aux"` to get commuters from out-of-state. The optional argument `use_cache = TRUE` stores downloaded LODES data in a cache directory on the user's computer; this is recommended to avoid having to re-download data for future analyses.

```
library(lehdr)
library(tidycensus)
library(sf)
library(tidyverse)
library(tigris)

wa_lodes_od <- grab_lodes(
  state = "wa",
  year = 2018,
  lodes_type = "od",
  agg_geo = "tract",
  state_part = "main",
  use_cache = TRUE
)
```

The result is a dataset that shows tract-to-tract commute flows (`S000`) and broken down by a variety of characteristics, referenced in the LODES documentation[14].

lehdr can be used for a variety of purposes including transportation and economic development planning. For example, the workflow below illustrates how to use LODES data to understand the origins of commuters to the Microsoft campus (represented by its Census tract) in Redmond, Washington. Commuters from LODES will be normalized by the total population age 18 and up, acquired with **tidycensus** for 2018 Census tracts in Seattle-area counties. The dataset `ms_commuters` will include Census tract geometries (obtained with `geometry = TRUE` in **tidycensus**) and an estimate of the number of Microsoft-area commuters per 1000 adults in that Census tract.

```
seattle_adults <- get_acs(
  geography = "tract",
```

---
[14] https://lehd.ces.census.gov/data/lodes/LODES7/LODESTechDoc7.5.pdf

11.3 Other US government datasets

TABLE 11.3 Origin-destination data acquired with the lehdr package

year	state	w_tract	h_tract	S000	SA01	SA02	SA03
2018	WA	53001950100	53001950100	515	109	241	165
2018	WA	53001950100	53001950200	75	12	40	23
2018	WA	53001950100	53001950300	23	2	20	1
2018	WA	53001950100	53001950400	13	3	6	4
2018	WA	53001950100	53001950500	27	5	18	4
2018	WA	53001950100	53003960100	1	0	0	1
2018	WA	53001950100	53003960300	1	0	0	1
2018	WA	53001950100	53003960400	1	1	0	0
2018	WA	53001950100	53003960500	1	0	1	0
2018	WA	53001950100	53005010201	1	1	0	0

```r
  variables = "S0101_C01_026",
  state = "WA",
  county = c("King", "Kitsap", "Pierce",
             "Snohomish"),
  year = 2018,
  geometry = TRUE
)

microsoft <- filter(wa_lodes_od, w_tract == "53033022803")

ms_commuters <- seattle_adults %>%
  left_join(microsoft, by = c("GEOID" = "h_tract")) %>%
  mutate(ms_per_1000 = 1000 * (S000 / estimate)) %>%
  st_transform(6596) %>%
  erase_water(area_threshold = 0.99)
```

The result can be visualized on a map with **ggplot2** as in Figure 11.15, or alternatively with any of the mapping tools introduced in Chapter 6. Note the use of the erase_water() function from the **tigris** package introduced in Section 7.5 with a high area threshold to remove large bodies of water like Lake Washington from the Census tract shapes.

```r
ggplot(ms_commuters, aes(fill = ms_per_1000)) +
  geom_sf(color = NA) +
  theme_void() +
  scale_fill_viridis_c(option = "cividis") +
  labs(title = "Microsoft commuters per 1000 adults",
       subtitle = "2018 LODES data, Seattle-area counties",
       fill = "Rate per 1000")
```

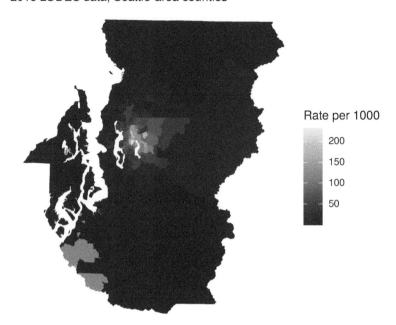

FIGURE 11.15 Map of commuter origins to the Microsoft Campus area, Seattle-area counties

11.3.3 Bureau of Labor Statistics data with blscrapeR

Another essential resource for data on US labor market characteristics is the Bureau of Labor Statistics[15], which makes their data available to users via an API. BLS data can be accessed from R with the **blscrapeR** package (Eberwein, 2021). Like other government APIs, the BLS API requires an API key, obtained from https://data.bls.gov/registrationEngine/. Sign up with an email address then validate the key using the link sent to your email, then return to R to get started with **blscrapeR**.

```
# remotes::install_github("keberwein/blscrapeR")
library(blscrapeR)

set_bls_key("YOUR KEY HERE")
```

blscrapeR includes a number of helper functions to assist users with some of the most commonly requested statistics. For example, the `get_bls_county()` function fetches employment data for all counties in the United States for a given month.

```
latest_employment <- get_bls_county("May 2021")
```

---

[15] https://www.bls.gov/

11.3 Other US government datasets

TABLE 11.4 May 2021 employment data from the BLS

area_code	fips_state	fips_county	area_title	period	labor_force	employed	unemployed
CN0100100000000	01	001	Autauga County, AL	2021-05-01	26257	25565	692
CN0100300000000	01	003	Baldwin County, AL	2021-05-01	100091	97283	2808
CN0100500000000	01	005	Barbour County, AL	2021-05-01	8148	7704	444
CN0100700000000	01	007	Bibb County, AL	2021-05-01	8554	8250	304
CN0100900000000	01	009	Blount County, AL	2021-05-01	24977	24421	556
CN0101100000000	01	011	Bullock County, AL	2021-05-01	4412	4239	173
CN0101300000000	01	013	Butler County, AL	2021-05-01	8739	8289	450
CN0101500000000	01	015	Calhoun County, AL	2021-05-01	46083	44232	1851
CN0101700000000	01	017	Chambers County, AL	2021-05-01	15639	15057	582
CN0101900000000	01	019	Cherokee County, AL	2021-05-01	12029	11741	288

TABLE 11.5 Series IDs for use in blscrapeR

series_title	series_id	seasonal	periodicity_code
(Seas) Population Level	LNS10000000	S	M
(Seas) Population Level	LNS10000000Q	S	Q
(Seas) Population Level – Men	LNS10000001	S	M
(Seas) Population Level – Men	LNS10000001Q	S	Q
(Seas) Population Level – Women	LNS10000002	S	M
(Seas) Population Level – Women	LNS10000002Q	S	Q

More complex queries are also possible with the `bls_api()` function; however, this requires knowing the BLS *series ID*. Series IDs are composite strings made up of several alphanumeric codes that can represent different datasets, geographies, and industries, among others. **blscrapeR** includes an internal dataset, `series_ids`, to help users view some of the most commonly requested IDs.

```
head(series_ids)
```

`series_ids` only contains indicators from the Current Population Survey and Local Area Unemployment Statistics databases. Other data series can be constructed by following the instructions on <https://www.bls.gov/help/hlpforma.htm>[16]. The example below illustrates this for data on the Current Employment Statistics[17], for which the series ID is formatted as follows[18]:

- Positions 1–2: The prefix (in this example, SA)

- Position 3: The seasonal adjustment code (either S, for seasonally adjusted, or U, for unadjusted)

- Positions 4–5: The two-digit state FIPS code

- Positions 6–10: The five-digit area code[19], which should be set to 00000 if an entire state is requested.

- Positions 11–18: The super sector/industry code[20]

---

[16] https://www.bls.gov/help/hlpforma.htm
[17] https://www.bls.gov/ces/
[18] https://www.bls.gov/help/hlpforma.htm#SA
[19] https://download.bls.gov/pub/time.series/sm/sm.area
[20] https://download.bls.gov/pub/time.series/sm/sm.industry

TABLE 11.6 Accommodations employment in Maui, Hawaii

year	period	periodName	latest	value	footnotes	seriesID
2022	M03	March	true	11.5	P Preliminary	SMU15279807072100001
2022	M02	February	NA	11.3		SMU15279807072100001
2022	M01	January	NA	11.1		SMU15279807072100001
2021	M12	December	NA	11.4		SMU15279807072100001
2021	M11	November	NA	10.9		SMU15279807072100001
2021	M10	October	NA	10.7		SMU15279807072100001

- Positions 19–20: The data type code[21]

We'll use this information to get unadjusted data on accommodation workers in Maui, Hawaii since the beginning of 2018. Note that the specific codes detailed above will vary from indicator to indicator, and that not all industries will be available for all geographies.

```
maui_accom <- bls_api(seriesid = "SMU15279807072100001",
                      startyear = 2018)
```

This time series can be visualized with **ggplot2**:

```
maui_accom %>%
  mutate(period_order = rev(1:nrow(.))) %>%
  ggplot(aes(x = period_order, y = value, group = 1)) +
  geom_line(color = "darkgreen") +
  theme_minimal() +
  scale_y_continuous(limits = c(0, max(maui_accom$value) + 1)) +
  scale_x_continuous(breaks = seq(1, nrow(maui_accom), 12),
                     labels = paste("Jan", 2018:2022)) +
  labs(x = "",
       y = "Number of jobs (in 1000s)",
       title = "Accommodation employment in Maui, Hawaii",
       subtitle = "Data source: BLS Current Employment Statistics")
```

The data suggest a stable accommodations sector prior to the onset of the COVID-19 pandemic, with recovery in the industry starting toward the end of 2020.

11.3.4 Working with agricultural data with tidyUSDA

Agriculture can be a difficult sector on which to collect market statistics, as it is not available in many data sources such as LODES or the BLS Current Employment Statistics. Dedicated statistics on US agriculture can be acquired with the **tidyUSDA** R package (Lindblad, 2021), which interacts with USDA data resources using tidyverse-centric workflows. Get an API key at https://quickstats.nass.usda.gov/api and use that to request data from the USDA QuickStats API. In this example, the API key is stored as a "USDA_KEY" environment variable.

---

[21]https://download.bls.gov/pub/time.series/sm/sm.data_type

11.3 Other US government datasets

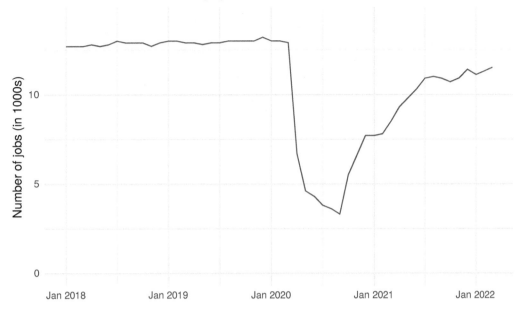

FIGURE 11.16 Time-series of accommodations employment in Maui, Hawaii

```
library(tidyUSDA)

usda_key <- Sys.getenv("USDA_KEY")
```

Now, let's see which US counties produce have the most acres devoted to cucumbers. To use the getQuickstat() function effectively, it is helpful to construct a query first at https://quickstats.nass.usda.gov/ and see what options are available, then bring those options as arguments into R.

```
cucumbers <- getQuickstat(
  key = usda_key,
  program = "CENSUS",
  data_item = "CUCUMBERS, FRESH MARKET - ACRES HARVESTED",
  sector = "CROPS",
  group = "VEGETABLES",
  commodity = "CUCUMBERS",
  category = "AREA HARVESTED",
  domain = "TOTAL",
  geographic_level = "COUNTY",
  year = "2017"
)
```

TABLE 11.7 Top cucumber-growing counties in the United States

state_name	county_name	Value
FLORIDA	MANATEE	2162
MICHIGAN	BERRIEN	1687
CALIFORNIA	SAN JOAQUIN	1558
GEORGIA	BROOKS	1495
NORTH CAROLINA	SAMPSON	1299
FLORIDA	HILLSBOROUGH	1070
MICHIGAN	BAY	453
NEW JERSEY	GLOUCESTER	438

This information allows us to look at the top cucumber-growing counties in the United States, shown in Table 11.7.

```
cucumbers %>%
  select(state_name, county_name, Value) %>%
  arrange(desc(Value))
```

11.4 Getting government data without R packages

The breath of R's developer ecosystem means that in many cases if you have a need for US government data, an enterprising developer has written some code that you can use. However, this won't be true for every data resource, especially as US government agencies continue to release new API endpoints to open up access to data. In this case, you may be interested in writing your own data access functions for this purpose. This section gives some pointers on how to make HTTP requests to data APIs using the httr package and process the data appropriately for use in R.

11.4.1 Making requests to APIs with httr

Most data API packages in R will rely on the **httr** R package (Wickham, 2020), which provides R functions for common HTTP requests such as GET, POST, and PUT, among others. In this example, we'll use the httr::GET() function to make a request to the new Department of Housing and Urban Development Comprehensive Affordable Housing Strategy (CHAS) API[22]. This will require getting a HUD API token and storing it as an environment variable as described earlier in this chapter.

Every API will be structured differently with respect to how it accepts queries and authenticates with API tokens. In this example, the documentation instructs users to pass the API token in the HTTP request with the Authentication: Bearer header. The example URL in the documentation, https://www.huduser.gov/hudapi/public/chas?type=3&year=2012-2016&stateId=51&entityId=59[23], is passed to GET(), and the add_headers() function in **httr** is used to assemble an appropriate string to send to the API.

---

[22]https://www.huduser.gov/portal/dataset/chas-api.html

[23]https://www.huduser.gov/hudapi/public/chas?type=3&year=2012-2016&stateId=51&entityId=59

11.4 Getting government data without R packages

TABLE 11.8 CHAS API request as a data frame

geoname	sumlevel	year	A1	A2	A3	A4	A5
Fairfax County, Virginia	County	2012-2016	12895.0	21805.0	34700.0	14165.0	17535.0

```
library(glue)
library(httr)
library(jsonlite)
library(tidyverse)

my_token <- Sys.getenv("HUD_TOKEN")

hud_chas_request <- GET(
  paste0("https://www.huduser.gov/hudapi/public/chas",
         "?type=3&year=2012-2016&stateId=51&entityId=59"),
  add_headers(Authorization = glue("Bearer {my_token}"))
)

hud_chas_request$status_code
```

```
## [1] 200
```

As we got back the HTTP status code 200, we know that our request to the API succeeded. To view the returned data, we use the `content()` function and translate the result to text with the `as = "text"` argument.

```
content(hud_chas_request, as = "text")
```

The data were returned as JavaScript Object Notation, or JSON[24], a very common data format for returning data from APIs. While the data shown above is moderately interpretable, it will require some additional translation for use in R. The **jsonlite** package (Ooms, 2014) includes a powerful set of tools for working with JSON objects in R; here, we'll use the `fromJSON()` function to convert JSON to a data frame.

```
hud_chas_request %>%
  content(as = "text") %>%
  fromJSON()
```

The data are now formatted as an R data frame, which can be embedded in your R-based workflows.

11.4.2 Writing your own data access functions

If you plan to use a particular API endpoint frequently, or for different areas and times, it is a good idea to write a data access function. API functions should ideally identify the components of the API request that can vary, and allow users to make modifications in ways that correspond to those components.

---

[24]https://www.json.org/json-en.html

The URL used in the example above can be separated into two components: the *API endpoint* https://www.huduser.gov/hudapi/public/chas[25], which will be common to all API requests, and the *query* ?type=3&year=2012-2016&stateId=51&entityId=59, which shows components of the request that can vary. The query will generally follow a question mark sign and be composed of key-value pairs, e.g. type=3, which in this example means county according to the HUD API documentation. These queries can be passed as a list to the query parameter of GET() and formatted appropriately in the HTTP request by **httr**.

The function below is an example you can use to get started writing your own API data request functions. The general components of the function are as follows:

- The function is defined by a name, get_hud_chas(), with a number of parameters to which the user will supply arguments. Default arguments can be specified (e.g. year = "2013-2017" to request that data without the user needing to specify it directly. If a default argument is NULL, it will be ignored by the GET() request.

- The function first checks to see if an existing HUD token exists in the user's environment referenced by the name HUD_TOKEN if the user hasn't supplied it directly.

- Next, the function constructs the GET() query, translating the function arguments to an HTTP query.

- If the request has failed, the function returns the error message from the API to help inform the user about what went wrong.

- If the request succeeds, the function converts the result to text, then the JSON to a data frame which will be pivoted to long form using **tidyr**'s pivot_longer() function. Finally, the result is returned. I use the return() function here, which isn't strictly necessary but I use it because it is helpful for me to see where the function exits in my code.

- You may notice a couple new notational differences from code used elsewhere in this book. Functions are referenced with the packagename::function() notation, and the new base R pipe |> is used instead of the magrittr pipe %>%. This allows the function to run without loading any external libraries with library(), which is generally a good idea when writing functions to avoid namespace conflicts.

```
get_hud_chas <- function(
  type,
  year = "2013-2017",
  state_id = NULL,
  entity_id = NULL,
  token = NULL
) {

  # Check to see if a token exists
  if (is.null(token)) {
    if (Sys.getenv("HUD_TOKEN") != "") {
      token <- Sys.getenv("HUD_TOKEN")
    }
  }

  # Specify the base URL
```

---

[25]https://www.huduser.gov/hudapi/public/chas

```
  base_url <- "https://www.huduser.gov/hudapi/public/chas"

  # Make the query
  hud_query <- httr::GET(
    base_url,
    httr::add_headers(Authorization = glue::glue("Bearer {token}")),
    query = list(
      type = type,
      year = year,
      stateId = state_id,
      entityId = entity_id)
  )

  # Return the HTTP error message if query failed
  if (hud_query$status_code != "200") {
    msg <- httr::content(hud_query, as = "text")
    return(msg)
  }

  # Return the content as text
  hud_content <- httr::content(hud_query, as = "text")

  # Convert the data to a long-form tibble
  hud_tibble <- hud_content |>
    jsonlite::fromJSON() |>
    dplyr::as_tibble() |>
    tidyr::pivot_longer(cols = !dplyr::all_of(c("geoname", "sumlevel", "year")),
                        names_to = "indicator",
                        values_to = "value")

  return(hud_tibble)

}
```

Now, let's try out the function for a different locality in Virginia:

```
get_hud_chas(type = 3, state_id = 51, entity_id = 510)
```

The function returns affordable data for Alexandria, Virginia (shown in Table 11.9) in a format that is more suitable for use with tidyverse tools.

11.5 Exercises

- Visit `https://ipums.org` and spend some time exploring NHGIS and/or IPUMS. Attempt to replicate the NHGIS workflow for a different variable and time period.

TABLE 11.9 Affordable housing data for Alexandria, Virginia using a custom-built API function

geoname	sumlevel	year	indicator	value
Alexandria city, Virginia	County	2013-2017	A1	1185.0
Alexandria city, Virginia	County	2013-2017	A2	6745.0
Alexandria city, Virginia	County	2013-2017	A3	7930.0
Alexandria city, Virginia	County	2013-2017	A4	1335.0
Alexandria city, Virginia	County	2013-2017	A5	5355.0
Alexandria city, Virginia	County	2013-2017	A6	6690.0
Alexandria city, Virginia	County	2013-2017	A7	1245.0
Alexandria city, Virginia	County	2013-2017	A8	3580.0
Alexandria city, Virginia	County	2013-2017	A9	4825.0
Alexandria city, Virginia	County	2013-2017	A10	1465.0

- Explore the **censusapi** package and create a request to a Census API endpoint using a dataset that you weren't previously familiar with. Join the data to shapes you've obtained with **tigris** and make a map of your data.

12

Working with Census data outside the United States

Although the methods presented in this book are generalizable, the examples given in the book to this point have focused on the United States. Readers of this book are most certainly not limited to the United States in their projects, and will be interested in how to apply the methods presented to examples around the world. International Census & demographic data is the focus of this chapter. The first section focuses on *global demographic indicators* for between-country comparisons and gives an overview of the **idbr** R package for accessing these indicators through the US Census Bureau's International Data Base[1]. The sections that follow cover country-specific packages from around the world that deliver Census data to R users in similar ways to the US-focused packages covered earlier in this book. Examples from Canada, Mexico, Brazil, and Kenya illustrate how to apply the methods readers have learned in this book to Census data analyses in other parts of the world.

12.1 The International Data Base and the idbr R package

The US Census Bureau's International Database (IDB) is a repository of dozens of demographic indicators for over 200 countries around the world. The IDB includes both historical information by year for most countries as well as population projections to 2100. In-place demographic characteristics are derived from a wide range of international Censuses and surveys, and future projections are estimated with the cohort-component method.

The Census Bureau makes these datasets available to researchers through an interactive data tool[2] and also through its API, allowing for programmatic data access. The **idbr** R package (Walker, 2021b), first released in 2016 and updated to version 1.0 in 2021, uses a simple R interface to help users gain access to and analyze data from the IDB.

To get started with **idbr**, users can install the package from CRAN and set their Census API keys with the `idb_api_key()` function. If a user's Census API key is already installed using **tidycensus**, **idbr** will pick it up making this step unnecessary.

```
library(idbr)
# Unnecessary if API key is installed with tidycensus
idb_api_key("YOUR KEY GOES HERE")
```

The core function implemented in **idbr** is `get_idb()`, which can access all datasets available in the IDB. There are two main datasets available: the 1-year-of-age population dataset, which allows for population data by country to be downloaded broken down by age and sex;

---

[1] https://www.census.gov/programs-surveys/international-programs/about/idb.html
[2] https://www.census.gov/data-tools/demo/idb/#/country?YR_ANIM=2021

TABLE 12.1 Nigeria population data acquired with idbr

code	year	name	pop
NG	1990	Nigeria	97041753
NG	1991	Nigeria	99406734
NG	1992	Nigeria	101837787
NG	1993	Nigeria	104387403
NG	1994	Nigeria	107053762
NG	1995	Nigeria	109838554
NG	1996	Nigeria	112744016
NG	1997	Nigeria	115773046
NG	1998	Nigeria	118927944
NG	1999	Nigeria	122200766

and the 5-year-of-age dataset, which gives access to population by 5-year age bands but also many other fertility, mortality, and migration indicators that are not organized by age bands.

A basic IDB query uses a country, specified as either an ISO-2 code or a country name in English, a vector of one or more variables, and a vector of one or more years. For example, we can fetch historical and projected population for Nigeria from 1990 to 2100:

```
nigeria_pop <- get_idb(
  country = "Nigeria",
  variables = "pop",
  year = 1990:2100
)
```

The result can then be visualized as a time series using **ggplot2**.

```
library(tidyverse)

ggplot(nigeria_pop, aes(x = year, y = pop)) +
  geom_line(color = "darkgreen") +
  theme_minimal() +
  scale_y_continuous(labels = scales::label_number_si()) +
  labs(title = "Population of Nigeria",
       subtitle = "1990 to 2100 (projected)",
       x = "Year",
       y = "Population at midyear")
```

idbr includes functionality to help users look up variable codes for use in their function calls. The function `idb_variables()` returns a data frame with all variables available in the IDB, along with an informative label; the `idb_concepts()` function prints out a list of *concepts* that can be supplied to the concept parameter and will return a group of variables that belong to the same concept (e.g. mortality rates, components of population growth).

For worldwide analyses, users can specify the argument `country = "all"` to return all available countries in the IDB. In concert with tidyverse tools, this makes global comparative

12.1 The International Data Base and the idbr R package

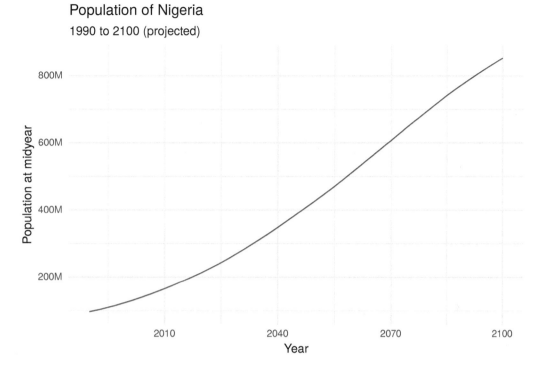

FIGURE 12.1 Historical and projected population of Nigeria

analyses straightforward. In the example below, we get data on life expectancy at birth for all countries in 2021, then view the top 10 and bottom 10 countries.

```
world_lex <- get_idb(
  country = "all",
  variables = "e0",
  year = 2021
)

world_lex %>%
  slice_max(e0, n = 10)
```

```
world_lex %>%
  slice_min(e0, n = 10)
```

12.1.1 Visualizing IDB data

idbr aims to return data from the IDB in a structure suitable for creative data visualizations. This includes population pyramids (like those introduced in Chapter 4) and world maps.

As mentioned earlier in this chapter, `get_idb()` can retrieve population data by age and sex in one–year age bands if the `age` or `sex` arguments are supplied but `variables` is left

TABLE 12.2 Countries with the longest life expectancies at birth, 2021

code	year	name	e0
MC	2021	Monaco	89.40
SG	2021	Singapore	86.19
MO	2021	Macau	84.81
JP	2021	Japan	84.65
SM	2021	San Marino	83.68
CA	2021	Canada	83.62
IS	2021	Iceland	83.45
HK	2021	Hong Kong	83.41
AD	2021	Andorra	83.23
IL	2021	Israel	83.15

TABLE 12.3 Countries with the shortest life expectancies at birth, 2021

code	year	name	e0
AF	2021	Afghanistan	53.25
CF	2021	Central African Republic	55.07
SO	2021	Somalia	55.32
MZ	2021	Mozambique	56.49
SL	2021	Sierra Leone	58.45
SS	2021	South Sudan	58.60
TD	2021	Chad	58.73
LS	2021	Lesotho	58.90
SZ	2021	Eswatini	59.13
NE	2021	Niger	59.70

blank. An illustration of this is found in the example below, where `get_idb()` is used to get 1-year-of-age population data for 2021 in Japan, broken down by sex.

```
japan_data <- get_idb(
  country = "Japan",
  year = 2021,
  age = 0:100,
  sex = c("male", "female")
)
```

Following the example illustrated in Section 4.5.2, a population pyramid can be created by changing data values for one sex to negative, then plotting the data as back-to-back horizontal bar charts.

```
japan_data %>%
  mutate(pop = ifelse(sex == "Male", pop * -1, pop)) %>%
  ggplot(aes(x = pop, y = as.factor(age), fill = sex)) +
  geom_col(width = 1) +
  theme_minimal(base_size = 12) +
  scale_x_continuous(labels = function(x) paste0(abs(x / 1000000), "m")) +
```

12.1 The International Data Base and the idbr R package

TABLE 12.4 1-year age band population data for Japan in 2021

code	year	name	pop	sex	age
JP	2021	Japan	450869	Male	0
JP	2021	Japan	458588	Male	1
JP	2021	Japan	478968	Male	2
JP	2021	Japan	503320	Male	3
JP	2021	Japan	519507	Male	4
JP	2021	Japan	535623	Male	5
JP	2021	Japan	544629	Male	6
JP	2021	Japan	553183	Male	7
JP	2021	Japan	563710	Male	8
JP	2021	Japan	567511	Male	9

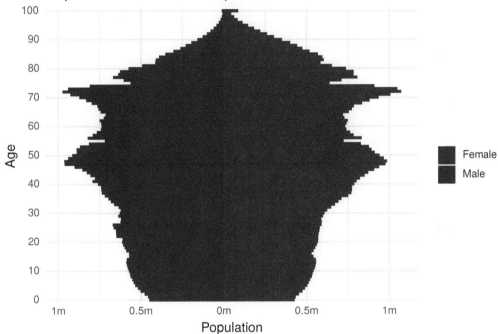

FIGURE 12.2 Population pyramid for Japan

```
scale_y_discrete(breaks = scales::pretty_breaks(n = 10)) +
scale_fill_manual(values = c("darkred", "navy")) +
labs(title = "Population structure of Japan in 2021",
     x = "Population",
     y = "Age",
     fill = "")
```

The aging structure of Japan's population is notable, especially the number of female centenarians (women aged 100 and up) that stand out in Figure 12.2, a number that is likely to grow in the years ahead.

`get_idb()` also includes a `geometry` parameter to help users make world maps of demographic indicators. By setting `geometry = TRUE`, `get_idb()` returns a simple features object with a geometry column attached to the demographic data. Here, we fetch data on total fertility rate by country in 2021.

```
library(idbr)
library(tidyverse)

fertility_data <- get_idb(
  country = "all",
  year = 2021,
  variables = "tfr",
  geometry = TRUE,
)
```

There are a couple points of note with respect to the data returned. The dataset leaves out a number of small countries that would not appear on a zoomed-out world map. To ensure that they are retained, use the argument `resolution = "high"` for a higher-resolution dataset (which you may also want for regional mapping). There are also rows returned as NA in the dataset; this is to fill in areas that should appear on the map (like Antarctica) but do not have data in the IDB.

The data returned can be mapped with `geom_sf()` from ggplot2 or any other mapping tool introduced in this book. Specifying a suitable map projection with `coord_sf()` for the whole world – like the Robinson projection used below – is recommended.

```
ggplot(fertility_data, aes(fill = tfr)) +
  theme_bw() +
  geom_sf() +
  coord_sf(crs = 'ESRI:54030') +
  scale_fill_viridis_c() +
  labs(fill = "Total fertility\nrate (2021)")
```

12.1.2 Interactive and animated visualization of global demographic data

In Chapter 6, readers learned how to create interactive maps of US Census data with the **tmap** and **leaflet** packages. While the same methods *can* be applied to the global data returned by `get_idb()`, the Web Mercator projection used by the Leaflet library is not ideal for world maps as it inflates the area of countries near the poles relative to countries near the Equator. This problem with Web Mercator was covered in Section 6.5.2 and is particularly important for global maps. In turn, it is preferable to turn to other interactive charting libraries such as **ggiraph**.

To get started, let's return to our global fertility data and create a new column that will store information we want to include in our tooltip that appears when a viewer hovers over a country.

12.1 The International Data Base and the idbr R package

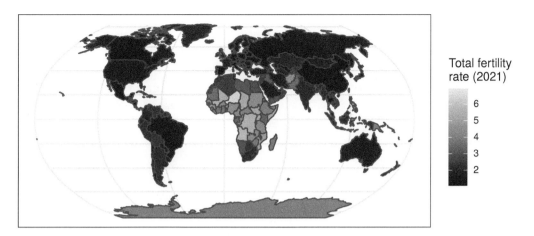

FIGURE 12.3 Map of total fertility rates by country in 2021

```
fertility_data$tooltip <- paste0(fertility_data$name,
                                 ": ",
                                 round(fertility_data$tfr, 2))
```

We now set up the visualization. Note that instead of using geom_sf(), we use **ggiraph**'s geom_sf_interactive(), which includes some new aesthetics. We map our new tooltip column to the tooltip aesthetic, which will generate a hover tooltip, and we map the code column to the data_id aesthetic, which will change the country's color when hovered over. The girafe() function then renders the interactive graphic; we can specify that we want map zooming with girafe_options().

```
library(ggiraph)

fertility_map <- ggplot(fertility_data, aes(fill = tfr)) +
  theme_bw() +
  geom_sf_interactive(aes(tooltip = tooltip, data_id = code), size = 0.1) +
  coord_sf(crs = 'ESRI:54030') +
  scale_fill_viridis_c() +
  labs(fill = "Total fertility\nrate (2021)")

girafe(ggobj = fertility_map) %>%
  girafe_options(opts_zoom(max = 10))
```

The time-series data availability in the IDB also works well for creating animated time-series graphics with help from the **gganimate** package (Pedersen and Robinson, 2020). **gganimate** extends **ggplot2** to animate sequences of plots using intuitive syntax to **ggplot2** users. Earlier in this chapter, we learned how to create a population pyramid from IDB data for a single year (2021). If data are requested instead for multiple years, **gganimate** can be used to transition from year to year in an animated population pyramid, showing how the population structure of a country has evolved (and is projected to evolve) over time.

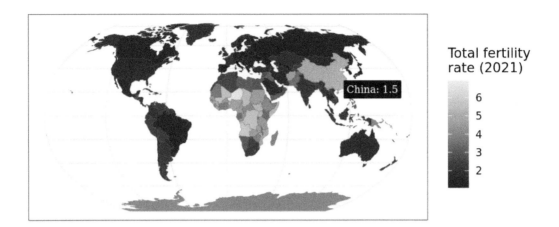

FIGURE 12.4 Interactive world map with ggiraph

```
library(idbr)
library(tidyverse)
library(gganimate)

mx_pyramid_data <- get_idb(
  country = "Mexico",
  sex = c("male", "female"),
  age = 0:100,
  year = 1990:2050
) %>%
  mutate(pop = ifelse(sex == "Male", pop * -1, pop))

mx_animation <- ggplot(mx_pyramid_data,
                      aes(x = pop,
                          y = as.factor(age),
                          fill = sex)) +
  geom_col(width = 1) +
  theme_minimal(base_size = 16) +
  scale_x_continuous(labels = function(x) paste0(abs(x / 1000000),
                                                 "m")) +
  scale_y_discrete(breaks = scales::pretty_breaks(n = 10)) +
  scale_fill_manual(values = c("darkred", "darkgreen")) +
  transition_states(year) +
  labs(title = "Population structure of Mexico in {closest_state}",
       x = "Population",
       y = "Age",
       fill = "")

animate(mx_animation, height = 500, width = 800)
```

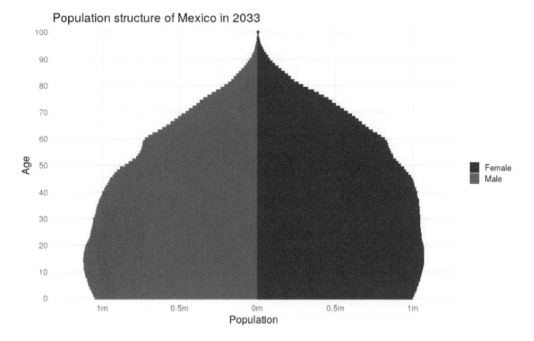

FIGURE 12.5 Animated time-series population pyramid of Mexico

The animation illustrates the aging of Mexico's population, both observed and projected, during the specified time period.

12.2 Country-specific Census data packages

Analysts around the world need access to high-quality tabular and spatial demographic data to make decisions much like the data that are available from the US Census Bureau. While Censuses are conducted by countries around the world, few countries expose their data by API in the way the US Census Bureau does. Nonetheless, there are a number of R packages that help analysts acquire and work with country-specific Census data. This section of the book covers a sampling of those packages, with examples given from Canada, Mexico, Kenya, and Brazil. These examples will illustrate how to apply many of the techniques introduced in this book to a variety of non-US examples.

12.2.1 Canada: cancensus

The **cancensus** R package (von Bergmann et al., 2021) grants comprehensive access to Canadian Census data through the CensusMapper APIs[3]. While Statistics Canada, the Canadian agency that distributes Canadian demographic data, does not maintain these APIs, package author Jens von Bergmann's CensusMapper product makes Canadian data accessible via a web interface and data API. Working with **cancensus** will feel familiar to **tidycensus** users, as **cancensus** integrates well within the tidyverse and includes an

---

[3] https://censusmapper.ca/

TABLE 12.5 List of Census vectors in cancensus

vector	type	label	units	parent_vector	aggregation	details
v_CA16_401	Total	Population, 2016	Number	NA	Additive	CA 2016 Census; Population and Dwellings; Population, 2016
v_CA16_402	Total	Population, 2011	Number	NA	Additive	CA 2016 Census; Population and Dwellings; Population, 2011
v_CA16_403	Total	Population percentage change, 2011 to 2016	Number	NA	Average of v_CA16_402	CA 2016 Census; Population and Dwellings; Population percentage change, 2011 to 2016
v_CA16_404	Total	Total private dwellings	Number	NA	Additive	CA 2016 Census; Population and Dwellings; Total private dwellings
v_CA16_405	Total	Private dwellings occupied by usual residents	Number	v_CA16_404	Additive	CA 2016 Census; Population and Dwellings; Total private dwellings; Private dwellings occupied by usual residents

option to return feature geometry along with demographic data for a wide range of Canadian geographies.

To get started with **cancensus**, obtain an API key from the CensusMapper website and store it in your .Renviron file as discussed in Section 11. Supplying this key to the cancensus.api_key option will allow the key to be picked up automatically by cancensus functions.

The list_census_vectors() function operates in a similar way to load_variables() in **tidycensus**. Specify a dataset to generate a browsable data frame of variable IDs, shown in Table 12.5, which can be used to query data from the CensusMapper API.

```
library(cancensus)
library(tidyverse)
options(cancensus.api_key = Sys.getenv("CANCENSUS_API_KEY"))

var_list <- list_census_vectors("CA16")
```

The data can be browsed with the View() function in RStudio much like it is recommended to do with **tidycensus** in Section 2.3. Two companion functions, list_census_regions() and list_census_datasets(), should also be used to identify region codes and dataset codes for which you'd like to request data.

Once determined, the appropriate codes can be passed as arguments to various parameters in the get_census() function, the main data access function used in **cancensus**. The example below fetches data on English speakers by Census tract in the Montreal area in this way.

```
montreal_english <- get_census(
  dataset = "CA16",
  regions = list(CMA = "24462"),
  vectors = "v_CA16_1364",
  level = "CT",
  geo_format = "sf",
  labels = "short"
)
```

The example above constructs the query to the CensusMapper API using the following arguments:

- dataset = "CA16" requests data from the 2016 Canadian Census;
- regions = list(CMA = "24462") gets data for the Montreal metropolitan area by using a named list, matching the region type CMA with the specific ID for the Montreal area;
- vectors = "v_CA16_1364" fetches data for the specific Census vector that represents English language speaking, as identified using list_census_vectors();
- level = "CT" gets data at the Census tract level;

12.2 Country-specific Census data packages

TABLE 12.6 Data on English speakers in Montreal by Census tract

Population	Households	GeoUID	Type	PR_UID	Shape Area	Dwellings	Adjusted Population (previous Census)	geometry
2638	1328	4620001.00	CT	24	0.46188	1452	2608	MULTIPOLYGON (((-73.50938 4...
3516	1762	4620002.00	CT	24	0.38927	1902	3039	MULTIPOLYGON (((-73.51732 4...
6373	2937	4620003.00	CT	24	0.74010	3103	6238	MULTIPOLYGON (((-73.52152 4...
3176	1603	4620004.00	CT	24	0.44828	1704	3261	MULTIPOLYGON (((-73.51911 4...
3060	1593	4620005.00	CT	24	0.56464	1749	3163	MULTIPOLYGON (((-73.50938 4...
4467	1993	4620006.00	CT	24	0.64815	2149	4458	MULTIPOLYGON (((-73.50779 4...

- `geo_format = "sf"` instructs `get_census()` to return simple feature geometry for Census tracts along with the Census data;
- `labels = "short"` returns simpler labels in the output dataset.

Once queried, our data appear as follows:

By default, cancensus returns a number of contextual variables along with the requested Census vector; total population, total households, total dwellings, and area can be used for normalization automatically. This information can be used to normalize data on English speakers by calculating a new column with `mutate()` named `pct_english`; familiar mapping tools from Chapter 6 are then used to visualize the data.

```
library(tmap)
tmap_mode("view")

montreal_pct <- montreal_english %>%
  mutate(pct_english = 100 * (v_CA16_1364 / Population))

tm_shape(montreal_pct) +
  tm_polygons(
    col = "pct_english",
    alpha = 0.5,
    palette = "viridis",
    style = "jenks",
    n = 7,
    title = "Percent speaking<br/>English at home"
  )
```

Exploring the data interactively reveals distinctive patterns of English language speaking in Montreal. While English is the most common language spoken at home in some central Census tracts and the southwestern portion of the Island of Montreal, percentages are much lower in suburban areas and the outer portions of the metropolitan region.

12.2.2 Kenya: rKenyaCensus

The **rKenyaCensus** package (Kariuki, 2020) makes indicators from the 2019 Kenya Population and Housing Census available to R users. This package is the result of a painstaking effort by the package author Shel Kariuki[4] to scrape data from PDFs containing the Census results – which is the way the original data were distributed – and convert them to R data frames. Kenyan Census tables install directly with the package and can be accessed by name. Let's take a look at table `V4_T2.30`, which contains information on religion by county in Kenya. We will assign this table to the variable `religion` for ease of reference.

---
[4] https://shelkariuki.netlify.app/

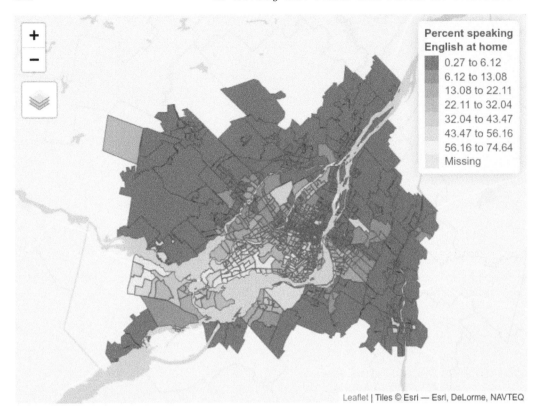

FIGURE 12.6 Interactive map of the percentage English language speakers at home in the Montreal area

```
# remotes::install_github("Shelmith-Kariuki/rKenyaCensus")
library(rKenyaCensus)
library(tidyverse)

religion <- V4_T2.30
```

The data are organized by county in Kenya with religious affiliations arranged by column. The first row represents data for the entirety of Kenya. Some additional data wrangling steps allow for the data to be readily plotted using familiar tooling. For example, the inclusion of the Total column makes calculation of proportions straightforward. In the example below, a new column prop_islam is computed and passed to a **ggplot2** bar chart.

TABLE 12.7 Religion data from the 2019 Kenyan Census

County	Total	Catholic	Protestant	Evangelical	AfricanInstituted	Orthodox	OtherChristian
KENYA	47213282	9726169	15777473	9648690	3292573	201263	1732911
MOMBASA	1190987	170797	241554	151939	46094	3104	100568
KWALE	858748	43624	116453	82176	22193	720	47892
KILIFI	1440958	95148	388893	307345	54980	2858	128298
TANA RIVER	314710	11306	22866	6791	4015	176	11148

12.2 Country-specific Census data packages

```
religion_prop <- religion %>%
  filter(County != "KENYA") %>%
  mutate(county_title = str_to_title(County),
         prop_islam = Islam / Total)

ggplot(religion_prop, aes(x = prop_islam,
                          y = reorder(county_title, prop_islam))) +
  geom_col(fill = "navy", alpha = 0.6) +
  theme_minimal(base_size = 12.5) +
  scale_x_continuous(labels = scales::percent) +
  labs(title = "Percent Muslim by County in Kenya",
       subtitle = "2019 Kenyan Census",
       x = "",
       y = "",
       caption = "Data source: rKenyaCensus R package")
```

The visualization in Figure 12.7 illustrates that most Kenyan counties have very small Muslim populations; however, a smaller number of counties have much larger Muslim populations, with percentages in some cases close to 100%. This raises related questions about the geography of these patterns, which can be mapped using additional functionality in **rKenyaCensus**.

The **rKenyaCensus** package includes a built-in county boundaries dataset to facilitate mapping of the various indicators in the Census, KenyaCounties_SHP. This object is of class SpatialPolygonsDataFrame, which will need to be converted to a simple features object with sf::st_as_sf().

```
library(sf)

kenya_counties_sf <- st_as_sf(KenyaCounties_SHP)

ggplot(kenya_counties_sf) +
  geom_sf() +
  theme_void()
```

With a little additional cleaning, the religion data can be joined to the county boundaries dataset, facilitating choropleth mapping with **ggplot2** and geom_sf().

```
kenya_islam_map <- kenya_counties_sf %>%
  mutate(County = str_remove(County, " CITY")) %>%
  left_join(religion_prop, by = "County")

ggplot(kenya_islam_map, aes(fill = prop_islam)) +
  geom_sf() +
  scale_fill_viridis_c(labels = scales::percent) +
  theme_void() +
  labs(fill = "% Muslim",
       title = "Percent Muslim by County in Kenya",
```

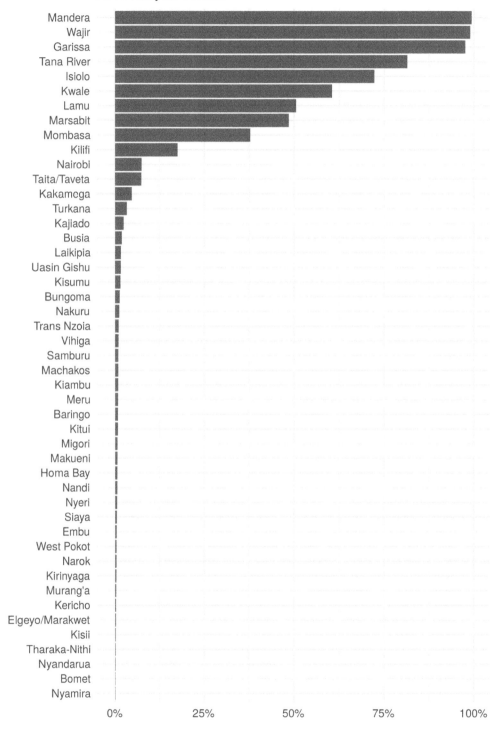

FIGURE 12.7 Bar chart of Islam prevalence by county in Kenya

12.2 Country-specific Census data packages   331

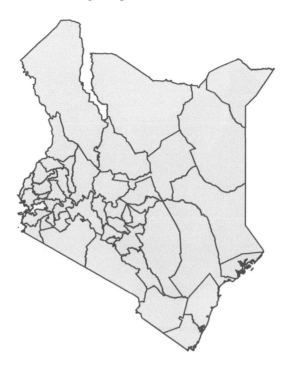

FIGURE 12.8 ggplot2 map of Kenya county boundaries

```
    subtitle = "2019 Kenyan Census",
    caption = "Data acquired with the rKenyaCensus R package")
```

The map in Figure 12.9 shows how the distinct religious divides in Kenya play out geographically in the country, with proportionally few Muslim residents in the western part of Kenya and much higher percentages in the eastern part of the country.

12.2.3 Mexico: combining mxmaps and inegiR

Mexico's national statistics office, the Instituto Nacional de Estadística y Geografía (INEGI)[5], offers an API for access to many of its datasets. Used together, the **mxmaps** (Valle-Jones, 2021) and **inegiR** (Flores, 2019) R packages allow for geographic analysis and visualization of Mexican Census data obtained from INEGI.

Like other APIs introduced in this book, the INEGI API requires an API token. This token can be requested from the INEGI website[6]. Once acquired, it can be used by both **mxmaps** and **inegiR** to request data; I'm saving my key as the environment variable INEGI_API_KEY.

The **mxmaps** package is based on the **choroplethr** R package (Lamstein, 2020), which offers a convenient interface for visualizing data from the US ACS. **mxmaps** wraps **inegiR**'s data access functions in the `choropleth_inegi()` function, which can request data for indicators from INEGI's data bank[7] and visualize them. The map below shows the percentage of

---

[5] https://www.inegi.org.mx/
[6] http://www3.inegi.org.mx//sistemas/api/indicadores/v1/tokenVerify.aspx
[7] https://en.www.inegi.org.mx/app/indicadores/

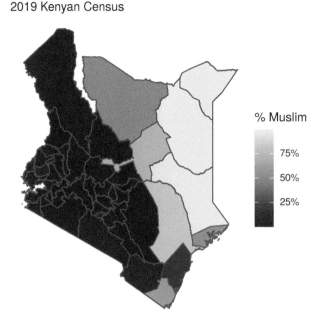

FIGURE 12.9 Choropleth map of Islam in Kenya

residents by state born in Mexico but outside their current state of residence using indicator code 304003001. For full functionality, the **inegiR** and **mxmaps** packages must both be installed from GitHub with `remotes::install_github()`.

```
# remotes::install_github("Eflores89/inegiR")
# remotes::install_github("diegovalle/mxmaps")
library(mxmaps)
token_inegi <- Sys.getenv("INEGI_API_KEY")

state_regions <- df_mxstate_2020$region
choropleth_inegi(token_inegi, state_regions,
                 indicator = "3104003001",
                 legend = "%",
                 title = "Percentage born outside\nstate of residence") +
  theme_void(base_size = 14) +
  labs(caption = "Data sources: INEGI, mxmaps R package")
```

Downloading data from the INEGI API

The map in Figure 12.10 illustrates that the states with the most internal migrants in Mexico include Baja California Sur, Quintana Roo, and Nuevo Leon. This information can also be accessed directly using the inegiR package. The `inegi_series()` function returns one state at a time, which can be combined into a country-wide dataset with `map_dfr()`.

12.2 Country-specific Census data packages

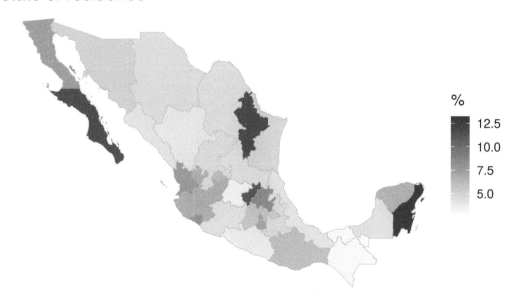

FIGURE 12.10 Choropleth map of internal migration in Mexico by state

```
library(inegiR)
token_inegi <- Sys.getenv("INEGI_API_KEY")
state_regions <- mxmaps::df_mxstate_2020$region

pct_migrants <- map_dfr(state_regions, ~{
  inegi_series(series_id = "3104003001",
               token = token_inegi,
               geography = .x,
               database = "BISE") %>%
    mutate(state_code = .x)
})
```

12.2.4 Brazil: aligning the geobr R package with raw Census data files for spatial analysis

The three country-specific examples shown above – Canada, Kenya, and Mexico – illustrate workflows for obtaining data directly from an R package and mapping the results. In many cases, however, data do not come directly with an R package, requiring analysts to research the raw data and work with raw files. The example below illustrates how to accomplish this for Brazil, where spatial data are found in a well-documented and widely used R package **geobr**, but demographic data will be downloaded and processed directly from the Instituto Brasileiro de Geografia e Estatística website using R tooling.

TABLE 12.8 Illustration of data acquired with inegiR

date	date_shortcut	values	notes	state_code
NA	Q1	6.182882		01
NA	Q1	7.725998		02
NA	Q1	11.723691		03
NA	Q1	4.443911		04
NA	Q1	4.581990		05
NA	Q1	8.135314		06
NA	Q1	3.030496		07
NA	Q1	4.665803		08
NA	Q1	7.377861		09
NA	Q1	4.055359		10

12.2.4.1 Spatial data for Brazil with geobr

The **geobr** (Pereira and Goncalves, 2021) package from IPEA[8] helps you load spatial datasets for many Brazilian geographies, including Census boundaries. These datasets are generously hosted by IPEA and downloaded into a user's R session when the corresponding function is called. The `code_tract` parameter can accept a 7-digit municipality code; below, the code for Rio de Janeiro is used. Municipality codes can be looked up with the corresponding `read_municipality()` function.

```
# Install first from GitHub:
# remotes::install_github("ipeaGIT/geobr", subdir = "r-package")
library(geobr)

rj_tracts <- read_census_tract(code_tract = 3304557)
```

By default, **geobr** retrieves a Census tract dataset with simplified geometries that is appropriate for small-scale demographic visualization but inappropriate for spatial analysis or large-scale visualization. A quick visualization in Figure 12.11 shows the Census tract geometries for Rio de Janeiro:

```
ggplot(rj_tracts) +
  geom_sf(lwd = 0.1) +
  theme_void()
```

12.2.4.2 Identifying demographic data resources for Brazil

For many demographic applications, users will want to identify Brazilian data that can be joined to the Census tract shapes. This makes for a useful exercise to illustrate an applied workflow for working with the raw tables. The example below will get data from the 2010 Brazilian Census and merge it to the geometries obtained with **geobr** for analysis.

Census data tables can be downloaded from the Instituto Brasileiro de Geografia e Estatística (IBGE) website's FTP page[9], which provides aggregated data at the Census tract level.

---

[8] https://www.ipea.gov.br/portal/
[9] https://ftp.ibge.gov.br/Censos/Censo_Demografico_2010/Resultados_do_Universo/Agregados_por_Setores_Censitarios/

12.2 Country-specific Census data packages

FIGURE 12.11 Basic ggplot2 map of Census tracts in Rio de Janeiro

These datasets are organized by state, so we'll need to identify the appropriate zipped folder to download and bring in the data we need into R. R's web access tools, discussed in Section 11, can be used to download this data. Our analysis will focus on Brasilia, the capital city of Brazil; in turn, we'll need data for the Distrito Federal (Federal District) in which Brasilia is located.

```
library(httr)

dir.create("brazil")

data_url <- paste0(
  "https://ftp.ibge.gov.br/Censos/",
  "Censo_Demografico_2010/Resultados_do_Universo/",
  "Agregados_por_Setores_Censitarios/DF_20171016.zip"
)

zip_name <- basename(data_url)

out_file <- file.path("brazil", zip_name)

GET(data_url, write_disk(out_file, overwrite = TRUE))

unzip(out_file, exdir = "brazil/brasilia_data")
```

The workflow above operates as follows:

- After loading in **httr**, we create a new subdirectory named `brazil` in our working directory. This will store the data we download from IBGE.
- We then identify the URL from which we'll be requesting data, and extract the basename from the URL with `basename()`; this will be the name of the file on disk.

TABLE 12.9 Data from the 2010 Brazilian Census

Cod_setor	Cod_Grandes Regiões	Nome_Grande_Regiao	Cod_UF	Nome_da_UF	Cod_meso	Nome_da_meso	Cod_micro
5.300108e+14	5	Região Centro-Oeste	53	Distrito Federal	5301	Distrito Federal	53001
5.300108e+14	5	Região Centro-Oeste	53	Distrito Federal	5301	Distrito Federal	53001
5.300108e+14	5	Região Centro-Oeste	53	Distrito Federal	5301	Distrito Federal	53001
5.300108e+14	5	Região Centro-Oeste	53	Distrito Federal	5301	Distrito Federal	53001
5.300108e+14	5	Região Centro-Oeste	53	Distrito Federal	5301	Distrito Federal	53001

- `file.path()` constructs the output location for the downloaded file, then `GET()` from httr makes the HTTP request, using `write_disk()` to write the downloaded file to the output location.
- Finally, `unzip()` unzips the folder. We can take a look at what we got back:

```
data_path <- "brazil/brasilia_data/DF/Base informa\x87oes setores2010
    universo DF/CSV"
```

```
list.files(data_path)
```

```
##  [1] "Basico_DF.csv"          "Domicilio01_DF.csv"
##  [3] "Domicilio02_DF.csv"     "DomicilioRenda_DF.csv"
##  [5] "Entorno01_DF.csv"       "Entorno02_DF.csv"
##  [7] "Entorno03_DF.csv"       "Entorno04_DF.csv"
##  [9] "Entorno05_DF.csv"       "Pessoa01_DF.csv"
## [11] "Pessoa02_DF.csv"        "Pessoa03_DF.csv"
## [13] "Pessoa04_DF.csv"        "Pessoa05_DF.csv"
## [15] "Pessoa06_DF.csv"        "Pessoa07_DF.csv"
## [17] "Pessoa08_DF.csv"        "Pessoa09_DF.csv"
## [19] "Pessoa10_DF.csv"        "Pessoa11_DF.csv"
## [21] "Pessoa12_DF.csv"        "Pessoa13_DF.csv"
## [23] "PessoaRenda_DF.csv"     "Responsavel01_DF.csv"
## [25] "Responsavel02_DF.csv"   "ResponsavelRenda_DF.csv"
```

The actual path, given that a subdirectory includes an accented character, may differ based on your operating system and/or your locale. Finally, we'll copy all the files from the nested subdirectory to the `brasilia_data` directory.

```
file.copy(data_path, "brazil/brasilia_data", recursive = TRUE)
```

12.2.4.3 Working with Brazilian demographic data

Our files include a series of CSV files that represent characteristics of Census tracts in the Federal District. Let's read in the first file, `Basico_DF.csv`, and see what we get back. An important step here is to use `read_csv2()` rather than `read_csv()`; this is because the file is semicolon-delimited and uses commas for decimal places, meaning that `read_csv()` will interpret it incorrectly. The use of accented characters in the raw data file also requires setting the encoding to Latin-1.

```
basico <- read_csv2("brazil/brasilia_data/CSV/Basico_DF.csv",
                    locale = locale(encoding = "latin1"))
```

12.2 Country-specific Census data packages

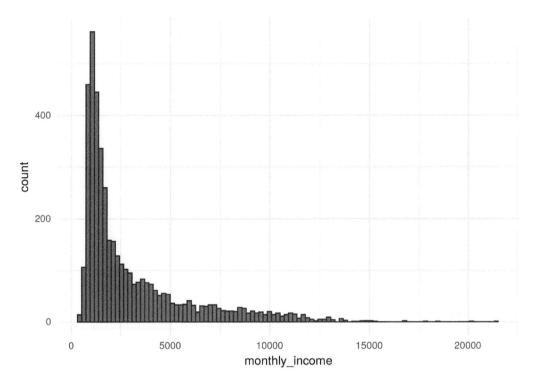

FIGURE 12.12 ggplot2 histogram of median monthly income by Census tract in Brasilia

The file includes a series of ID codes then a series of variables associated with each Census tract. The meaning of the variables in the dataset can be looked up from a PDF data dictionary available for download from the IBGE website[10]. We'll focus here on variable V007, which represents the average monthly income in each Census tract for income-earning households.

```
brasilia_income <- basico %>%
  mutate(code_tract = as.character(Cod_setor)) %>%
  select(code_tract, monthly_income = V007)

ggplot(brasilia_income,
       aes(x = monthly_income)) +
  geom_histogram(bins = 100,
                 alpha = 0.5,
                 color = "navy",
                 fill = "navy") +
  theme_minimal()
```

Typical of neighborhood income distributions, the histogram in Figure 12.12 is heavily right-skewed with a large cluster of Census tracts with low earnings and a long tail of wealthier areas.

---

[10]https://ftp.ibge.gov.br/Censos/Censo_Demografico_2010/Resultados_do_Universo/Agregados_por_Setores
_Censitarios/Documentacao_Agregado_dos_Setores_20180416.zip

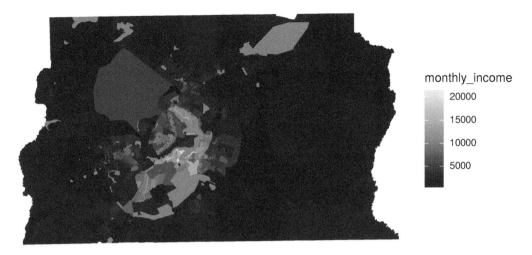

FIGURE 12.13 Choropleth map of median monthly income in Brasilia

12.2.4.4 Spatial analysis of Brazilian Census data

This information can be used in concert with the **geobr** package to analyze the geographic distribution of earnings using techniques covered elsewhere in this book. The first step of the analysis involves returning to **geobr** to get Census tracts for Brasilia/the Federal District using Brasilia's municipality code. The argument `simplified = FALSE` is used given the need for higher-quality tract boundaries for spatial analysis. Next, the income data is merged to the tract shapes, then transformed to an appropriate coordinate reference system for Brasilia.

```
library(sf)
library(geobr)

brasilia_tracts <- read_census_tract(
  code_tract = 5300108,
  simplified = FALSE
) %>%
  select(code_tract)

brasilia_income_geo <- brasilia_tracts %>%
  left_join(brasilia_income, by = "code_tract") %>%
  st_transform(22523)
```

After joining the data, the income information can be visualized on a map using familiar mapping tools like `geom_sf()`.

```
ggplot(brasilia_income_geo, aes(fill = monthly_income)) +
  geom_sf(color = NA) +
  theme_void() +
  scale_fill_viridis_c()
```

12.2 Country-specific Census data packages

The map in Figure 12.13 highlights income disparities between central Brasilia and the more rural portions of the Federal District, and identifies the Lago Sul area as home to the highest-earning households, on average.

As the spatial data have been prepared appropriately, methods from Section 7.7 can be used to explore patterns of spatial autocorrelation by monthly income in the Brasilia area. The example below uses the local form of the Moran's I statistic, described in more detail in Section 7.7.3. The steps used are described in brief below:

- A queens-case contiguity-based spatial weights matrix is created with functions from the **spdep** package, using the argument zero.policy = TRUE as a small number of tracts in the dataset have no neighbors after omitting NA values;
- The monthly_income column is scaled to a z-score;
- The LISA statistic is computed with the localmoran_perm() function, using the permutation-based method to compute statistical significance;
- The LISA results are attached to the spatial dataset and recoded into cluster groups for visualization.

```
library(spdep)
set.seed(1983)

# Omit NAs and scale the income variable
brasilia_input <- brasilia_income_geo %>%
  mutate(scaled_income = as.numeric(scale(monthly_income))) %>%
  na.omit()

# Generate contiguity-based weights with zero.policy = TRUE
brasilia_weights <- brasilia_input %>%
  poly2nb() %>%
  nb2listw(zero.policy = TRUE)

# Compute the LISA and convert to a tibble
brasilia_lisa <- localmoran_perm(brasilia_input$scaled_income,
                                 brasilia_weights,
                                 nsim = 999L,
                                 alternative = "two.sided",
                                 na.action = na.pass) %>%
  as_tibble() %>%
  set_names(c("local_i", "exp_i", "var_i", "z_i", "p_i",
              "p_i_sim", "pi_sim_folded", "skewness", "kurtosis"))

# Add the LISA columns to the spatial dataset
# and recode them into cluster values
brasilia_lisa_clusters <- brasilia_input %>%
  select(code_tract, scaled_income) %>%
  bind_cols(brasilia_lisa) %>%
  mutate(lisa_cluster = case_when(
    p_i >= 0.05 ~ "Not significant",
    scaled_income > 0 & local_i > 0 ~ "High-high",
```

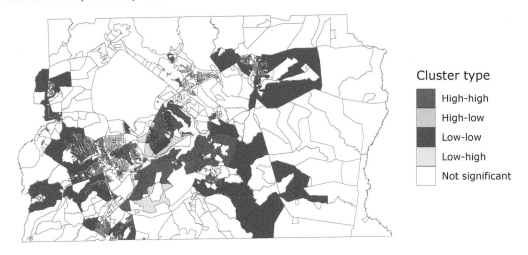

Data sources: 2010 Brazilian Census via IBGE; geobr R package

FIGURE 12.14 LISA cluster map of median monthly income by Census tract in Brasilia

```
    scaled_income > 0 & local_i < 0 ~ "High-low",
    scaled_income < 0 & local_i > 0 ~ "Low-low",
    scaled_income < 0 & local_i < 0 ~ "Low-high"
))
```

Once computed, the results can be visualized on a map.

```
color_values <- c(`High-high` = "red",
                  `High-low` = "pink",
                  `Low-low` = "blue",
                  `Low-high` = "lightblue",
                  `Not significant` = "white")

ggplot(brasilia_lisa_clusters, aes(fill = lisa_cluster)) +
  geom_sf(size = 0.1) +
  theme_void(base_family = "Verdana") +
  scale_fill_manual(values = color_values) +
  labs(title = "LISA clusters of median monthly income",
       subtitle = "Census tracts, Brasilia/Federal District",
       fill = "Cluster type",
       caption = "Data sources: 2010 Brazilian Census via IBGE; geobr R package")
```

The map in Figure 12.14 illustrates strong clustering of high monthly incomes in central Brasilia with low-high spatial outliers scattered among the higher-earning areas. Low-low cluster census tracts tend to be found on the edges of the urbanized portions of the Federal District, representing more sparsely populated rural areas.

12.3 Other international data resources

Many other international data resources in R exist, and a robust R ecosystem is emerging to standardize these data source. In Europe, the **nomisr** R package (Odell, 2018) provides an interface to official statistics for the United Kingdom, and the **insee** package (Leclerc, 2021) does the same for France. Europe-wide data can be acquired using the **eurostat** package (Lahti et al., 2017), which interacts with European Union data sources. In Africa, the afrimapr[11] project is an effort to disseminate Africa-wide datasets to R users. It includes a wide variety of spatial datasets for countries across Africa.

Microdata users will be interested in the IPUMS International project[12], which makes historical microdata for over 100 countries available to researchers. Like US IPUMS data, the **ipumsr** R package can help users work with international microdata in R. A companion project from IPUMS is the International Historical Geographic Information System (IHGIS) project, which aims to disseminate international spatial and demographic data in much the same way that the NHGIS project (covered in Section 11.1.1 does for historical Census data for the US.

12.4 Exercises

- Use **idbr** to create a map of a global demographic indicator as shown above. Use **ggiraph** and **ggplot2** to make both static and interactive versions of the map.
- Choose one of the four countries highlighted in this chapter (Canada, Kenya, Mexico, or Brazil) and explore their corresponding R packages further. Explore the available demographic indicators for those countries, and try making a map or a chart for that indicator.

---

[11] https://afrimapr.github.io/afrimapr.website/
[12] https://international.ipums.org/international/

Conclusion

As its title suggests, this book is an overview of methods, maps, and models that can be used to complete applied social research projects with R and Census data. As the breadth of approaches and packages covered in the book suggests, R's ecosystems for working on these topics are large and constantly changing. While I have chosen to focus on packages that I have developed (**tidycensus, tigris, mapboxapi, crsuggest**, and **idbr**) and frameworks like **tidyverse** and **sf** that integrate with those packages, there are many other directions you could go to learn more.

Census data analysts may be interested in delving deeper into the packages introduced in Chapter 11, such as **censusapi**, **ipumsr**, and **lehdr**. Geospatial analysts will want to read *Geocomputation with R* (Lovelace et al., 2019) to gain a strong command of R's capabilities for spatial data. Those interested in machine learning and prediction, which is not a major focus of this book, should read Ken Steif's *Public Policy Analytics* (Steif, 2021), which uses some of the skills learned in this book in applied machine learning workflows for public policy. Readers who want to learn more about modeling, especially within a tidyverse framework, should keep an eye on *Tidy Modeling with R* (Kuhn and Silge, 2021), which is not yet complete at the time of this writing but when finished will offer a comprehensive overview of how to model data in a tidyverse-friendly way.

My own packages outlined in this book will also be updated as new data are made available. The top development priority for **tidycensus** is making access to 2020 Census data seamless as they are released to data.census.gov and the Census API, allowing users to work with 2020 Census data in the exact same way they've been working with other datasets in the package. Other feature suggestions are welcome on the **tidycensus** GitHub issues page[13], as are user contributions!

My future development work on these packages will certainly be done with recent threats to Census data quality in mind. While we are very fortunate in the R community to have such a breadth of data resources at our fingertips, it is important to be aware of how ephemeral that can be. The Census Bureau did not release its typical 2020 1-year ACS estimates due to data collection problems during the COVID-19 pandemic, breaking time series workflows that use 1-year ACS data. While the 5-year 2016-2020 ACS dataset was still released, larger margins of error due to COVID-19 data collection problems will be present in ACS samples for the next several years.

The Census data community has also been grappling with the tension between preserving respondent privacy and maintaining data quality. The 2020 decennial Census data are released using *differential privacy*, an approach that introduces errors into datasets in an attempt to preserve overall population characteristics while limiting the possibility of re-identifying individuals in the data (Abowd, 2018). Advocates for differential privacy argue that disclosure avoidance techniques are necessary to prevent re-identification attacks which are important for the Census Bureau given modern database reconstruction technologies

---

[13]https://github.com/walkerke/tidycensus/issues

(Hawes, 2020). Critics of this approach, however, have argued that differential privacy will have a disastrous impact on Census data quality (Ruggles et al., 2019), potentially making microdata and block-level aggregate data unusable. The loss in data quality also threatens to disproportionately impact rural populations and racial & ethnic minorities. This will make it difficult to evaluate racial health disparities (Santos-Lozada et al., 2020), accurately estimate COVID-19 mortality rates (Hauer and Santos-Lozada, 2021), and complete countless other analyses that require access to high-quality small-area Census data. More broadly, Ruggles and Van Riper (2021) argue that re-identification attacks with database reconstruction are little different than what is produced with a random number generator, and that the actual threats to the population from Census data re-identification are already publicly available on the internet.

Census data have also faced threats from higher levels of authority. A major initiative of the Trump Administration was to introduce a citizenship question on the decennial Census, which is currently only asked on the American Community Survey. This move, which was ultimately rejected by the US Supreme Court, was widely interpreted as an effort to suppress responses in nonwhite and immigrant communities (Frey, 2019). It also may have been a precursor to an effort to make Congressional apportionment contingent only on the *citizen* population rather than the entire population as is mandated in the US Constitution. Although this effort did not succeed, the Trump Administration engaged in other efforts to underfund and undermine the success of the Census, which was exacerbated by data collection difficulties during the COVID-19 pandemic (Bahrampour et al., 2021).

This discussion is not intended to conclude the book on a cynical note. Rather, it is to reinforce one of the book's central take-aways: the critical importance of **high-quality, open data** and **free and open source software** to analyze that data. Census data are a democratizing force in many ways. They allow communities to analyze their own characteristics and use that information to solve problems that may otherwise be overlooked by higher levels of government. They allow analysts to independently evaluate re-districting scenarios and call them out if they are disenfranchising local residents. They help us "see" latent inequalities that manifest themselves within societies and propose solutions to rectify them.

These initiatives are facilitated not only by open data, but by open tools to analyze them. As discussed earlier in this book, even open government datasets have traditionally been hard to work with. They may have been stored in bulky datasets that require navigation of complex file systems, and required expensive commercial software platforms to process and analyze them. In contrast, resources like R and the Census API put Census data queries at the fingertips of the user, and integrate with data analysis tools in ways that help analysts get to insights faster. This is not to dismiss the learning curve of R and the methods and skills necessary to generate and understand those insights. Rather, it is to stress that open data and open source software reduce financial and logistical barriers to access, creating a larger and more diverse user community that can generate unique insights and help solve problems. Contributing to this effort is one of my main reasons for writing this book and publishing it as an open website, and I very much appreciate all of you who have taken the time to read through it.

Bibliography

Abowd, J. (2018). The U.S. Census Bureau adopts differential privacy. In *KDD '18: The 24th ACM SIGKDD International Conference on Knowledge Discovery and Data Mining.* ACM.

Agafonkin, V. (2020). Leaflet. Version 1.9.1. https://leafletjs.com

American Community Survey Office, U. C. B. (2020). *Instructions for Applying Statistical Testing to American Community Survey Data.*

American Community Survey Office, U. C. B. (2021). *American Community Survey 2015-2019 ACS 5-year PUMS files ReadMe.*

Anselin, L. (1995). Local indicators of spatial association-lisa. *Geographical Analysis*, 27(2):93–115.

Anselin, L., Bera, A., Florax, R., and Yoon, M. (1996). Simple diagnostic tests for spatial dependence. *Regional Science and Urban Economics*, 26(1):77–104.

Appelhans, T., Detsch, F., Reudenbach, C., and Woellauer, S. (2020). *mapview: Interactive Viewing of Spatial Data in R.* R package version 2.9.0.

Assunção, R., Neves, M., Câmara, G., and Da Costa Freitas, C. (2006). Efficient regionalization techniques for socio-economic geographical units using minimum spanning trees. *International Journal of Geographical Information Science*, 20(7):797–811.

Bache, S. M. and Wickham, H. (2020). *magrittr: A Forward-Pipe Operator for R.* R package version 2.0.1.

Bahrampour, T., Rabinowitz, K., and Mellnik, T. (2021). Lower-than-expected state population totals stoke concerns about the 2020 census. *The Washington Post.* https://www.washingtonpost.com/dc-md-va/2021/04/27/2020-census-undercount/

Battersby, S., Finn, M., Usery, E., and Yamamoto, K. (2014). Implications of web mercator and its use in online mapping. *Cartographica: The International Journal for Geographic Information and Geovisualization*, 49(2):85–101.

Bivand, R. and Wong, D. (2018). Comparing implementations of global and local indicators of spatial association. *TEST*, 27(3):716–748.

Bivand, R. and Yu, D. (2020). *spgwr: Geographically Weighted Regression.* R package version 0.6-34.

Bivand, R. S., Pebesma, E., and Gomez-Rubio, V. (2013). *Applied spatial data analysis with R, Second edition.* Springer, NY.

Boehmke, B. and Greenwell, B. (2019). *Hands-On Machine Learning with R.* Chapman and Hall/CRC.

Brewer, C. (2016). *Designing Better Maps: A Guide for GIS Users*. Esri Press, New York NY.

Brewer, C., Hatchard, G., and Harrower, M. (2003). Colorbrewer in print: A catalog of color schemes for maps. *Cartography and Geographic Information Science*, 30(1):5–32.

Brunsdon, C., Fotheringham, A., and Charlton, M. (1996). Geographically weighted regression: A method for exploring spatial nonstationarity. *Geographical Analysis*, 28(4):281–298.

Chang, W., Cheng, J., Allaire, J., Sievert, C., Schloerke, B., Xie, Y., Allen, J., McPherson, J., Dipert, A., and Borges, B. (2021). *shiny: Web Application Framework for R*. R package version 1.6.0.

Cheng, J., Karambelkar, B., and Xie, Y. (2021). *leaflet: Create Interactive Web Maps with the JavaScript 'Leaflet' Library*. R package version 2.0.4.1.

Clarke, E. and Sherrill-Mix, S. (2017). *ggbeeswarm: Categorical Scatter (Violin Point) Plots*. R package version 0.6.0.

Cooley, D. (2020). *mapdeck: Interactive Maps Using 'Mapbox GL JS' and 'Deck.gl'*. R package version 0.3.4.

Dowle, M. and Srinivasan, A. (2021). *data.table: Extension of 'data.frame'*. R package version 1.14.0.

Dunnington, D. (2019). *rosm: Plot Raster Map Tiles from Open Street Map and Other Sources*. R package version 0.2.5.

Dunnington, D., Pebesma, E., and Rubak, E. (2021). *s2: Spherical Geometry Operators Using the S2 Geometry Library*. R package version 1.0.6.

Eberwein, K. (2021). *blscrapeR: An API Wrapper for the Bureau of Labor Statistics (BLS)*. R package version 3.2.0.

Elbers, B. (2021). A method for studying differences in segregation across time and space. *Sociological Methods and Research*, 10.1177/00491241219862.

Ellis, G. and Burk, D. (2020). *ipumsr: Read 'IPUMS' Extract Files*. R package version 0.4.5.

Evergreen, S. (2020). *Effective Data Visualization: The Right Chart for the Right Data*. SAGE, Thousand Oaks, CA.

Flores, E. (2019). *inegiR: Integrate INEGI's (Mexican Stats Office) API with R*. R package version 3.0.0.

Fox, J. and Weisberg, S. (2019). *An R Companion to Applied Regression*. Sage, Thousand Oaks, CA, third edition.

Freedman Ellis, G. and Schneider, B. (2021). *srvyr: 'dplyr'-Like Syntax for Summary Statistics of Survey Data*. R package version 1.0.1.

Frey, W. H. (2019). America wins as trump abandons the citizenship question from the 2020 census.

Garnier, S., Ross, N., BoB Rudis, Filipovic-Pierucci, A., Galili, T., Timelyportfolio, Greenwell, B., Sievert, C., Harris, D., and Chen, J. J. (2021). *sjmgarnier/viridis: viridis 0.6.0 (pre-CRAN release)*. Zenodo.

Getis, A. and Ord, J. (1992). The analysis of spatial association by use of distance statistics. *Geographical Analysis*, 24(3):189–206.

Glenn, E. H. (2019). *acs: Download, Manipulate, and Present American Community Survey and Decennial Data from the US Census*. R package version 2.1.4.

Gohel, D. and Skintzos, P. (2021). *ggiraph: Make 'ggplot2' Graphics Interactive*. R package version 0.7.10.

Gollini, I., Lu, B., Charlton, M., Brunsdon, C., and Harris, P. (2015). Gwmodel: Anrpackage for exploring spatial heterogeneity using geographically weighted models. *Journal of Statistical Software*, 63(17): 1–50.

Goss, J. (1995). "we know who you are and we know where you live": The instrumental rationality of geodemographic systems. *Economic Geography*, 71(2):171.

Green, J., Wang, L., and Mahmoudi, D. (2021). *lehdr: Grab Longitudinal Employer-Household Dynamics (LEHD) Flat Files*. R package version 0.2.4.

Hafen, R. (2020). *geofacet: 'ggplot2' Faceting Utilities for Geographical Data*. R package version 0.2.0.

Hauer, M. and Santos-Lozada, A. (2021). Differential privacy in the 2020 census will distort covid-19 rates. *Socius: Sociological Research for a Dynamic World*, 7:237802312199401.

Hawes, M. (2020). Implementing differential privacy: Seven lessons from the 2020 United States Census. *Harvard Data Science Review*.

Healy, K. (2019). *Data Visualization: A Practical Introduction*. Princeton University Press, Princeton, NJ.

Henry, L. and Wickham, H. (2020). *purrr: Functional Programming Tools*. R package version 0.3.4.

Hester, J., Csárdi, G., Wickham, H., Chang, W., Morgan, M., and Tenenbaum, D. (2021). *remotes: R Package Installation from Remote Repositories, Including 'GitHub'*. R package version 2.3.0.

James, G., Witten, D., Hastie, T., and Tibshirani, R. (2013). *An Introduction to Statistical Learning*. Springer, New York.

Jenks, G. (1967). The data model concept in statistical mapping. *International Yearbook of Cartography*, 7:186–190.

Kariuki, S. (2020). *rKenyaCensus: 2019 Kenya Population and Housing Census Results*. R package version 0.0.2.

Kenny, C. T., McCartan, C., Fifield, B., and Imai, K. (2021). redist: Simulation methods for legislative redistricting. Available at The Comprehensive R Archive Network (CRAN).

Knaflic, C. (2015). Storytelling with Data: A Data Visualization Guide for Business Professionals. Hoboken, NJ: John Wiley & Sons.

Kuhn, M., Jackson, S., and Cimentada, J. (2020). *corrr: Correlations in R*. R package version 0.4.3.

Kuhn, M. and Silge, J. (2021). Tidy Modeling with R: A Framework for Modeling in the Tidyverse. Sebastopol, CA: O'Reilly Media.

Lahti, L., Huovari, J., Kainu, M., and Biecek, P. (2017). Retrieval and analysis of eurostat open data with the eurostat package. *The R Journal*, 9(1):385–392.

Lamstein, A. (2020). *choroplethr: Simplify the Creation of Choropleth Maps in R*. R package version 3.7.0.

Lamstein, A. and Powell, L. (2018). A guide to working with us census data in R. Technical report, The R Consortium.

Leclerc, H. (2021). *insee: Tools to Easily Download Data from INSEE BDM Database*. R package version 1.1.0.

Li, G. (2021). *totalcensus: Extract Decennial Census and American Community Survey Data*. R package version 0.6.6.

Lindblad, B. (2021). *tidyUSDA: A Minimal Tool Set for Gathering USDA Quick Stat Data for Analysis and Visualization*. R package version 0.3.1.

Lovelace, R., Nowosad, J., and Muenchow, J. (2019). *Geocomputation with R*. CRC Press.

Lu, B., Harris, P., Charlton, M., and Brunsdon, C. (2014). The gwmodel r package: further topics for exploring spatial heterogeneity using geographically weighted models. *Geo-spatial Information Science*, 17(2):85–101.

Lucchesi, L. and Wikle, C. (2017). Visualizing uncertainty in areal data with bivariate choropleth maps, map pixelation and glyph rotation. *Stat*, 6(1):292–302.

Lumley, T. (2010). *Complex Surveys: A Guide to Analysis Using R*. John Wiley and Sons.

Manson, S., Schroeder, J., Van Riper, D., Kugler, T., and Ruggles, S. (2021). National historical geographic information system: Version 16.0.

Matloff, N. (2017). *Statistical Regression and Classification*. Chapman and Hall/CRC.

McCartan, C. and Kenny, C. T. (2021). *PL94171: Tabulate P.L. 94-171 Redistricting Data Summary Files*. R package version 0.3.2.

Mora, R. and Ruiz-Castillo, J. (2011). Entropy-based segregation indices. *Sociological Methodology*, 41(1):159–194.

Munzner, T. (2014). *Visualization Analysis and Design*. A K Peters/CRC Press.

Müller, K. and Wickham, H. (2021). *tibble: Simple Data Frames*. R package version 3.1.3.

Odell, E. (2018). nomisr: Access nomis uk labour market data with r. *The Journal of Open Source Software*, 3(27):859.

Ooms, J. (2014). The jsonlite package: A practical and consistent mapping between json data and r objects. *arXiv:1403.2805 [stat.CO]*.

Pattani, A., Recht, H., and Grey, J. (2021). In appalachia and the mississippi delta, millions face long drives to stroke care. *KHN*.

Pebesma, E. (2018). Simple features for R: Standardized support for spatial vector data. *The R Journal*, 10(1):439.

Pedersen, T. L. (2020). *patchwork: The Composer of Plots*. R package version 1.1.1.

Pedersen, T. L., Ooms, J., and Govett, D. (2021). *systemfonts: System Native Font Finding*. R package version 1.0.2.

Pedersen, T. L. and Robinson, D. (2020). *gganimate: A Grammar of Animated Graphics*. R package version 1.0.7.

Pereira, R. H. M. and Goncalves, C. N. (2021). *geobr: Download Official Spatial Data Sets of Brazil*. R package version 1.6.1.

Peterson, G. (2020). *GIS Cartography*. CRC Press.

R Core Team (2021). *R: A Language and Environment for Statistical Computing*. R Foundation for Statistical Computing, Vienna, Austria.

R Special Interest Group on Databases (R-SIG-DB), Wickham, H., and Müller, K. (2021). *DBI: R Database Interface*. R package version 1.1.1.

Ratnakumar, S., Mick, T., and Davis, T. (2021). *rappdirs: Application Directories: Determine Where to Save Data, Caches, and Logs*. R package version 0.3.3.

Recht, H. (2021). *censusapi: Retrieve Data from the Census APIs*. R package version 0.7.2.

Rey, S., Arribas-Bel, D., and Wolf, L. (2020). *Geographic Data Science with Python* https://geographicdata.science/book/intro.html.

RStudio Team (2021). *RStudio: Integrated Development Environment for R*. RStudio, PBC, Boston, MA.

Ruggles, S., Fitch, C., Magnuson, D., and Schroeder, J. (2019). Differential privacy and census data: Implications for social and economic research. *AEA Papers and Proceedings*, 109:403–408.

Ruggles, S., Flood, S., Goeken, R., Grover, J., Meyer, E., Pacas, J., and Sobek, M. (2020). Ipums USA: Version 10.0.

Ruggles, S. and Van Riper, D. (2022). The role of chance in the census bureau database reconstruction experiment. *Population Research and Policy Review* 41, pp. 781–788.

Santos-Lozada, A., Howard, J., and Verdery, A. (2020). How differential privacy will affect our understanding of health disparities in the United States. *Proceedings of the National Academy of Sciences*, 117(24):13405–13412.

Schroeder, J. and Van Riper, D. (2013). Because muncie's densities are not manhattan's: Using geographical weighting in the expectation-maximization algorithm for areal interpolation. *Geographical Analysis*, 45(3):216–237.

Shannon, J. (2020). Dollar stores, retailer redlining, and the metropolitan geographies of precarious consumption. *Annals of the American Association of Geographers*, 111(4):1200–1218.

Sievert, C. (2020). *Interactive Web-Based Data Visualization with R, plotly, and shiny*. Chapman and Hall/CRC.

Silge, J. (2021). Pca and umap with tidymodels and tidytuesday cocktail recipes.

Singleton, A. and Spielman, S. (2013). The past, present, and future of geodemographic research in the United States and United Kingdom. *The Professional Geographer*, 66(4):558–567.

Spielman, S. and Singleton, A. (2015). Studying neighborhoods using uncertain data from the american community survey: A contextual approach. *Annals of the Association of American Geographers*, 105(5):1003–1025.

Steif, K. (2021). *Public Policy Analytics*. CRC Press.

Tennekes, M. (2018). tmap: Thematic maps in R. *Journal of Statistical Software*, 84(6), pp. 1–39.

Tobler, W. (1970). A computer movie simulating urban growth in the detroit region. *Economic Geography*, 46:234.

Vaidyanathan, R., Xie, Y., Allaire, J., Cheng, J., Sievert, C., and Russell, K. (2020). *htmlwidgets: HTML Widgets for R*. R package version 1.5.3.

Valle-Jones, D. (2021). *mxmaps: Create Maps of Mexico*. https://www.diegovalle.net/mxmaps/, https://github.com/diegovalle/mxmaps.

Vicino, T., Hanlon, B., and Short, J. (2011). A typology of urban immigrant neighborhoods. *Urban Geography*, 32(3):383–405.

von Bergmann, J., Shkolnik, D., and Jacobs, A. (2021). *cancensus: R package to access, retrieve, and work with Canadian Census data and geography*. R package version 0.4.2.

Walker, K. (2016a). Locating neighbourhood diversity in the american metropolis. *Urban Studies*, 55(1):116–132.

Walker, K. (2016b). tigris: An R package to access and work with geographic data from the us census bureau. *The R Journal*, 8(2):231.

Walker, K. (2018). Scaling the interactive dot map. *Cartographica: The International Journal for Geographic Information and Geovisualization*, 53(3):171–184.

Walker, K. (2021a). *crsuggest: Obtain Suggested Coordinate Reference System Information for Spatial Data*. R package version 0.3.1.

Walker, K. (2021b). *idbr: R Interface to the US Census Bureau International Data Base API*. R package version 1.0.

Walker, K. (2021c). *mapboxapi: R Interface to 'Mapbox' Web Services*. R package version 0.2.1.9000.

Walker, K. and Herman, M. (2021). *tidycensus: Load US Census Boundary and Attribute Data as 'tidyverse' and 'sf' -Ready Data Frames*. R package version 1.0.0.9000.

Wickham, H. (2011). The split-apply-combine strategy for data analysis. *Journal of Statistical Software*, 40(1).

Wickham, H. (2014). Tidy data. *Journal of Statistical Software*, 59(10).

Wickham, H. (2016). *ggplot2: Elegant Graphics for Data Analysis*. Springer-Verlag, New York.

Wickham, H. (2019). *Advanced R*. Chapman and Hall/CRC.

Wickham, H. (2019). *stringr: Simple, Consistent Wrappers for Common String Operations*. R package version 1.4.0.

Wickham, H. (2020). *httr: Tools for Working with URLs and HTTP*. R package version 1.4.2.

Wickham, H. (2021a). *forcats: Tools for Working with Categorical Variables (Factors)*. R package version 0.5.1.

Wickham, H. (2021b). *tidyr: Tidy Messy Data*. R package version 1.1.3.

Wickham, H., Averick, M., Bryan, J., Chang, W., McGowan, L., François, R., Grolemund, G., Hayes, A., Henry, L., Hester, J., Kuhn, M., Pedersen, T., Miller, E., Bache, S., Müller, K., Ooms, J., Robinson, D., Seidel, D., Spinu, V., Takahashi, K., Vaughan, D., Wilke, C., Woo, K., and Yutani, H. (2019). Welcome to the tidyverse. *Journal of Open Source Software*, 4(43):1686.

Wickham, H. and Bryan, J. (2021). *usethis: Automate Package and Project Setup*. R package version 2.0.1.

Wickham, H., François, R., Henry, L., and Müller, K. (2021a). *dplyr: A Grammar of Data Manipulation*. R package version 1.0.7.

Wickham, H., Girlich, M., and Ruiz, E. (2021b). *dbplyr: A 'dplyr' Back End for Databases*. R package version 2.1.1.

Wickham, H. and Grolemund, G. (2017). *R for Data Science*. O'Reilly, Sebastopol, CA.

Wickham, H. and Hester, J. (2021). *readr: Read Rectangular Text Data*. R package version 2.0.0.

Wickham, H., Ooms, J., and Müller, K. (2021c). *RPostgres: 'Rcpp' Interface to 'PostgreSQL'*. R package version 1.3.2.

Wickham, H. and Seidel, D. (2022). *scales: Scale Functions for Visualization*. R package version 1.2.0.

Wilke, C. (2019). *Fundamentals of Data Visualization*. O'Reilly Media, Sebastopol, CA.

Wilke, C. O. (2021). *ggridges: Ridgeline Plots in 'ggplot2'*. R package version 0.5.3.

Wong, D. and Sun, M. (2013). Handling data quality information of survey data in gis: A case of using the american community survey data. *Spatial Demography*, 1(1):3–16.

Çetinkaya-Rundel, M. and Hardin, J. (2021). *Introduction to Modern Statistics*. OpenIntro, first edition.

Index

Note: Locators in *italics* represent figures and **bold** indicate tables in the text.

A

Accessibility, 187, 189
ACS *see* American Community Survey (ACS)
ACS PUMS data, 256, **258,** 260
    available variables, 261
    recoding variables, 261–262, *262*
    variables filters, 262, **263**
Agricultural data, 310–312, *311,* **312**
ALARM project, 12–13
American Community Survey (ACS), 2, 5, 20–22, **22,** 269
    classification, 247–250, *250*
    comparing estimates, 49–56
    Comparison Profile Tables, 52–53, **53**
    data, 269, 272
    Data Profile, 174
    estimates over time, 76–78, **77,** *78*
    iterating with tidyverse tools, 53–56, **56**
    margins of error, 56–60
    Migration Flows API, 155
    Public Use Microdata Sample APIs, 17
    spatial clustering, 250–253, *251, 253*
    time-series analysis, 50–52, **51–52**
American FactFinder, 5
Analytic pipelines, 45
API functions, 313–315
Apportionment, 1
Areal interpolation, 107, 181, 194–195, **195**
Area-weighted areal interpolation, 182, *183*
arrange(), 42, 216
as_dot_density(), 141, 291
Assignment operator, 8–9

B

basename(), 335
beeswarm plots, 87–88
Block group, 2
bls_api(), 309
blscraper package, 308–310, **309–310**
Box-and-whisker plot, 62, 64, *64*

Buffers, 192–194, *194*
Bureau of Labor Statistics data, 308–310, **309–310**

C

cancensus package, Canada, 325–327, **326,** *327*
Cartographic techniques, 129, 161–165, 196–198
case_when(), 48, 205–206, 281–282
Categorical predictors, 282
Census API, 5–7, *6,* 256, 259
censusapi package, 285, 302–305, *303,* **304,** *305*
Census Bureau, 123, 255, 317
    data, downloads, 5, *5*
    third-party data distributors, 7–8
Census data
    analysts, 343
    hierarchies, 2–3, *3*
    resources, 302–305, *303,* **304,** *305*
    visualization, 15–16
Census polygons, 196–198
Census Reporter, 7
Census tracts, 2, *4,* **26,** *104, 131, 172,* **176,** *180,* 181, *181,* 291, *292, 335*
Census visualization, ggplot2, 61–62
    getting started, 62–64, *63–64*
    multivariate relationships, scatter plots, 64–66, *65, 66*
Choropleth maps
    ggplot2 package, 126–127, *127*
    tmap, 129–133, *130–133*
choroplethr package, 11, 331
Clustering algorithm, 248
Collinearity, 221, 227, 247
ColorBrewer palette, 128
Color-coded scatterplot, 249, *250*
Comparison Profile Tables, 52–53, **53**
Comprehensive R Archive Network (CRAN), 8, 10

353

Contiguity-based neighbors, 199
Contiguity-based spatial weights matrix, 339
Cook Political Report, 142
Coordinate reference systems (CRS), 109–110, 168–169
coord_sf(), 113–115, *114–115*
Correlation matrix, 227
corrr package, 227
count() function, 270
Country-specific Census data packages
    cancensus R package, Canada, 325–327, **326, 327**
    geobr R package, Brazil, 333–340
    INEGI, Mexico, 331–333, *332, 333,* **334**
    rKenyaCensus package, Kenya, 327–331, **328,** *328, 330–331*
COVID-19 pandemic, 2, 12, 343–344
COVID-19 Vaccine Tracker, 12, *14*
crsuggest package, 110–113, **111,** 290

D
Dasymetric dot-density map, 141
Dasymetric dot-density methodology, 16
data.census.gov, 5
Data, US Census, 3
    Census API, 5–7, *6*
    downloads from Census Bureau, 5, *5*
    third-party data distributors, 7–8
dbConnect() function, 298
DBeaver database, 297, *298, 299*
dbplyr package, 300
Decennial census, 18–20, **20**
Demographic data resources, 334–336, *335,* **336**
Density plot, 84, *85*
Derived margins of error, 57–59
Differential privacy, 19, 343
Dimension reduction, 230–234, **231,** *232,* 248
dissimilarity() function, 215–216, **216, 217**
Distance analysis, 187–188
    catchment areas, 192–194, *194*
    demographic estimates, areal interpolation, 194–195, **195**
    distances, calculating, 188–190, *189–190*
    travel times, calculating, 190–192, *192*
Diverging color palettes, 136
Diversity gradient, 218–220, *221*
Dot-density maps, 140–141, *141*
dplyr package, 39, 48, 216, 269, 291

E
Ecological fallacy, 176
Economic Census API, 304
Edge effects, 247
Enumeration units, 2
    hierarchies, 2–3, *3*
erase_water(), 197, 307
error bars, 74–76, **75,** *76*
Exploratory data analysis, 222–223
Exploratory spatial data analysis (ESDA), 198–199

F
Faceted maps, 138–140, *139*
Faceted plots, 82, 102
facet_wrap(), 84
"Feature engineering," 225
Federal Information Processing Series (FIPS), 31
Federal Statistical Research Data Centers (FSRDCs), 255
filter(), 42, 46, 58, 278, 282, 288
5-year ACS, 2
Free and open source software (FOSS), 11

G
GDAL ecosystem, 97
geobr R package, 333
    demographic data resources, 334–336, *335,* **336**
    spatial analysis, 338–340, *340*
    spatial data, 334
    working, Brazilian demographic data, 336–337, *338*
GeoDa-style LISA quadrant plot, 208–209
Geodemographic classification, 248–250, *250*
Geofaceted plots, 88–89, *89–91*
Geographically weighted regression (GWR), 241–242
    bandwidth for, 242–243
    fitting and evaluating, 243–246, *245–246*
    limitations of, 246–247
Geographic coordinate system, 109
Geographic features, 93, 96
Geographic information science, 167
Geographic information system (GIS), 93, 167
    layers, *168*
Geographic space, 249

Index

Geographic visualization, 156–157
Geography
    subsets, 24–26, **25–26**
    variables, 22–24, **22–24**
GEOIDs, 31–32
Geometries
    multipolygon geometries, 119–122
    polygons, converting to points, 117–119, *119*
    shifting and rescaling, 115–117, *116–117*
geom_sf(), 101–103, *103*, 113, 117, 209, 223, 244, 249, 252, 291, 292, 322, 329, 338
    choropleth mapping, 126–127, *127*
    customizing maps, 127–128, *127–128*
    manual color palette, 143
geom_sf_interactive(), 323
GEOS ecosystem, 97
Geospatial analysts, 343
Gerrymandering, 12–13
get_acs(), 17, 20, 21, 22, 24, 32, 44, 47, 51, 52, 55, 67, 164, 174, 214, 269, 271
get_decennial(), 17, 18, 22, 32, 44
get_estimates(), 17, 33–35, **34–35**, 79, 123
get_flows(), 17, 35, 155
get_idb(), 317, 319–320, 322
Getis-Ord local G statistic, 204
get_pums(), 17, 257–260, **258, 260,** 261, 266, 269, 276–277, 280
gganimate package, 323
ggbeeswarm package, 87–88
ggiraph package, 153, *158*, 323
ggplot2 package, 39, 61, 62, 67, 68, 101–103, *102, 103,* 125, 209, 231, 244, 246, 304–305, 307, 318, 323, 328
    ACS estimates over time, 76–78
    advanced visualization, 86–92
    Census visualization, 61–66
    charts, 69–71, *71*
    choropleth mapping, 126–127, *127*
    customizing maps, 127–128, *127–128*
    customizing visualization, 66–72
    geom_bar(), 62
    geom_errorbar(), 75
    geom_histogram(), 62
    geom_line(), 77
    geom_ribbon(), 77
    group-wise comparisons, 82–86
    histogram, *235, 337*
    manual color palette, 143
    margins of error, 72–76
    population pyramids, 78–82
    scale_fill_distiller(), 127
    scale_fill_manual(), 143
    scale_x_continuous(), 71
ggridges package, 86–87, *87*
girafe(), 323
girafe_options(), 323
GitHub repository, 12, 15, 142
Global demographic data, 322–325, *323–325*
Global demographic indicators, 317
glue(), 107
Government data without R packages
    data access functions, 313–315
    requesting to APIs with httr, 312–313, **313**
grab_lodes(), 306
Graduated symbol map, 137–138, *138*
Graph-based neighbors, 199
Graticule, 113
group_by(), 46–47, 300
Grouped filter, 46
group_modify(), 216, 219
Group-wise Census data analysis, 45–46
    group-wise comparisons, 46–47, **47**
    tabulating new groups, 47–49, **49**
Group-wise data tabulation, 272–274, **273**
Group-wise data visualization, 176–180, **178,** *179*
Group-wise margins of error, 59–60
Group-wise visualization, 82–86, **83–84,** *85–86*
GWmodel package, 242

H

Harmonized variables, 294–295
Health resource access, 12
"High" clustering, 206
Histogram, *131*
    Census visualization, 62, *63*
    ggplot2, *235, 337*
    minimum distances, *190*
Housing unit, 260
htmlwidgets package, 91–92, 162
httr::GET(), 312
httr package, 312–313, **313**

I

idbr package, core function, 317–318; *see also* International Database (IDB)
idb_variables(), 318
include.self(), 205

Individual-level microdata, 269
INEGI *see* Instituto Nacional de Estadística y Geografía (INEGI)
inegi package, 331
inegi_series(), 332
Inference, 247
inner_join(), 218, 219
Inner spatial join, 215
Instituto Brasileiro de Geografia e Estatística (IBGE), 334–335, 337
Instituto Nacional de Estadística y Geografía (INEGI), 331–333, *332, 333,* **334**
Interactive mapping
    alternative approaches, 151–154, *152–154*
    Leaflet, 147–151, *148–150*
Interactive viewing, 103–105, *104*
Interactive visualization, 91–92
International Database (IDB), 317–319
    animated visualization, 322–325
    interactive visualization, 322–325
    visualizing data, 319–322, **321**, *321*
interpolate_pw(), 184, 195
IPUMS *see* Minnesota Population Center (IPUMS)
ipums_collect(), 301
ipumsr package, 11, 285, 296, 301, 341
    ipumsr::read_ipums_micro(), 296
Isochrones, 192–194, *194*

J
JavaScript Object Notation (JSON), 6–7, 313
Jenks natural-breaks method, 132
jsonlite package, 313

K
k-means, 248

L
Labor market, analyses, 306–308, **307**, *308*
labs(), 68
Lagrange multiplier tests, 240
Layered grammar of graphics approach, 61
leaflet(), 150
Leaflet JavaScript library, 147
leaflet package, 125, 147–151, *148–150,* 151–152, 193
left_join(), 142, 143, 185
Legal entities, 93, 96
lehdr package, 285, 306–308, **307**, *308*

Lines, 98, *99*
LISA *see* local indicators of spatial association (LISA)
load_variables(), 27, 58, 326
localG(), 205
Local indicators of spatial association (LISA), 206–210, *209–210*
localmoran(), 207
localmoran_perm(), 207, 208
Local multicollinearity, 247
Local regression coefficient, 242
Local spatial autocorrelation, 204–206, *206*
Longitudinal and Employer-Household Dynamics (LEHD), 306
    Origin-Destination Employment Statistics (LODES), 306–308, **307**, *308*

M
magrittr package, 45, 67
mapboxapi package, 134, 190–191
Mapbox basemaps, 134
mapdeck(), 155–156
mapdeck R package, 155
map_dfr(), 54–55, 109, 332
Map types
    dot-density maps, 140–141, *141*
    faceted maps, 138–140, *139*
    ggplot2 package (*see* ggplot2 package)
    graduated symbol map, 137–138, *138*
    interactive mapping, 147–154
    tmap (*see* tmap)
mapview R package, 103–105, *104*, 148, 162
Margins of error (MOEs), 39, 56–60, 157, **279**, 279–280, *281*
    data setup, 72–74, **73**, *74*
    moe_product(), 59
    moe_prop(), 59
    moe_ratio(), 59
    moe_sum(), 59
    visualizing, 72–76, *76*
Mastering Shiny (Wickham), 159
mb_isochrone(), 193
mb_matrix(), 190–191
Microdata
    big Census microdata in R, 300–302
    Census API, 256
    definition, 255
    getting from IPUMS, 294–296, *295, 296*
    harmonized, 256
    IPUMS, 256

loading to database, 296–297
PUMAs (*see* Public Use Microdata
Areas (PUMAs))
PUMS (*see* Public Use Microdata Series
(PUMS))
tidycensus, 257–260
Migration flows, 155–156, *156*
Minimum spanning trees, 251
Minnesota Population Center (IPUMS), 285,
292–294
accessing microdata database with R,
297–300, *298, 299*
big Census microdata in R, 300–302
data browser, *294*
data cart, *295*
display options, 295–296
extract request screen, 296, *296*
getting microdata, 294–296, *295, 296*
harmonized variables, 294–295
home page, *256*
loading microdata to database, 296–297
microdata, 256
sample selection, 294
Model R-squared drops, 229
modify(), 219
Moran's *I* test, 203–204, *204*, 206–207,
235–236, *236*, 239–241
Multi-group segregation indices, 217–218,
218, *219*
mutate(), 44, 291
Mutual Information Index *M*, 217
mxmaps package, 331

N

National election mapping, 142–143
National Historical Geographic Information
System (NHGIS), 7, 285–286
browser interface, *286*
datasets, 286
geographic levels, 286
getting started, 286–287, *287*
mapping in R, 289–292, **290**, *290, 292*
R-based data analysis workflows,
287–289, **289**, *289*
Select Data menu, 286–287
topics, 286
years, 286
nb2listw(), 201–202
Neighborhood, 199
NHGIS *see* National Historical Geographic
Information System (NHGIS)

nomisr package, 341
Non-Census data, cartographic workflows,
141–142
national election mapping, 142–143
working with ZCTAs, *144*, 144–147,
146, *146, 147, 148*
Non-spatial model, 238
Non-standard evaluation, 43
Normalizing, count data, 43
Null hypothesis, 227

O

Obama Administration's Digital
Government Strategy, 5–6
1-year ACS, 2
opentripplanner package, 192
osrm package, 192
Other government data resources
blscrapeR package, 308–310, **309–310**
censusapi package, 285, 302–305,
303,**304**,*305*
government data without R packages,
312–315, **313**
IPUMS, 292–302
lehdr package, 306–308, **307**,*308*
NHGIS, 285–292
tidyUSDA package, 310–312, *311*,**312**
Outcome variable, 220, 223–225, *224*, **225**
Outlier Analysis geoprocessing tool, 207

P

patchwork package, 102, 105, 107
PL94171 package, 11, 12
places(), 117
plot(), 95–97, 125, 289–290, **290**
Plot legibility, 67–69, *68–69*
plotly library, 91–92
plotly package, 61
Point-in-polygon spatial join, 172–176, **173**,
174, 175, **176**
Points, 97–98, *98*
polygons, converting to, 117–119, *119*
poly2nb(), 200
Polygon-in-polygon spatial joins, 172
Polygons, 99–100, *100*
converting to points, 117–119, *119*
Population Estimates Program (PEP), 33–34
Population pyramids
designing and styling, 80–82, *81–82*
Population Estimates API, 78–80, *79*,
80

Population-weighted areal interpolation, 183–185, *184–185*
PostgreSQL8 database, 297
prcomp(), 230
predict(), 231
Prediction, 247
Predictors, 221
Principal components analysis, 230–234, **231**, *232*
Principal components regression, 233–234
PROJ ecosystem, 97
Proximity analysis, 187–188
    catchment areas, 192–194, *194*
    demographic estimates, areal interpolation, 194–195, **195**
    distances, calculating, 188–190, *189–190*
    travel times, calculating, 190–192, *192*
Proximity-based neighbors, 199
Public Policy Analytics (Steif), 343
Public Use Microdata Areas (PUMAs)
    definition, 263
    geographies, 263–265, *264, 265*, 274, *274*
    working in PUMS data, 265–267, **266, 267**
Public Use Microdata Series (PUMS), 255; *see also* PUMS data
    datasets, 269
    working with PUMAs, 265–267, **266, 267**
PUMAs *see* Public Use Microdata Areas (PUMAs)
PUMS *see* Public Use Microdata Series (PUMS)
PUMS data
    group-wise data tabulation, 272–274, **273**
    mapping, *274*, 274–275
    margins of error, **279**, 280, *281*
    modeling, 280–284, *281*
    preparation, 281–282
    tabulation of weights, 269–272, **270**
pums_variables, 261
purrr package, 39, 54
p-value, 204, 227

Q

QGIS, 163–165, *164*
Qualitative palettes, 137
Quantitative data, 136

R

r5r package, 192
Reactive mapping, 158–161
readr package, 39
redist package, 13
Redistricting, 12–13, *14*
Redistricting summary file, 19
Regionalization, 250–253, *251, 253*
Regression modeling, 220–222
    dimension reduction, 230–234, **231**, *232*
    exploratory data analysis, 222–223
    "feature engineering," 225
    linear model, 225–230
    outcome variable, visualization, 223–225, *224*, **225**
    spatial regression, 234–241
Regular expressions, 68–69
Replicate weights, 269, 275–277, *276*
rKenyaCensus package, 327–331, **328**, *328, 330–331*
rosm package, 134
Row-standardized spatial weights, 202
R package; *see also individual entries*
    analyses, 11–13, *14–15*
    Census data packages, 11–12
    data structures, 8–9
    definition, 8
    exporting data visualizations, 71–72
    exporting maps, 161–163
    functions, 9–10
    getting started, 8
    package ecosystems, 10–11
    packages, 9–10
    visualization software, 163–165, *164, 165*
RPostgres package, 298
RStudio, 8

S

s2 package, 110
saveWidget(), 162
Scatter plots, 64–66, *65, 66*
    ggsave(), 72
    ggsave(). tmap_save(), 162
Segregation and diversity indices, 213
    data setup, spatial analysis, 213–215
    dissimilarity index, 215–217, **216, 217**
    diversity gradient, 218–220, *221*

Index

multi-group segregation indices, 217–218, **218**, *219*
segregation package, 213, 215, 217, 218, 219
Sequential color palettes, 136–137, *136–137*
sf objects, 125
sf package, 110, 188
    st_as_sf(), 329
    st_buffer(), 192
    st_centroid(), 118
    st_crs(), 109
    st_drop_geometry(), 288–289
    st_filter(), 171, 188
    st_interpolate_aw(), 182, 183
    st_intersects(), 176
    st_join(), *172*, 176, 178, 215
    st_read(), 187
    st_transform(), 113, 169, 173, 175
    st_union(), 291
    st_within(), 170
shapefile, 163
shift_geometry(), 116, 117, 126, 142
Shiny, 158–161, *159*, 210, *210*
Simple feature geometry, 123
SKATER algorithm, 250–251
Sliver polygons, 197–198
Small Area Health Insurance Estimates API, 304–305
Small area time-series analysis, 180–182, *181*
    areal interpolation, 181
    comparisons, 185–187
    distance analysis, 187–195
    population-weighted areal interpolation, 183–185, *184–185*
    proximity analysis, 187–195
Small multiples, 82
Social Explorer, 7
Spatial analysis, 167
    data setup, segregation and diversity indices, 213–215
    distance and proximity analysis, 187–195
    joins, 172–180
    overlay, 167–171
    small area time-series analysis, 180–187
    spatial autocorrelation, 202–210
    spatial neighborhoods, 199–202
    spatial weights matrices, 199–202
Spatial autocorrelation, 202–210, 221, 234
Spatial clustering, 202, 250–253, *251, 253*
Spatial clusters, 207
Spatial econometrics, 236

Spatial error models, 238–239
Spatial joins, 171–172, *172*
    group-wise data visualization, 176–180, **178**, *179*
    point-in-polygon spatial join, 172–176, **173**, *174, 175*, **176**
Spatial lags, 203–204, *204*, 236–238
Spatially clustered, 204
Spatial neighborhoods, 198–202
Spatial non-stationarity, 241
Spatial outliers, 207
Spatial overlay, 167–168
    cartography, 196–198
    coordinate reference systems, 168–169
    core-based statistical areas, 169
    geometries within metropolitan area, 169–170
    spatial subsetting, 170–171
Spatial predicate, 170
Spatial randomness, 202
spatialreg package, 237
Spatial regression, 234–236, *235–236*
    alternative approaches, 239–241
    spatial error models, 238–239
    spatial lag models, 236–238
Spatial spillover effects, 236–237
Spatial subsetting, 170–171
Spatial uniformity, 202
Spatial weights, 201
Spatial weights matrices, 198–202
spdep package, 201, 202, 205, 251, 339
spgwr package, 246–247
Split-apply-combine model, data analysis, 45–46
sp package, 167
srvyr package, 269, 275, 278–279
    srvyr::filter(), 278
State Data Centers (SDCs), 263
states(), 93–94, 117, 142
Static maps, 147
Statistical entities, 2, 93, 96
Statistics of Income (SOI) data, 144
stringr package, 68
str_remove(), 73
Structured Query Language (SQL), 297–298
Subpopulation, 277, 278
Successive difference replication method, 275
    standard errors, 276
suggest_crs(), 169
summarize(), 47, 55, 179, 273, 300
Survey design

errors with srvyr, 278–279, **279**
getting replicate weights, 275–277, *276*
standard errors to margins of error, **279**, 279–280
survey object, creation, 277–278
survey package, 275, 277, 282
survey::subset(), 278

T

Theil's Entropy Index *H*, 217
Third-party data distributors, 7–8
tibble, 39, 40
tidycensus package, 7, 10, 123, 196, 269, 277, 285, 291
ACS, 20–22, **22**
aims, 17
color palettes, 137
core functions, 17
data structure, 29–30, **30**
debugging errors, 36–37
decennial census, 18–20, **20**
dot-density visualization, 140–141
GEOIDs, 31–32
geometry parameter, 123–125
get_estimates(), 33–35, **34–35**
get_flows(), 35
getting started, 17–22
group-wise comparisons, 83
maps and charts, linking, 156–158, *158*
margin of error, 56
microdata, 257–260
migration flows, mapping, 155–156, *156*
population-weighted areal interpolation, 183
regression modeling, 221, **222**
renaming variable IDs, 32–33, **33**
spatial datasets, 168–169
variables, searching for, 27–29, **28**
"Tidy" data, 29
tidyr package, 39
tidyUSDA package, 310–312, *311*, **312**
tidyverse ecosystem, 11
tidyverse package, 39
ACS, comparing estimates, 49–56
ACS iterating, 53–56, **56**
exploring Census data, 40–45
filtering data, 40–43, **41**
group-wise Census data analysis, 45–49
margins of error (MOEs), 39, 56–60
sorting data, 40–43, **41**
summary variables, 43–45, **44–45**
tidyverse tools
group-wise data tabulation, 272–274, **273**
tabulation of weights, 269–272, **270**
TIGER/Line database, 93
cartographic boundary shapefiles, 105–106, *106*
data types, 93
yearly differences, 107–108, *108*
tigris package, 93, 195, 196–197, 274, 304–305, 307
basic usage, 93–97, *95–97*
caching data, 106–107
coordinate reference systems, 109–115
data availability, 100–101, **100–101**
datasets, combining, 108–109
features, 97–101
geometries, working with, 115–122
national election mapping, 142–143
plotting geographic data, 101–105
population-weighted areal interpolation, 183
spatial datasets, 168–169
tigris::shift_geometry(), 153
workflows, 105–109
Tiled mapping, 151
Time-series analysis, 50–52, **51–52**
tmap package, 125
adding reference elements, map, 133–136, *135–136*
choropleth maps, 129–133, *130–133*
color palettes, 136–137, *136–137*
Leaflet maps, 149
map-making, 128–129, *129*
tmap_mode(), 149
tmap_save(), 162, 163
tmap::tm_dots(), 292
tmaptools::palette_explorer(), 137
tm_bubbles(), 137–138, 146–147
tm_compass(), 134–135
tm_credits(), 135
tm_dots() function, 140
tm_facets(), 138
tm_fill(), 130, 132, 138–139
tm_layout(), 133, 139
tm_polygons(), 129–130, 133, 138–139
tm_scale_bar(), 134–135
to_survey(), 277
transmute(), 215, 291

Index

U
"Unknown variable" error message, 36
Unsupervised machine learning, 247
US Census, 1
    data, 3–8
US Census Bureau, 1, 2, 17, 46, 269
    Slack Community, 7
US Constitution, 1
US Department of Commerce, 1
usethis package, 303

V
variables_filter, 262, 266, 281
Variance inflation factor (VIF), 228–230

View() command, 27, 261, 288, 326
vif() function, 228
viridis package, 136
Visualization software, 163–165, *164, 165*

W
Web Mercator, 153, *153*
Weighted block point approach, 184
Weighted samples, 269

Z
zctas(), 145
Zip Code Tabulation Areas (ZCTAs), 3, *4, 144,* 144–147, **146,** *146, 147, 148*